Complex Behavior of Grid-Connected Power Electronics Systems

Jingxi Yang • Chi Kong Tse

Complex Behavior of Grid-Connected Power Electronics Systems

Springer

Jingxi Yang
City University of Hong Kong
Kowloon, Hong Kong

Chi Kong Tse
City University of Hong Kong
Kowloon, Hong Kong

ISBN 978-3-031-69339-7 ISBN 978-3-031-69340-3 (eBook)
https://doi.org/10.1007/978-3-031-69340-3

This Springer imprint is published by the registered company Springer Nature Switzerland AG
The registered company address is: Gewerbestrasse 11, 6330 Cham, Switzerland

Preface

Grid-connected power electronic converters are devices that are used to convert electrical power between the power grid and other sources or loads, such as renewable energy sources, electric vehicles, or industrial machinery. These converters are essential for integrating renewable energy sources into the power grid and for improving the efficiency and flexibility of the grid's operation. However, the behavior of grid-connected power electronic converters is complex and nonlinear, which can lead to instability and other issues that can affect the reliable operation of the power grid. For example, the interaction between multiple converters and the grid can lead to oscillations and other dynamic behavior that can cause voltage and frequency fluctuations, compromising the quality and reliability of the power supply. Therefore, it is necessary to have a thorough understanding of the complex behavior and stability issues of grid-connected power electronic converters in order to design and control these devices effectively. This requires a multidisciplinary approach that combines expertise in power electronics, control theory, and power systems. Researchers need to develop advanced control strategies that can adapt to changes in the grid's operating conditions and mitigate the effects of nonlinear behavior and other stability issues. They also need to consider the impact of grid-connected power converters on the overall operation of the power grid and ensure that these devices are integrated in a way that ensures the safe, reliable, and efficient operation of the grid.

Thus, a thorough understanding of the complex behavior and stability issues of grid-connected power electronic converters is essential for the successful integration of renewable energy sources into the power grid and for improving the efficiency and flexibility of the grid's operation. It enables researchers to design and control these devices effectively, develop advanced control strategies that can adapt to changes in the grid's operating conditions, and mitigate the effects of nonlinear behavior and other stability issues. It also ensures that these devices are integrated in a way that ensures the safe, reliable, and efficient operation of the power grid, ultimately leading to a more sustainable and reliable energy future.

Complex Behavior of Grid-Connected Power Electronics Systems is a comprehensive resource on the latest research in the field of power electronics and systems.

This book provides a detailed analysis of the complex behavior and stability issues of grid-connected power converters, which are essential for integrating renewable energy sources into the power grid and improving the efficiency and flexibility of the grid's operation. The book offers a multidisciplinary approach that combines expertise in circuit modeling, control theory, nonlinear system analysis, and power electronics system design. It covers topics such as bifurcation, nonlinear oscillations, synchronization, and stability of grid-connected power converters. Additionally, it highlights the latest research in these areas, including the development of advanced control strategies that can adapt to changes in the grid's operating conditions and mitigate the effects of nonlinear behavior and other stability issues. The book is an essential resource for researchers, engineers, and students working in the field of power electronics and power systems. It provides valuable insights into the design and control of grid-connected power converters for a more sustainable and reliable energy future. It offers a platform for researchers to share their latest findings and exchange ideas on how to address the challenges associated with grid-connected power converters. This fosters collaboration and innovation in the field, ultimately leading to the development of more efficient and reliable grid-connected power converters.

With its comprehensive coverage of the latest research in the field, this book serves as an indispensable resource for professionals engaged in power electronics and power systems. Our aim is for readers to find this book valuable not only as a comprehensive source of information but also as a technical guide for the development and implementation of efficient and reliable grid-connected power converters. We hope that it will provide the necessary knowledge and insights for professionals to navigate this dynamic field with confidence and success.

During the development of this book, we have had the privilege of being inspired by numerous individuals, to whom we extend our heartfelt gratitude. In particular, we would like to express our sincere appreciation to the members of our research group at City University of Hong Kong, with whom we have engaged in many fruitful discussions. We are especially grateful to Dong Liu and Charles Liu, whose valuable contributions have greatly influenced the ideas presented in this book. We would also like to extend special thanks to Meng Huang and his students at Wuhan University. Their assistance in conducting analyses, performing experimental work, and providing unwavering support to our research has been invaluable.

Last, but certainly not least, we would like to express our gratitude to all our research collaborators and students. Over the years, they have been a constant source of inspiration and support for our research endeavors.

Kowloon, Hong Kong — Jingxi Yang
Kowloon, Hong Kong — Chi Kong Tse
June 2024

Contents

Acronyms

AC	Alternating Current
CCM	Constant Current Mode
CFCT	Critical Fault Clearing Time
CPM	Constant Power Mode
DC	Direct Current
DVC	DC-Link Voltage Control
EQP	Equilibrium Point
ESR	Equivalent Series Resistance
GFLC	Grid-Following Converter
GFMC	Grid-Forming Converter
IBR	Inverter-Based Resource
LHP	Left-Half Plane
PCC	Point of Common Coupling
PLL	Phase-Locked Loop
PV	Photovoltaic
PWM	Pulse-Width Modulation
RoCoF	Rate of Change of Frequency
SCF	Short-Circuit Fault
SEP	Stable Equilibrium Point
SNB	Saddle-Node Bifurcation
SNPO	Saddle-Node Bifurcation of Periodic Orbits
SPO	Stable Periodic Orbit
UEP	Unstable Equilibrium Point
UPO	Unstable Periodic Orbit
VSC	Voltage Source Converter
ZOH	Zero-Order-Hold

Chapter 1
Introduction

To achieve net-zero emissions and combat global warming, the integration of renewable energy sources and power electronics into modern power systems is progressing rapidly. However, this transition poses two significant challenges. The first challenge arises from the intermittent nature of renewable energy sources, while the second stems from the inherent instability caused by the complex nonlinear behavior of power electronic converters [1–6]. In this book, our focus lies on understanding the intricate behavior of grid-connected converters and their impact on power system stability.

In this introductory chapter, we will begin by providing a brief overview of the roles of power electronics in modern power systems. Subsequently, we will delve into the fundamental control strategies employed by grid-forming converters, grid-following converters, active loads, and smart loads. Building upon this foundation, we will explore the control strategies used in multi-converter microgrid systems. Additionally, we will introduce the stability framework for converter-dominated systems and the averaging model for grid-connected converters and briefly discuss the relationship between the converters' complex behavior and power system stability. Finally, we will outline the structure and content of this book.

1.1 Roles of Power Electronics in Modern Power Systems

The electric power system stands as the largest and most intricate self-synchronized system in modern society. It encompasses multiple electricity generating stations, long-distance medium-/high-voltage transmission lines, and distribution systems that efficiently transfer power to individual consumers. In the past, fossil fuels served as the primary sources of energy. These fuels were converted into mechanical energy by prime movers, which then underwent further transformation into electrical energy through synchronous generators in the generating stations. However,

J. Yang, C. K. Tse, *Complex Behavior of Grid-Connected Power Electronics Systems*, https://doi.org/10.1007/978-3-031-69340-3_1

the urgent need to address the climate change crisis has prompted the widespread deployment of renewable energy sources, such as wind power and solar energy. These sources are now integrated into the power grids using power electronic converters as intermediaries. The rapid integration of renewable energy sources and power electronics has significantly reshaped the structure, fundamental characteristics, and dynamic behavior of our power systems. The importance of power electronics in modern power systems can be understood in three key aspects.

1.1.1 Electricity Generating Stations

For large-scale wind power generation systems, such as wind farms, there are two basic structures used to integrate wind power into medium- to high-voltage grids. The choice of structure depends on the type of generator employed. If a doubly fed induction generator is used, a partial-scale power converter with a back-to-back structure can be utilized. This converter, consisting of AC/DC and DC/AC conversion stages, has a power capacity of approximately 30% of the generator. In this setup, as depicted in Fig. 1.1a, the stator windings of the generator are directly connected to the grid, while the rotor windings are connected to the partial-scale power converter. This converter regulates the frequency and current of the generator rotor to maximize energy output. Additionally, to compensate for limited reactive power capability, a centralized static synchronous compensator is typically installed in wind farms. On the other hand, if a permanent magnet synchronous generator is employed, a full-scale power converter with a back-to-back structure can be used. In this configuration, as shown in Fig. 1.1b, the stator windings of the generator are connected to the grid through the full-scale power converter. This structure offers several advantages, including the avoidance of slip rings, simplified or eliminated gearbox, enhanced power and speed controllability, and increased flexibility in generating reactive power to support the grid. Consequently, the need for additional reactive power compensators is eliminated in this structure of wind farms. Moreover, due to the increasing power capacity of generators (expected to be 15 MW per generator in 2024 [7]), the multilevel technique, multiple paralleled converters, and multicell converter compatible with multi-winding generator may be adopted.

In constructing large-scale solar photovoltaic (PV) power generation systems, such as solar PV farms, it is common practice to connect a significant number of PV panels in series and parallel. This configuration aims to generate higher output voltage and power. There are two primary structures commonly used in solar PV farms. In Fig. 1.2a, we can observe the first typical structure. Here, multiple PV panels are connected in series to form what is known as a PV string. Each PV string is then linked to a local low-voltage DC bus through an individual DC/DC converter. This converter plays a crucial role in optimizing the energy generation of its respective PV string. Finally, a central DC/AC converter is employed to connect the local DC bus to the main AC power grid. In Fig. 1.2b, we can see the second

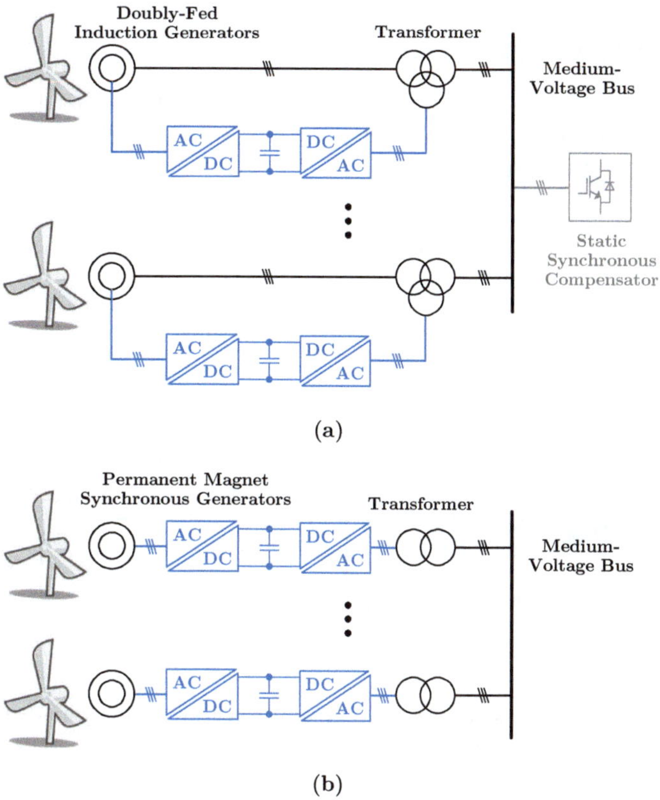

Fig. 1.1 Structures of large-scale wind power generation systems for (**a**) doubly fed induction generators and (**b**) permanent magnet synchronous generators

typical structure. In this case, numerous PV strings are connected in parallel directly to the local DC bus, without employing string converters. Subsequently, a central DC/AC converter is utilized to establish a connection between the local bus and the main power grid. This structure is commonly used in large open spaces to avoid mismatches in panel characteristics and to maximize the amount of harvested energy.

In order to address the unpredictability and fluctuation of wind and solar energy sources, it is advisable to incorporate additional energy storage equipment [7]. There are various techniques commonly used for energy storage, including pumped hydro storage, solar thermal power (utilizing thermal hot water or melted salts), hydrogen fuel cells, supercapacitors, and batteries (such as flow batteries, lead-acid batteries, and lithium-ion batteries). These energy storage devices can be strategically placed within wind and solar farms to mitigate the fluctuations in electricity generation. Figure 1.3a illustrates the presence of a shared DC bus within these farms, to which the energy storage devices are connected via DC/DC converters. Additionally, these

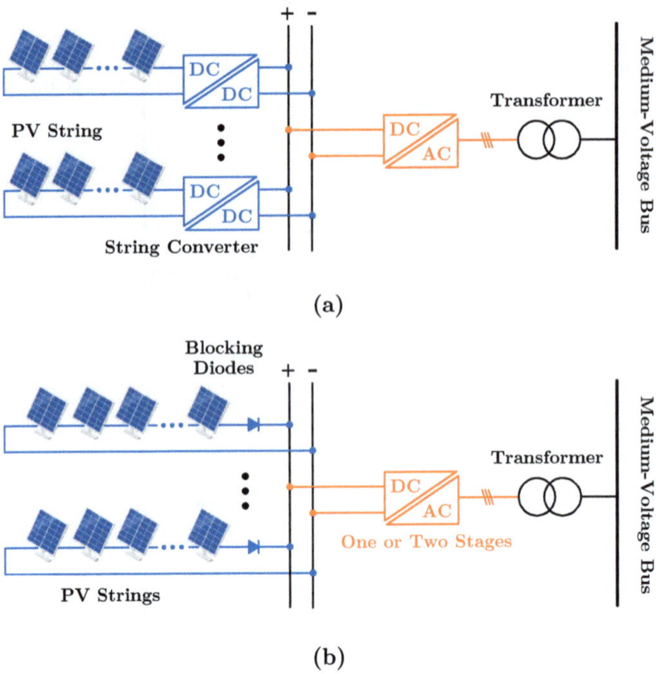

Fig. 1.2 Structures of large-scale solar power generation systems (**a**) using string converters and (**b**) without using string converters

devices can be organized into a centralized energy storage system, functioning as an independent power station to compensate for the variable wind and solar power within the grid. To achieve this, DC/DC and DC/AC converters are employed to link the energy storage devices to the primary AC power grid, as depicted in Fig. 1.3b.

1.1.2 Transmission Network

In high-voltage AC transmission systems, power electronic converters are employed within the AC transmission networks to regulate power flow, mitigate the negative effects of voltage dips, and compensate for harmonics and voltage imbalances. These converters include static synchronous compensators, static synchronous series compensators, and unified power flow controllers.

For high-voltage DC systems, terminal converters such as line-commutated converters (using thyristors) and modular multilevel converters are utilized. This technology is suitable for long-distance power transmission, offshore wind energy transportation, and interconnecting AC systems with varying frequencies [7, 8]. Figure 1.4 illustrates a typical high-voltage DC transmission system.

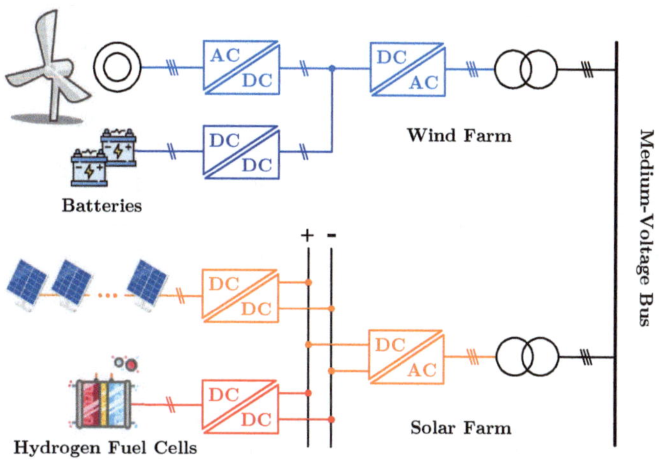

(a)

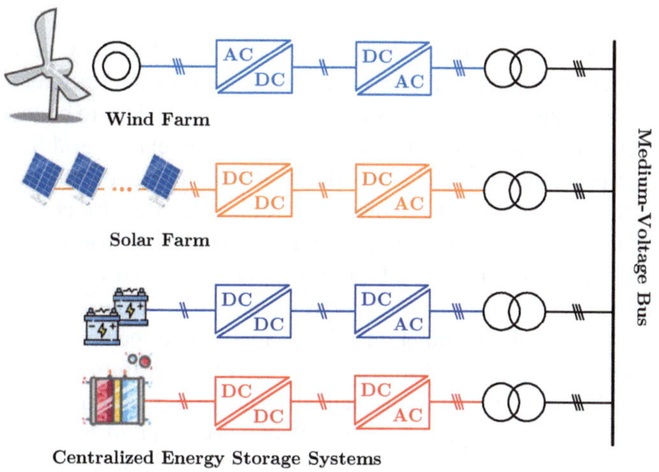

(b)

Fig. 1.3 Integration architecture for energy storage equipment in (**a**) wind/solar farms and (**b**) centralized energy storage systems

1.1.3 Distributed Systems

In the traditional distributed system, electricity is solely provided by the utility grid, and most power consumers consist of passive loads like electric heating systems, incandescent lamps, and induction motors. However, in the modern distributed system, there is a significant increase in inverter-based resources, electric vehicles, active loads, smart loads, and solid-state transformers. This shift completely

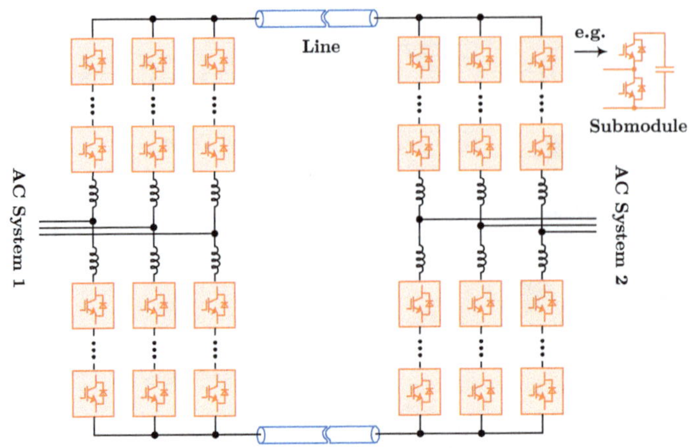

Fig. 1.4 High-voltage DC transmission system using modular multilevel converters

transforms the structure and topology of the distributed system, as depicted in Fig. 1.5. Inverter-based resources, for instance, serve as the interface for small-scale wind/solar power generation systems or energy storage devices to connect with the power grid [9]. These inverters can also play a role in regulating grid frequency and voltage while providing the necessary virtual inertia to support the grid. Electric vehicles, on the other hand, utilize power electronic converters for charging and can even contribute to power flow regulation through vehicle-to-grid techniques [10]. Active loads encompass a range of examples such as electric appliances, solid-state lighting, data centers, electrical drives, and smart manufacturing facilities. They employ pulse-width modulation (PWM) rectifiers or power-factor-correction converters as intermediaries to absorb constant active power from the grid [1]. Smart loads, comprising non-critical loads and electric springs, are able to effectively regulate bus voltage or frequency despite fluctuations caused by the intermittent nature of wind/solar power thanks to the electric spring, which is a power electronic converter [11, 12].

In addition, the interface converters mentioned earlier can be organized into clusters, forming individual AC or DC microgrids. These microgrids can then be interconnected to create networked microgrids. They have the capability to operate in either a grid-connected mode or an islanded mode [13, 14]. Furthermore, the integration of solid-state transformers offers a promising solution for connecting medium-voltage distribution power grids with low-voltage AC or DC microgrids. These transformers can also be utilized as efficient fast-charging equipment for electric vehicles. Typically, a solid-state transformer consists of a three-phase cascaded H-bridge AC/DC converter, multiple output-parallel DC/DC converters equipped with high-frequency transformers, and a low-voltage DC/AC converter [8].

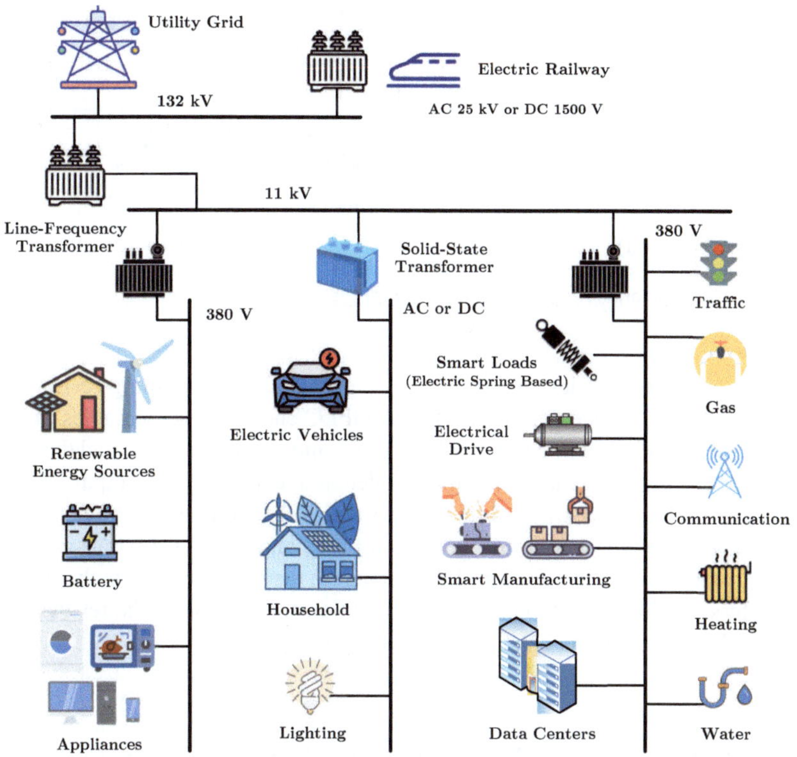

Fig. 1.5 Modern distribution systems. The typical voltage of transmission network is 132 kV, and the typical voltage of distribution network is 11 kV

1.2 Control of Grid-Connected Converters

On the one hand, power electronic converters play a crucial role in connecting renewable energy sources and energy storage devices to the power grid. When it comes to synchronization control, these converters can be classified into two main categories: grid-following converters (GFLCs) and grid-forming converters (GFMCs). On the other hand, interface power electronic converters are typically controlled as grid-following rectifiers to transmit power to loads. Each type of converter may deploy several control strategies with various structures and applicability, which will be thoroughly reviewed in this section.

Figure 1.6a and b illustrates the typical circuit topologies of the GFMC and GFLC, respectively. In the case of the GFMC, an LC-type filter is utilized, consisting of the inductance L_f, capacitance C_f, and series resistance r_f. Additionally, the GFMC incorporates various variables such as the DC-link voltage v_{dc}, PWM voltage v_o, current of L_f denoted as i_f, voltage of C_f represented by v_f, and the output current i_l of the GFMC. Moreover, the GFLC employs an LCL-type filter, which

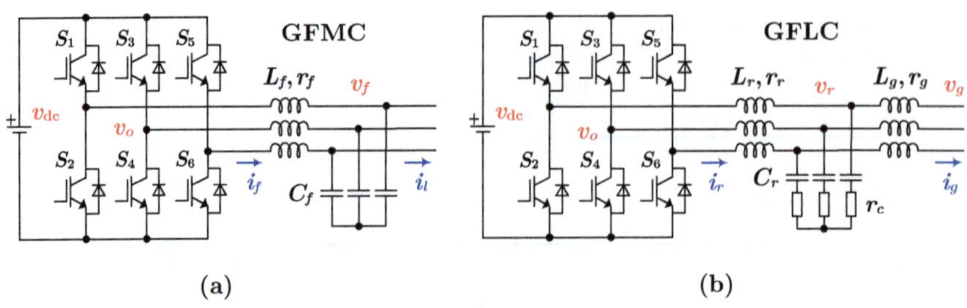

Fig. 1.6 Circuit topology of (**a**) GFMC and (**b**) GFLC

includes the converter-side inductance L_r, capacitance C_r, damping resistance r_c, and grid-side inductance L_g. Similarly, the GFLC encompasses variables like DC-link voltage v_{dc}, PWM voltage v_o, voltage of C_r and r_c labeled as v_r, and the current i_r flowing through L_r. The grid voltage is denoted as v_g, and the output current of the GFLC is represented by i_g. Both GFMCs and GFLCs rely on the filter capacitors to attenuate the PWM harmonics generated by v_o and prevent the transmission of these harmonics into the power grid. Moreover, in the case of GFLCs, the inclusion of r_c offers an effective and cost-efficient solution for mitigating the resonant peak of the LCL-type filter and enhancing system stability [15].

1.2.1 Inverter-Based Resources: Grid-Forming Control

1.2.1.1 Scenario 1: Constant DC-Link Voltage

The DC-link voltage of the GFMC can be kept constant by connecting it to a battery. As a result, the GFMC's primary task is to regulate the active and reactive powers on the AC side in order to maintain power balance within the AC-side network. Four common control schemes can be employed here.

(1) Droop Control
From Fig. 1.7a, we can observe the active power (p) and reactive power (q) outputs of the GFMC. Additionally, we have the output frequency (ω), phase angle (θ), reference voltage ($v_{f,\text{ref}}$), reference magnitude of voltage ($v_{fm,\text{ref}}$), and reference current ($i_{f,\text{ref}}$). Furthermore, P_0, Q_0, ω_0, and V_0 represent the nominal values of p, q, ω, and $v_{fm,\text{ref}}$, respectively. Moreover, ω_p and ω_q are the cutoff frequencies of the low-pass filters, while m_p and n_q are the droop coefficients. From Fig. 1.7a,

$$\omega = \omega_0 + \frac{\omega_p}{s + \omega_p} \times m_p(P_0 - p), \quad \theta = \frac{1}{s} \times \omega \tag{1.1}$$

$$v_{fm,\text{ref}} = V_0 + \frac{\omega_q}{s + \omega_q} \times n_q(Q_0 - q) \tag{1.2}$$

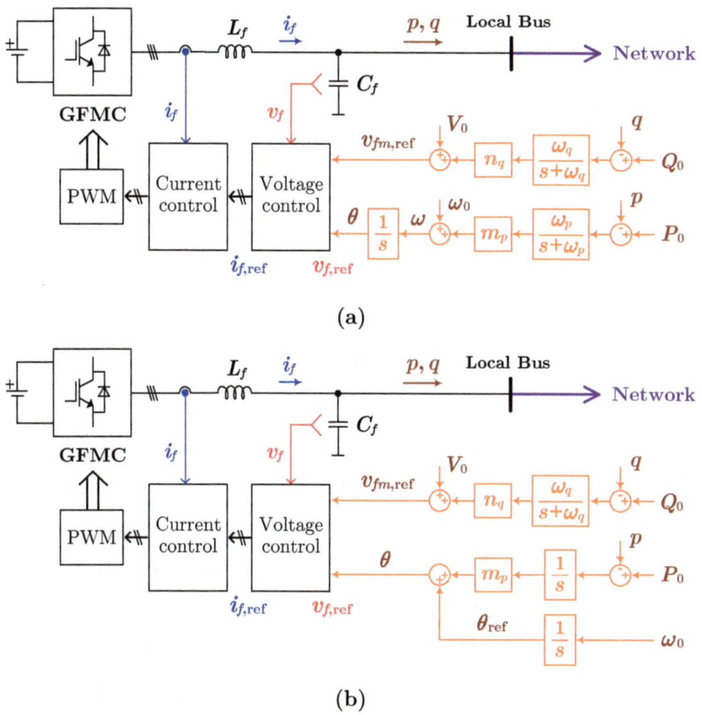

Fig. 1.7 Grid-forming control strategies: (**a**) droop control and (**b**) power synchronization control. (**c**) Virtual synchronous generator control, and (**d**) virtual oscillator control

In this case, the low-pass filter that corresponds to ω_p serves to slow down the rate of change of ω and mitigate the deviation of ω during the transient process. This effectively creates a virtual inertia for the GFMC. Thus, a smaller value for ω_p indicates a larger virtual inertia.

(2) Power Synchronization Control

The power synchronization control was initially developed as an additional control mechanism to guarantee the stable functioning of grid-connected converters, particularly in high-voltage DC applications under weak grid conditions [16]. Figure 1.7b shows the block diagram of the power synchronization control. From Fig. 1.7b,

$$\theta = \frac{1}{s} \times m_p(P_0 - p) + \theta_{\text{ref}}, \quad \theta_{\text{ref}} = \frac{1}{s} \times \omega_0 \tag{1.3}$$

and the generation of $v_{fm,\text{ref}}$ follows (1.2). Comparing (1.1) with (1.3), we observe that no low-pass filter is present in (1.3). In other words, the power synchronization control cannot provide the virtual inertia.

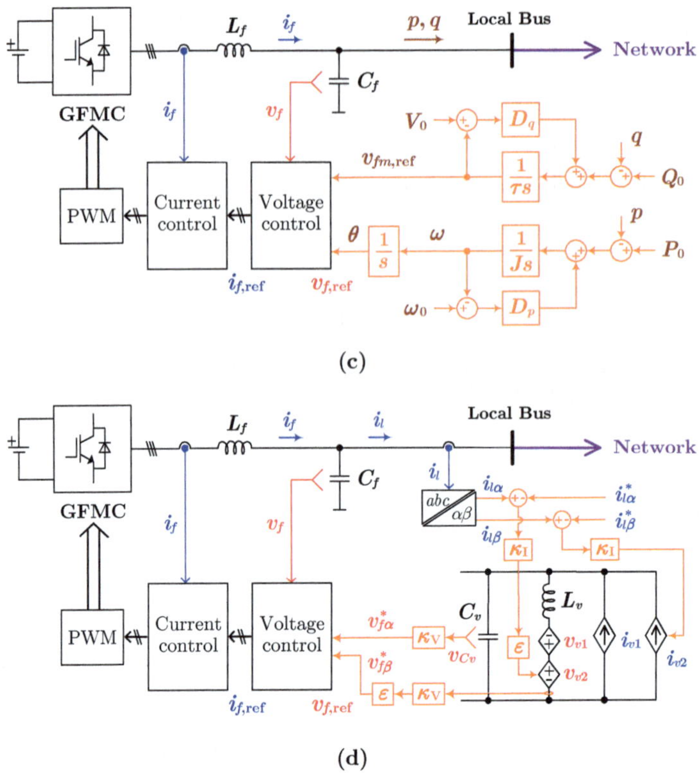

Fig. 1.7 (continued)

(3) Virtual Synchronous Generator Control

The concept of virtual synchronous generator control was introduced to replicate the characteristics of synchronous generators [17, 18]. In Fig. 1.7c, J denotes the inertia of the rotor in the synchronous generator, D_p represents the damping factor, τ signifies the coefficient of the integrator in the reactive power regulation loop, and D_q represents the voltage-drooping coefficient. From Fig. 1.7c,

$$\omega = \frac{D_p/J}{s + D_p/J}\omega_0 + \frac{D_p/J}{s + D_p/J} \times \frac{1}{D_p}(P_0 - p), \quad \theta = \frac{1}{s} \times \omega \tag{1.4}$$

$$v_{fm,\mathrm{ref}} = \frac{D_q/\tau}{s + D_q/\tau}V_0 + \frac{D_q/\tau}{s + D_q/\tau} \times \frac{1}{D_q}(Q_0 - q) \tag{1.5}$$

Comparing (1.1) with (1.4), and comparing (1.2) with (1.5), we get

$$\omega_p = \frac{D_p}{J}, \quad m_p = \frac{1}{D_p}, \quad \omega_q = \frac{D_q}{\tau}, \quad n_q = \frac{1}{D_q} \tag{1.6}$$

Therefore, the virtual synchronous generator control and the droop control are essentially the same.

(4) Virtual Oscillator Control

The concept of virtual oscillator control was introduced with the aim of enabling the GFMC to function as a nonlinear oscillator. This allows for the simulation of various types of oscillators, such as the dead-zone oscillator, Van der Pol oscillator, and Andronov-Hopf oscillator. These ideas are rooted in the theory of coupled nonlinear oscillators and the associated phenomena of synchrony [19, 20]. To illustrate the control mechanism of the virtual oscillator, we present the corresponding block diagram based on an Andronov-Hopf oscillator in Fig. 1.7d. The oscillator comprises an inductor denoted as L_v and a capacitor denoted as C_v. The parameters $\omega_0 = 1/\sqrt{L_v C_v}$ and $\varepsilon = \sqrt{L_v/C_v}$ characterize the behavior of the oscillator. Additionally, the current flowing through L_v is represented by i_{Lv}, while the voltage across C_v is denoted as v_{Cv}. Moreover, the oscillator incorporates two dependent voltage sources and two dependent current sources to complete its functionality, i.e.,

$$v_{v1} = \frac{\xi}{\omega_0}(2X_{\mathrm{osc,n}}^2 - \mathbf{x_{osc}}^2)\varepsilon i_{Lv} \tag{1.7}$$

$$i_{v1} = \frac{\xi}{\varepsilon\omega_0}(2X_{\mathrm{osc,n}}^2 - \mathbf{x_{osc}}^2)v_{Cv} \tag{1.8}$$

where ξ is a control parameter related to the convergence speed of the oscillator, $\mathbf{x_{osc}} = \sqrt{v_{Cv}^2 + (\varepsilon i_{Lv})^2}$, and $\sqrt{2}X_{\mathrm{osc,n}}$ is the nominal value of $\mathbf{x_{osc}}$.

In Fig. 1.7d, i_l is the output current of the GFMC, and its corresponding variables in $\alpha\beta$-frame are $i_{l\alpha}$ and $i_{l\beta}$. The current references of $i_{l\alpha}$ and $i_{l\beta}$ are

$$\begin{bmatrix} i_{l\alpha}^* \\ i_{l\beta}^* \end{bmatrix} = \frac{2}{3(v_{f\alpha}^2 + v_{f\beta}^2)} \begin{bmatrix} v_{f\alpha} & v_{f\beta} \\ v_{f\beta} & -v_{f\alpha} \end{bmatrix} \begin{bmatrix} P_0 \\ Q_0 \end{bmatrix} \tag{1.9}$$

where $v_{f\alpha}$ and $v_{f\beta}$ are the corresponding variables of v_f in $\alpha\beta$-frame. Therefore, the other dependent voltage and current sources are

$$v_{v2} = \varepsilon\kappa_{\mathrm{I}}(i_{l\alpha} - i_{l\alpha}^*) \tag{1.10}$$

$$i_{v2} = \kappa_{\mathrm{I}}(i_{l\beta} - i_{l\beta}^*) \tag{1.11}$$

and the generated voltage references are

$$v_{f\alpha}^* = \kappa_{\mathrm{V}} v_{Cv} \tag{1.12}$$

$$v_{f\beta}^* = \kappa_{\mathrm{V}}\varepsilon i_{Lv} \tag{1.13}$$

where κ_{I} and κ_{V} are the control parameters. It is assumed that $v_{f\alpha} \approx v_{f\alpha}^*$, and $v_{f\beta} \approx v_{f\beta}^*$. Combining (1.7) to (1.13), we get

$$\omega = \omega_0 + \frac{\kappa_{\mathrm{I}}\kappa_{\mathrm{V}}}{3C_v v_{fm}^2} \times (P_0 - p) \tag{1.14}$$

$$\frac{\mathrm{d}v_{fm,\mathrm{ref}}}{\mathrm{d}t} \approx \frac{\mathrm{d}v_{fm}}{\mathrm{d}t} = \frac{\xi}{\kappa_{\mathrm{V}}^2} v_{fm}(2V_{\mathrm{nom}}^2 - 2v_{fm}^2) + \frac{\kappa_{\mathrm{I}}\kappa_{\mathrm{V}}}{3C_v v_{fm}} \times (Q_0 - q) \tag{1.15}$$

where $V_{\mathrm{nom}} = \kappa_{\mathrm{V}} \times X_{\mathrm{osc,n}}$.

Comparing (1.1) with (1.14), we see that

$$m_p = \frac{\kappa_{\mathrm{I}}\kappa_{\mathrm{V}}}{3C_v v_{fm}^2} \tag{1.16}$$

implying that in terms of the active power sharing, the virtual oscillator control is similar to the droop control. However, for the virtual oscillator control, there is no low-pass filter, implying that the virtual oscillator control cannot provide any virtual inertia. To address this problem, a proportional and quasi-resonant controller was proposed to regulate $(i_{l\alpha} - i_{l\alpha}^*)$ and $(i_{l\beta} - i_{l\beta}^*)$, effectively supplying the virtual inertia to the grid [21]. For (1.15), in the steady state,

$$v_{fm,\mathrm{ref}} \approx v_{fm} = \sqrt{\frac{1}{2}\left(V_{\mathrm{nom}}^2 + \sqrt{V_{\mathrm{nom}}^4 + \frac{2\kappa_{\mathrm{I}}\kappa_{\mathrm{V}}^3}{3C_v\xi}(Q_0 - q)}\right)} \tag{1.17}$$

which is totally different from (1.2). Actually, the slope of the reactive power-voltage droop in (1.17) is not constant, but varies nonlinearly with q.

1.2.1.2 Scenario 2: Variable DC-Link Voltage

If the DC-link of the GFMC is connected to the wind/solar generation systems, we must consider the power balance between the GFMC and the converters that process the wind/solar power [22]. The control scheme is illustrated in Fig. 1.8. We have multiple wind/solar generation systems with their output terminals connected in parallel. Let us denote the output power of the ℓ-th wind/solar generation system by $p_{\mathrm{dc}\ell}$, where $\ell = 1, 2, \cdots, N$. By utilizing maximum power point tracking algorithms, we can determine the maximum available power output of the ℓ-th wind/solar generation system as $P_{\mathrm{max}\ell}$. This value serves as the upper limit for $p_{\mathrm{dc}\ell}$. For the ℓ-th wind/solar generation system,

$$p_{\mathrm{dc}\ell} = \min\{P_{\mathrm{max}\ell},\ P_{d0\ell} + D_{d\ell}(V_{d0\ell} - v_{\mathrm{dc}\ell})\} \tag{1.18}$$

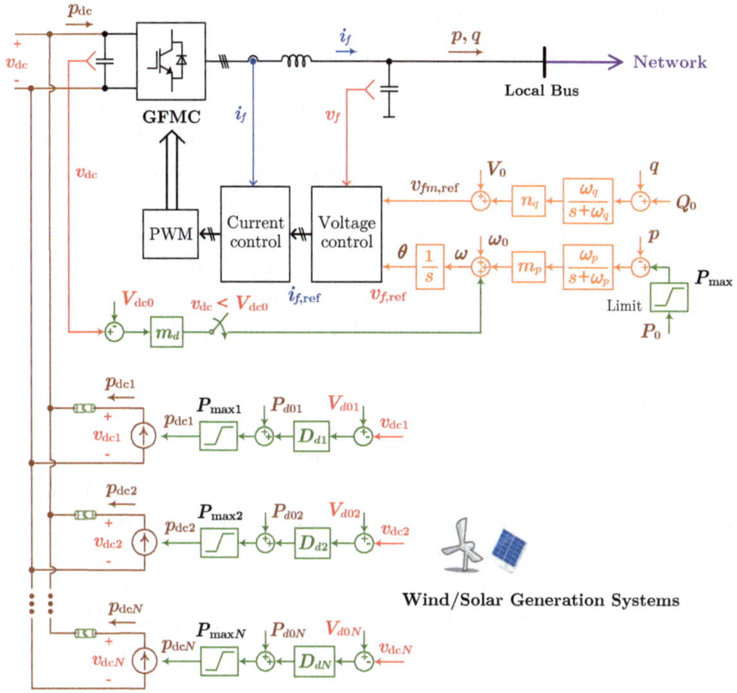

Fig. 1.8 Grid-forming control strategy considering the uncertainty of renewable energy sources

where $v_{dc\ell}$ is the output dc voltage, $D_{d\ell}$ is the droop coefficient, and $P_{d0\ell}$ and $V_{d0\ell}$ are the nominal values of $p_{dc\ell}$ and $v_{dc\ell}$, respectively.

For the GFMC,

$$p_{dc} = \sum_{k=1}^{N} p_{dck}, \quad P_{max} = \sum_{k=1}^{N} P_{maxk}$$

Here, P_{max} serves as the upper limit of P_0. From Fig. 1.8,

$$\omega = \omega_0 + \frac{\omega_p}{s+\omega_p} \times m_p(\min\{P_{max}, P_0\} - p) + m_d \times \min\{0, (v_{dc} - V_{dc0})\} \tag{1.19}$$

where m_d is the droop coefficient.

If p is greater than p_{dc}, both v_{dc} and $v_{dc\ell}$ will decrease and may become smaller than V_{dc0} and $V_{d0\ell}$. In the case of the ℓ-th wind/solar generation system, $p_{dc\ell}$ will increase according to (1.18), resulting in an increase in p_{dc}. However, for the GFMC, p will decrease according to (1.19). Consequently, a balance will be achieved between p and p_{dc}.

1.2.1.3 Extra Problem: Limiting Output Power

For the basic grid-forming control schemes illustrated in Fig. 1.7, the GFMC's output active and reactive powers are determined by the actual load demand. However, during load power fluctuations or short-circuit faults (SCFs), the GFMC's output power may exceed the converter's rated capacity, whether in transient or steady state.

To restrict the output power during transient processes, we can employ adaptive virtual impedance, which is integrated into the generation of voltage-control-loop references. This virtual impedance becomes significantly large when the output power surpasses a predetermined threshold. Consequently, the output power of the GFMC is confined to an acceptable range [23, 24].

On the other hand, to restrict the output power during the steady state, we can switch the control mode of the grid-forming system to the grid-following mode. In this configuration, the active power-frequency droop is replaced by the phase-locked loop (PLL), and the voltage closed-loop feedback control is replaced by power closed-loop feedback control. Both of these mechanisms generate the current references for the current control [25].

In practical applications, when the output power reaches the first threshold, the adaptive virtual impedance is activated to limit the output power. However, if the output power continues to increase and reaches the second threshold, the grid-forming control mode is transitioned to the grid-following control mode.

1.2.2 Inverter-Based Resources: Grid-Following Control

1.2.2.1 Scenario 1: Constant DC-Link Voltage

In this particular scenario, the DC-link voltage of the GFLC remains constant. As a result, there is no requirement for DC-link voltage control. The primary focus of the GFLC lies in regulating the AC-side variables. To maintain synchronization, there are two commonly employed control strategies, namely, vector-current control and direct-power control.

(1) Vector-Current Control

As depicted in Fig. 1.9a, the variables i_{gd} and i_{gq} correspond to i_g in the dq-frame, while v_{gd} and v_{gq} correspond to v_g. Additionally, p^* represents the reference for the output active power, q^* represents the reference for the output reactive power, i^*_{gd} is the reference for i_{gd}, and i^*_{gq} is the reference for i_{gq}. In the context of the PLL, ω denotes the output frequency, θ represents the output phase angle, ω_0 corresponds to the nominal value of ω, and ω_γ represents the cutoff frequency of the low-pass filter, which is applied to v_{gd}. Furthermore, κ_p and κ_i are the proportional and integral coefficients of the PLL, respectively, while β_p and β_i are the proportional and integral coefficients of the current loop control, respectively.

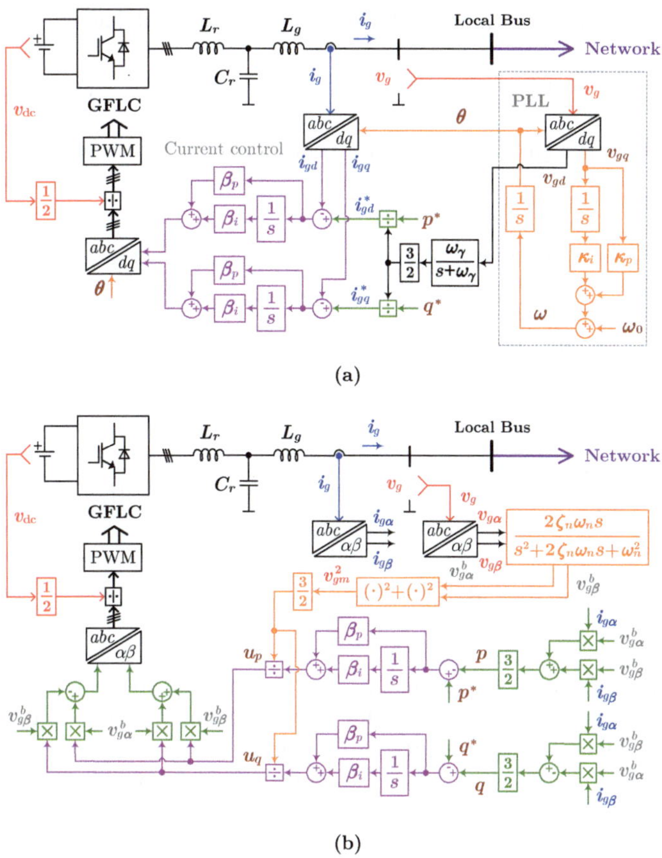

Fig. 1.9 Grid-following control strategies: (**a**) PLL-based vector-current control (**b**) direct-power control without using PLL

Referring to Fig. 1.9a, the PLL is utilized to determine the phase angle θ, enabling the transformation between the dq-frame and the abc-frame. Consequently, the current i_g can synchronize with v_g, ensuring that the output powers conform to the references p^* and q^*.

(2) Direct-Power Control

In contrast to vector-current control, direct-power control does not involve a PLL. Additionally, the active power (p) and reactive power (q) are calculated using current and voltage variables in the $\alpha\beta$-frame [26]. As shown in Fig. 1.9b, the $\alpha\beta$-frame variables for v_g are $v_{g\alpha}$ and $v_{g\beta}$, while for i_g, they are $i_{g\alpha}$ and $i_{g\beta}$. Furthermore, $v^b_{g\alpha}$ and $v^b_{g\beta}$ are obtained by applying a second-order low-pass filter to $v_{g\alpha}$ and $v_{g\beta}$, respectively. The cutoff frequency and damping coefficient of the filter are denoted as ω_n and ζ_n, respectively. Additionally, v_{gm} represents the magnitude of v_g. The relationship between vector-current control and direct-power control is elucidated as follows. To simplify the analysis, it is assumed that

$$v_{g\alpha}^b = v_{gm}\cos\theta, \quad v_{g\beta}^b = v_{gm}\sin\theta, \quad i_{g\alpha} = i_{gm}\cos(\theta+\phi), \quad i_{g\beta} = i_{gm}\sin(\theta+\phi)$$

where i_{gm} is the magnitude of i_g, and ϕ is the phase difference between i_g and v_g. Therefore,

$$p = \frac{3}{2}(v_{g\alpha}^b i_{g\alpha} + v_{g\beta}^b i_{g\beta}) = \frac{3}{2}v_{gm}i_{gm}\cos\phi \tag{1.20}$$

$$q = \frac{3}{2}(-v_{g\alpha}^b i_{g\beta} + v_{g\beta}^b i_{g\alpha}) = -\frac{3}{2}v_{gm}i_{gm}\sin\phi \tag{1.21}$$

From Fig. 1.9b,

$$\frac{u_p}{1.5v_{gm}^2} = \frac{1}{1.5v_{gm}^2}(\beta_p + \beta_i\frac{1}{s})(p^* - p) = \frac{1}{v_{gm}}(\beta_p + \beta_i\frac{1}{s})(i_{gd}^* - i_{gd}) \tag{1.22}$$

$$\frac{u_q}{1.5v_{gm}^2} = \frac{1}{1.5v_{gm}^2}(\beta_p + \beta_i\frac{1}{s})(-q^* + q) = \frac{1}{v_{gm}}(\beta_p + \beta_i\frac{1}{s})(i_{gq}^* - i_{gq}) \tag{1.23}$$

Here, the term $1/v_{gm}$ is used to nominalize $v_{g\alpha}^b$ and $v_{g\beta}^b$, so that $u_p/(1.5v_{gm})$ and $u_q/(1.5v_{gm})$ can be transformed to the corresponding variables in $\alpha\beta$-frame. In other words, there is an equivalent transformation from dq-frame to $\alpha\beta$-frame, and $u_p/(1.5v_{gm})$ and $u_q/(1.5v_{gm})$ are in the equivalent dq-frame, i.e.,

$$\frac{u_p}{1.5v_{gm}} = (\beta_p + \beta_i\frac{1}{s})(i_{gd}^* - i_{gd}) \tag{1.24}$$

$$\frac{u_q}{1.5v_{gm}} = (\beta_p + \beta_i\frac{1}{s})(i_{gq}^* - i_{gq}) \tag{1.25}$$

which are the same control principle as the vector-current control, as shown in Fig. 1.9a. However, due to the absence of PLL in the direct-power control, the GFLC may show more complicated nonlinear dynamics, compared to the vector-current controlled GFLC [27].

(3) Frequency-Active Power Droop Control

In the aforementioned two control schemes, both active power p and its reference remain constant, even when the grid frequency experiences a dip. In order to enhance the grid robustness, the GFLC should participate in the regulation of the grid frequency. Therefore, the frequency-active power droop control can be incorporated in the aforementioned control schemes [28]. For example, a vector-current control using the frequency-active power droop is shown in Fig. 1.10. Here, the grid frequency ω is estimated through the PLL and used for the frequency-active

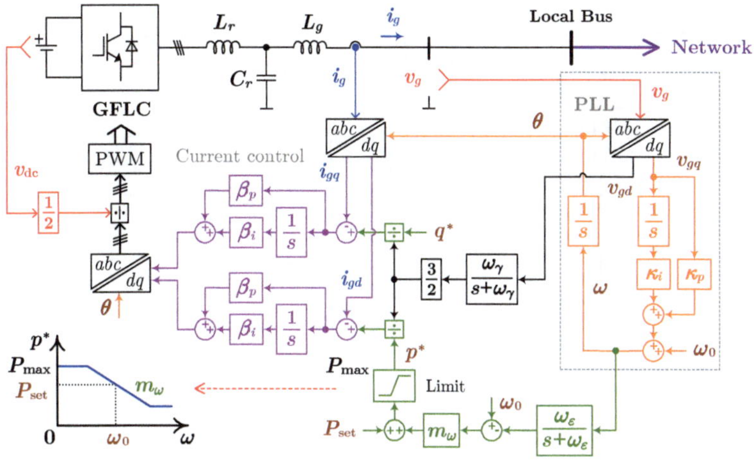

Fig. 1.10 Grid-following control scheme based on the vector-current control and the frequency-active power droop control

power droop after applying low-pass filtering. The cutoff frequency of the low-pass filter is denoted as ω_ε. Besides, m_ω is the droop coefficient, and P_{set} is the nominal value of p^*. From Fig. 1.10,

$$p^* = \min\left\{P_{max},\ P_{set} + m_\omega\left(\omega_0 - \frac{\omega_\varepsilon}{s+\omega_\varepsilon}\omega\right)\right\} \tag{1.26}$$

where P_{max} is the upper limit of p^*. Therefore, $p^* = P_{set}$ in the steady state, but p^* may significantly increase when ω exhibits a dip in the transient process.

1.2.2.2 Scenario 2: Variable DC-Link Voltage

In this scenario, the DC-link voltage is not constant, and the DC-link of GFLC is connected to one or several DC power sources (e.g., wind/solar generation systems). Depending on the control objectives for the DC power sources, there are two distinct control schemes for the GFLC.

(1) DC-Link-Capacitor-Based Virtual Inertia Control

As shown in Fig. 1.11, constant DC power p_{dc} is injected into the DC-link of the GFLC. In practice, p_{dc} may be generated from the wind/solar generation systems using a maximum power point tracking algorithm. The DC-link voltage v_{dc} is regulated by the GFLC to follow its reference v_{dc}^* [29]. Moreover, the frequency deviation $(\omega - \omega_0)$ is used as an additional term of v_{dc}^* after applying low-pass filtering and multiplying the coefficient k_ω. From Fig. 1.11, we have

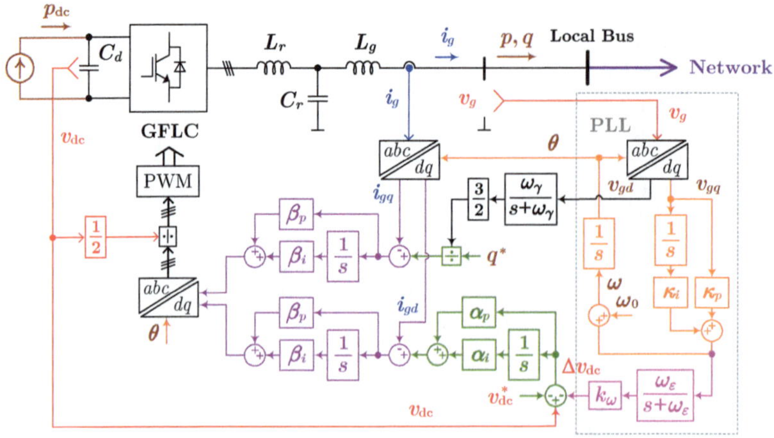

Fig. 1.11 GFLC using the DC-link-capacitor-based virtual inertia control

$$\Delta v_{dc} = v_{dc} - v_{dc}^* - \frac{\omega_\varepsilon}{s + \omega_\varepsilon} k_\omega(\omega - \omega_0) \tag{1.27}$$

which represents the input variable of the voltage controller. The proportional and integral gains of the voltage controller are α_p and α_i, respectively. From (1.27), $\omega = \omega_0$ in the steady state. Therefore, $v_{dc} = v_{dc}^*$ eventually.

Furthermore, assuming that the GFLC is powerless, the DC-link equation is

$$C_d v_{dc} \frac{dv_{dc}}{dt} = p_{dc} - p \tag{1.28}$$

where C_d is the DC-link capacitance. Suppose the voltage control loop operates perfectly. Then, according to (1.27),

$$v_{dc} = v_{dc}^* + k_\omega(\omega - \omega_0). \tag{1.29}$$

Combining (1.28) and (1.29) gives

$$C_d v_{dc}^* k_\omega \frac{d\omega}{dt} = p_{dc} - p. \tag{1.30}$$

Thus, the GFLC is able to participate in regulating the grid frequency and to provide a virtual inertia, which is $C_d v_{dc}^* k_\omega$. The physical meaning is that when the grid frequency ω exhibits a dip, v_{dc} decreases based on the control principle shown in (1.27). Therefore, $dv_{dc}/dt < 0$, leading to energy release from the DC-link capacitor. As a result, p increases according to (1.28), which is beneficial to preventing further drop of ω. For this control scheme, large C_d and/or v_{dc}^* and/or k_ω implies large virtual inertia. In practice, a supercapacitor may be used to guarantee sufficient virtual inertia.

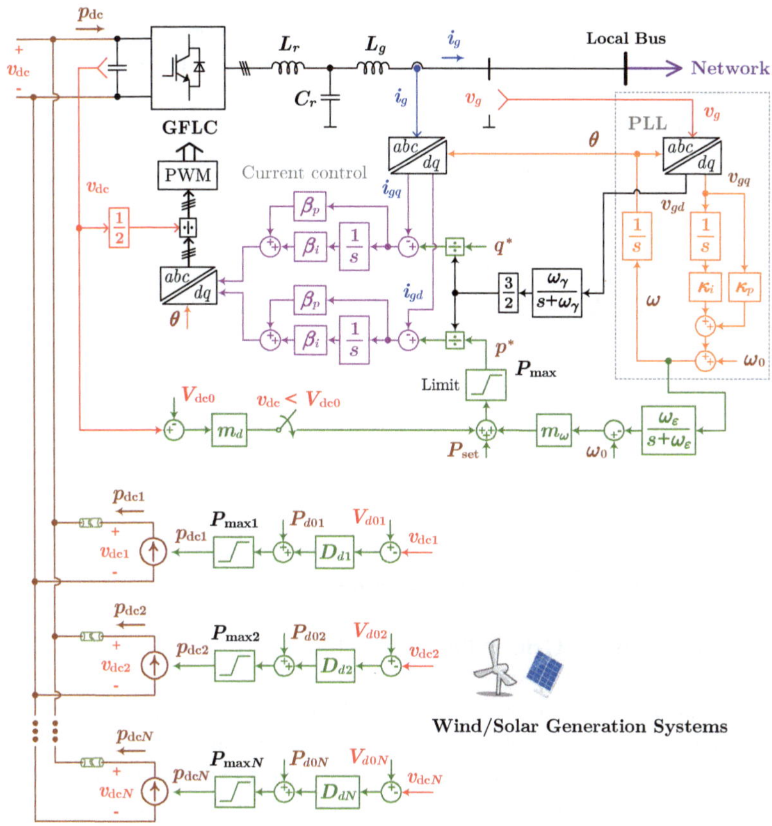

Fig. 1.12 GFLC using the frequency-active power and voltage-active power droop controls

(2) Frequency-Active Power and Voltage-Active Power Droop Controls
As shown in Fig. 1.12, the DC power sources (e.g., wind/solar generation systems) do not inject maximal power to the DC-link of GFLC, but an amount of power according to the voltage-power droop, as given in (1.18). For the GFLC, another DC-link voltage-active power droop control is adopted. These DC-link droop controls are aimed to maintain the DC-link power balance. Moreover, for the GFLC, the grid frequency-active power droop control is used to provide active power support in the event of a grid frequency dip. Referring to Fig. 1.12,

$$p^* = \min\left\{P_{\max},\ P_{\text{set}} + m_\omega(\omega_0 - \frac{\omega_\varepsilon}{s+\omega_\varepsilon}\omega) + m_d \times \min\{0, (v_{\text{dc}} - V_{\text{dc0}})\}\right\} \tag{1.31}$$

Contrary to the previous control scheme shown in Fig. 1.11, extra active power to support the grid is generated directly from the DC power sources.

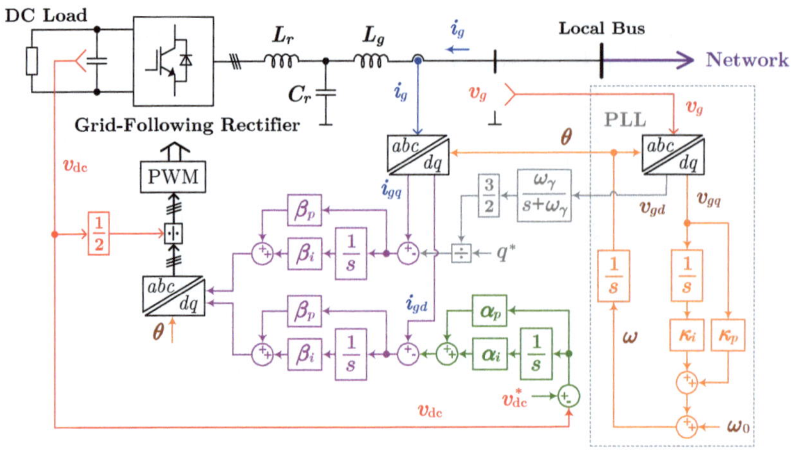

Fig. 1.13 Grid-following rectifier

1.2.3 Active Loads and Smart Loads

1.2.3.1 Active Loads: Grid-Following Rectifiers

In modern power distribution systems, there are numerous active loads. Unlike conventional passive loads, active loads use power electronic converters as intermediaries to connect to the grid. In practice, active loads include various appliances, solid-state lighting systems, electric vehicle chargers, etc. When a distributed energy storage device is charged, it is also an active load. For active loads, the interface converters are usually controlled as grid-following rectifiers [30], as shown in Fig. 1.13. Usually, such grid-following rectifiers do not participate in the reactive power regulation, as it may increase the power capacity of the rectifier. The DC-link voltage is regulated to follow its reference through a closed-loop feedback control. In this case, constant active power is absorbed by the grid-following rectifier.

1.2.3.2 Smart Loads: Electric Springs

A smart load is made up of an electric spring and a non-critical load (e.g., voltage-dependent load), as shown in Fig. 1.14. Here, the electric spring is used to participate in the regulation of the grid frequency [11, 12]. For a voltage-dependent load such as a constant resistive load, the active power consumption is

$$p_L = \frac{3}{2}\frac{v_{Lm}^2}{R_{\mathrm{VDL}}} \tag{1.32}$$

where v_{Lm} is the voltage across the load and R_{VDL} is the load resistance.

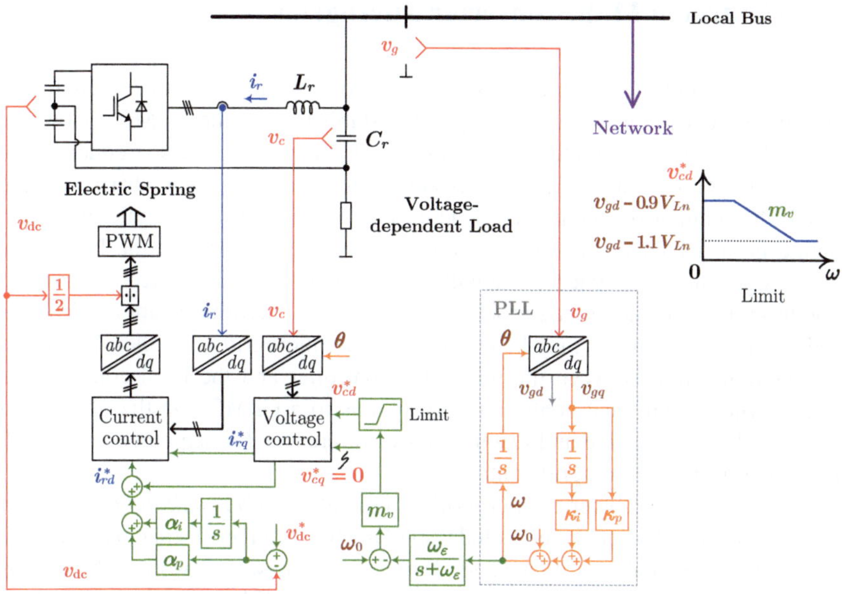

Fig. 1.14 Electric spring embedded with voltage-dependent load

For the electric spring, its output voltage magnitude is

$$v_{cm} \approx v_{cd}^* = m_v(\omega_0 - \omega) \tag{1.33}$$

where m_v is the droop coefficient, v_c is the voltage of C_r, v_{cm} is the magnitude of v_c, and v_{cd}^* and v_{cq}^* are the references of the d-axis and q-axis components of v_c, respectively. Besides, for (1.33), the upper limit of v_{cd}^* is $(v_{gd} - 0.9V_{Ln})$, whereas the lower limit of v_{cd}^* is $(v_{gd} - 1.1V_{Ln})$, where v_{gd} is the d-axis component of v_g, and V_{Ln} is the nominal value of v_{Lm}. From Fig. 1.14, we get

$$v_{Lm}^2 + v_{cm}^2 = v_{gm}^2 \tag{1.34}$$

where v_{gm} is the magnitude of v_g, and $v_{gm} \approx v_{gd}$.

Combining (1.32), (1.33), and (1.34), we get

$$p_L = \frac{3}{2R_{\text{VDL}}}(v_{gm}^2 - (m_v(\omega_0 - \omega))^2) \tag{1.35}$$

Thus, when ω declines, p_L decreases to reduce the power demand, which mitigates the drop of ω.

1.3 Control of Multi-converter Microgrids

A number of GFMCs and GFLCs may cluster together to form a microgrid. The microgrid's frequency and voltage are established by the GFMCs. However, due to the intrinsic characteristics of the local droop controllers, frequency and voltage deviations may exist [13]. Besides, the reactive power sharing among the GFMCs cannot be automatically achieved [31]. Moreover, the microgrid may operate in an islanded mode or a grid-connected mode [13]. For the latter case, the microgrid needs to synchronize with the utility grid before connecting. In order to address the aforementioned problems, extra control algorithms are needed. Part or all of these algorithms may be executed in the microgrid controller. Therefore, the communication between the microgrid controller and the local controllers is needed, where each local controller corresponds to a GFMC. On the other hand, for the GFLCs, the active and reactive power references (or setpoints) are usually dispatched from the microgrid controller.

Hence, the multi-converter microgrid is basically a *cyber-multitudinal-physical system* [32]. The physical layer includes (1) GFMCs and their local controllers, (2) GFLCs and their local controllers, and (3) microgrid's power network and the loads. The cyber layer includes the microgrid controller, and the communication links between different controllers (i.e., between different local controllers, and between the microgrid controller and all/specific local controllers). The word "multitudinal" indicates that the nodes/components in this system are heterogeneous, i.e., the types, parameters, and dynamic characteristics of the components are distinctly different. Such systems may display some new features, comparing with pure physical systems with homogeneous components.

Furthermore, a number of microgrids may cluster into a networked-microgrid system [14]. Each microgrid can be regarded as an independent unit using a grid-forming or grid-following control. From this point of view, the local-level and higher-level controls of the microgrids are similar to those of GFMCs and GFLCs in a microgrid.

1.3.1 Hierarchical Control Structure

The hierarchical control structure of a multi-converter microgrid is shown in Fig. 1.15 [13]. There are three levels from the bottom to the top:

1. Primary control level: The grid-forming and grid-following controls presented in the last section belong to this level. All of the control algorithms in this level are executed in the local controllers. There is no communication between different controllers in this level.
2. Secondary control level: In this level, the frequency and voltage of the microgrid are restored to their reference values, and the reactive power sharing is achieved among different GFMCs. Moreover, the overall generation cost of the microgrid

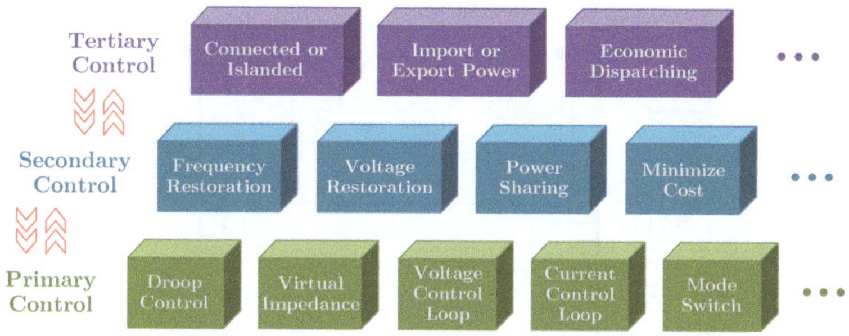

Fig. 1.15 Hierarchical control levels of a multi-converter microgrid

may be minimized by adjusting the active-power droop coefficients to achieve the optimal power distribution. In this case, an active power sharing control may be also needed.

3. Tertiary control level: There are several tasks in this level. The first task is to dispatch the active and reactive power references/setpoints and select the operation mode of specific converters (e.g., grid-forming control mode when the battery is discharged, and grid-following control mode when the battery is charged). The second task is to determine the operating mode of the microgrid (i.e., grid-connected mode or islanded mode) and proceed with seamless transitions between these modes. The third task is to control the power import/export of the microgrid.

1.3.2 Secondary Control

Depending on the topologies of the communication networks, the secondary control structures can be classified as decentralized control structure, centralized control structure, and distributed control structure [33]. The decentralized control structure is shown in Fig. 1.16a. In this structure, all of the secondary control algorithms are executed in the local controllers, and there is no communication between different controllers. The centralized control structure is shown in Fig. 1.16b. In this structure, all local controllers receive the corresponding control signals from the microgrid controller through the communication links. However, there is no communication link between different local controllers. The distributed control structure is shown in Fig. 1.16c. In this structure, each local controller communicates with its neighbor local controllers, and only a small fraction of local controllers is selected to communicate with the microgrid controller.

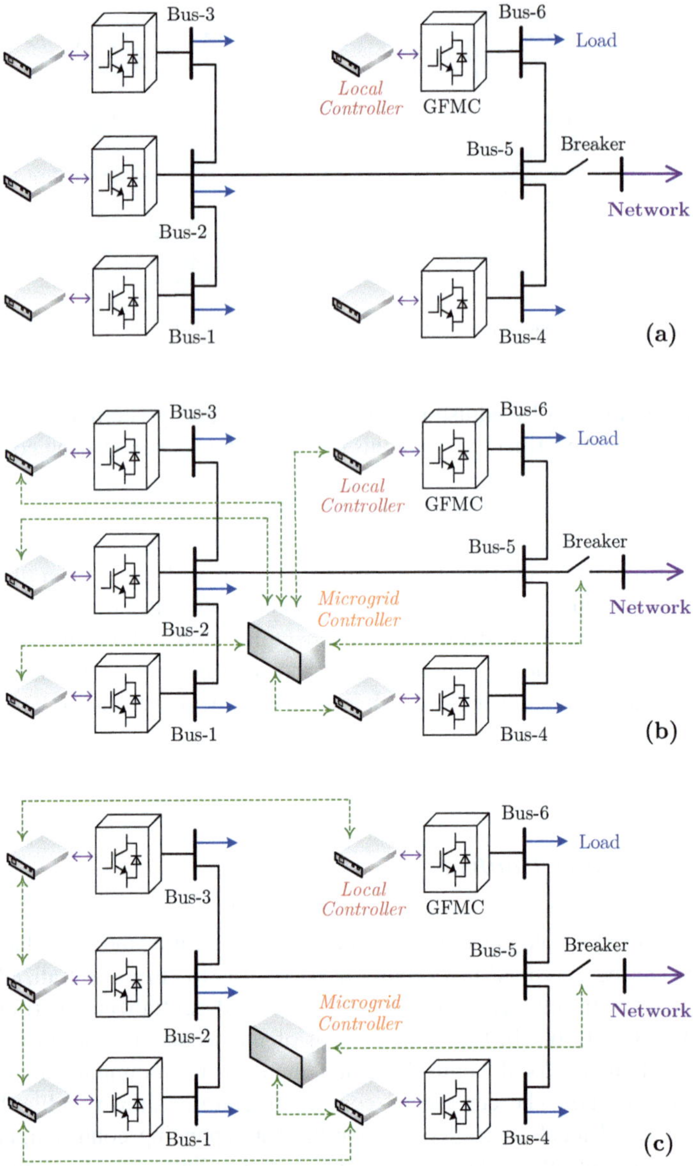

Fig. 1.16 Secondary control structure: (**a**) decentralized control, (**b**) centralized control, and (**c**) distributed control

1.3.2.1 Decentralized Control Structure

The basic principle of the decentralized control for offsetting deviations in grid frequency and voltage, and achieving reactive power sharing, is shown in Fig. 1.17.

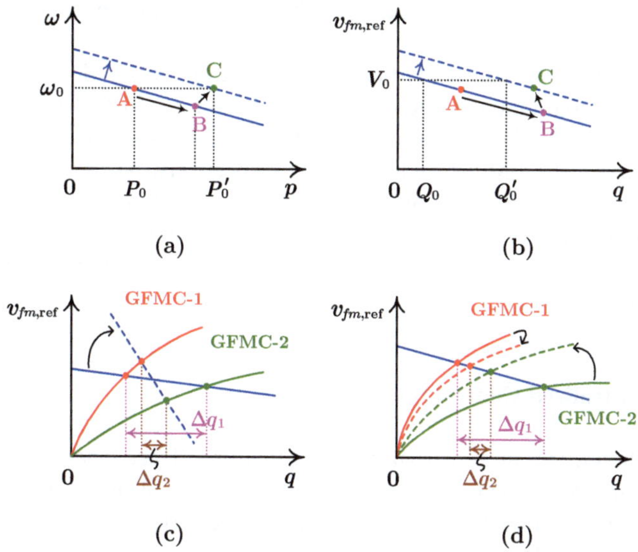

Fig. 1.17 Basic principle of the decentralized control: (**a**) grid frequency restoration, (**b**) grid voltage restoration, (**c**) reactive power sharing by adjusting the reactive-power droop coefficients, and (**d**) reactive power sharing by adjusting the virtual impedances

Specifically, in Fig. 1.17a, the solid blue curve represents the original active power-frequency droop. The system operates at point A initially, but moves to point B when p increases, leading to a frequency deviation. In this case, the setpoint is adjusted to move the droop curve from the solid blue one to the dashed blue one, and the operating point of the system may move to point C to restore the frequency [33].

In Fig. 1.17b, the solid blue curve represents the original reactive power-voltage droop. The system operates at point A initially, but moves to point B when q increases, leading to a voltage deviation. In order to offset this deviation, the setpoint is altered as $(Q_0^{'}, V_0)$ from (Q_0, V_0), where $Q_0^{'} > Q_0$. The dashed blue curve becomes the new reactive power-voltage droop curve, and the operating point of the system moves to point C to mitigate the voltage deviation [33].

There are two different methods to achieve reactive power sharing. Figure 1.17c shows the first method, while Fig. 1.17d shows the second method.

In the first method, the droop coefficient is adjusted. As shown in Fig. 1.17c, the solid red and green curves are the output q-$v_{fm,\mathrm{ref}}$ curves of GFMC-1 and GFMC-2, respectively. These curves are related to the connection-line impedances and the virtual impedances. The solid blue curve is the original reactive power-voltage droop curve, for both GFMC-1 and GFMC-2. The intersection point between the solid blue curve and the solid red/green curve is the operating point of the GFMC. From Fig. 1.17c, the discrepancy of these two GFMCs' output reactive powers is Δq_1. In this case, the droop coefficient is adjusted, and the dashed blue curve becomes the new droop curve. Then, the discrepancy of the reactive powers becomes Δq_2, where $\Delta q_2 \ll \Delta q_1$.

In the second method, the virtual impedances of the GFMCs are adaptively adjusted, e.g., large output reactive power leads to large virtual impedance [34]. As shown in Fig. 1.17d, the solid red and green curves are the original output q-$v_{fm,\text{ref}}$ curves of GFMC-1 and GFMC-2, respectively, whereas the dashed red and green curves are the new output q-$v_{fm,\text{ref}}$ curves of GFMC-1 and GFMC-2, respectively. Therefore, the discrepancy of the reactive powers is Δq_1 initially, but becomes Δq_2 at last. Here, $\Delta q_2 \ll \Delta q_1$, implying that the reactive power difference is significantly reduced.

1.3.2.2 Centralized Control Structure

For a specific GFMC in a microgrid, the virtual impedance, and/or the additional terms $\Delta v_{0\ell}$ and $\Delta\omega_{0\ell}$ can be adjusted to achieve the control targets of the secondary control [35, 36], as shown in Fig. 1.18. Here, we use $\Delta v_{0\ell}$ and $\Delta\omega_{0\ell}$ to explain the basic control principle.

For the centralized control, a specific bus (e.g., Bus-5 shown in Fig. 1.16) is selected to estimate the microgrid's frequency and voltage, i.e., ω_{MG} and v_{MG}, as shown in Fig. 1.19. Then, applying proportional-integral (PI) control, ω_{MG} and v_{MG}

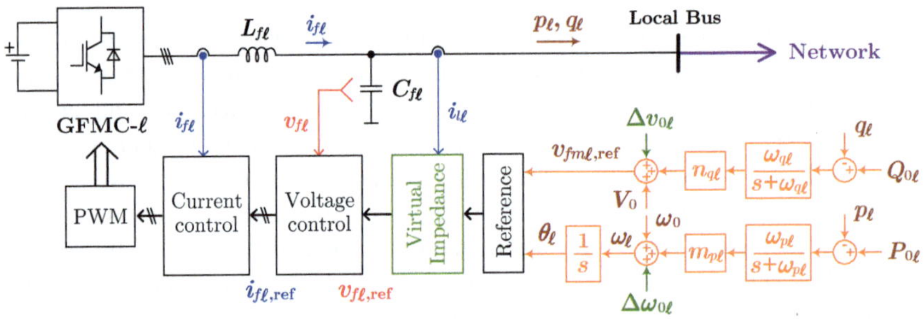

Fig. 1.18 Primary control strategy for the ℓ-th GFMC, where $\ell = 1, 2, \cdots, N$. In the microgrid, there are N GFMCs. Here, the virtual impedance and/or the additional terms $\Delta v_{0\ell}$ and $\Delta\omega_{0\ell}$ are adjusted according to the secondary control

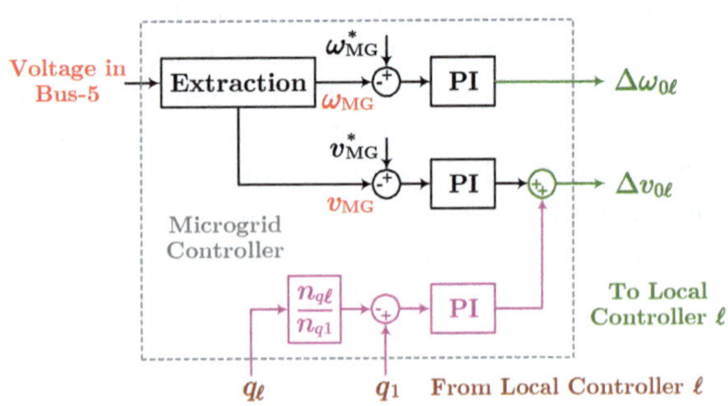

Fig. 1.19 Centralized control. As an example, only the information exchange between the microgrid controller and the ℓ-th local controller is exhibited

will follow their references ω_{MG}^* and v_{MG}^*, respectively. Besides, the information of all GFMCs' reactive powers is transmitted to the microgrid controller, and $(n_{q\ell}/n_{q1})q_\ell$ will follow q_1 through the PI control. Combining the aforementioned control, the control variables $\Delta v_{0\ell}$ and $\Delta\omega_{0\ell}$ are generated and sent to the ℓ-th local controller. Therefore, in the steady state, we have

$$\omega_{\mathrm{MG}} = \omega_{\mathrm{MG}}^*, \quad v_{\mathrm{MG}} = v_{\mathrm{MG}}^*, \quad n_{q1}q_1 = \cdots = n_{q\ell}q_\ell = \cdots = n_{qN}q_N \tag{1.36}$$

where $\ell = 1, 2, \cdots, N$, and N is the number of GFMCs in the microgrid.

1.3.2.3 Distributed Control Structure

For the distributed control, most of the control algorithms are executed in the local controllers [37], as shown in Fig. 1.20. In the microgrid controller, $\chi_{mc}(\omega_{\mathrm{MG}}^* - \omega_{\mathrm{MG}})$ and $\gamma_{mc}(v_{\mathrm{MG}}^* - v_{\mathrm{MG}})$ are calculated and sent to a small fraction of local controllers. In the ℓ-th local controller, $\Delta\omega_{0\ell}$ and $\Delta v_{0\ell}$ are derived through

$$\frac{\mathrm{d}\Delta\omega_{0\ell}}{\mathrm{d}t} = g_\ell \chi_{mc}(\omega_{\mathrm{MG}}^* - \omega_{\mathrm{MG}}) + \chi_{lc}\sum_{k=1}^{N} a_{\ell k}(\Delta\omega_{0k} - \Delta\omega_{0\ell}) \tag{1.37}$$

$$\frac{\mathrm{d}\Delta v_{0\ell}^1}{\mathrm{d}t} = g_\ell \gamma_{mc}(v_{\mathrm{MG}}^* - v_{\mathrm{MG}}) + \zeta_{lc}\sum_{k=1}^{N} a_{\ell k}(\Delta v_{0k}^1 - \Delta v_{0\ell}^1) \tag{1.38}$$

$$\frac{\mathrm{d}\Delta v_{0\ell}^2}{\mathrm{d}t} = \gamma_{lc}\sum_{k=1}^{N} a_{\ell k}(q_k - q_\ell) \tag{1.39}$$

$$\Delta v_{0\ell} = \Delta v_{0\ell}^1 + \Delta v_{0\ell}^2 \tag{1.40}$$

where $g_\ell = 1$ if there is a communication link between the ℓ-th local controller and the microgrid controller and $g_\ell = 0$ otherwise; $a_{\ell k} = 1$ if there is a communication link between the ℓ-th local controller and the k-th local controller, and $a_{\ell k} = 0$ otherwise.

Various distributed control structures have been proposed in the past [38], including the structure shown in Fig. 1.20. Chapter 9 will revisit these structures and summarize the similarities and differences among them.

1.3.3 Tertiary Control

In the tertiary control level, the microgrid needs to synchronize with the utility grid or other microgrids and determines its control scheme to import/export the active and reactive powers. These control algorithms are all executed in the microgrid controller.

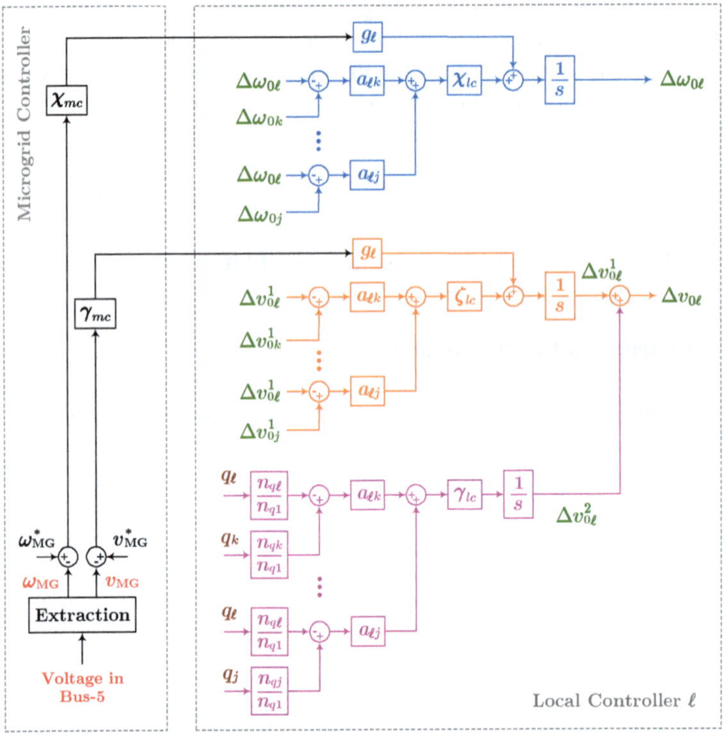

Fig. 1.20 Distributed control. As an example, only the information exchange between the microgrid controller and the ℓ-th local controller is exhibited. Here, $\ell, k, j = 1, 2, \cdots, N$, and $\ell \neq k \neq j$

As shown in Fig. 1.21, the breaker is open when the microgrid operates at the islanded mode. In this case, the microgrid does not synchronize with the external network, which may induce overcurrent and/or destabilize the microgrid if the breaker is closed directly. To address this problem, an auxiliary control loop is needed. The voltages before and after application of the breaker (i.e., v_{MG} and v_{NW}) are sampled, and the corresponding components in $\alpha\beta$-frame are derived, i.e., $v_{MG\alpha}$, $v_{MG\beta}$, $v_{NW\alpha}$, and $v_{NW\beta}$. A low-pass filtering (with the cutoff frequency of ω_{syn}) is applied to variable $v_{MG\alpha}v_{NW\beta}$, and the filtering result is used for the PI control to generate the synchronization term $\Delta\omega_s$, which is appended to ω^*_{MG}. From Fig. 1.21, switch S_0 determines if the aforementioned auxiliary control loop works or not. Here, $S_0 = 2$ only when the microgrid is planned to connect to the external network, whereas $S_0 = 1$ when the microgrid is islanded or has already connected to the external network. For $S_0 = 2$, after a period of time, $v_{MG\alpha}v_{NW\beta} \approx 0$, implying that v_{MG} is phase-locked to v_{NW}. Therefore, the microgrid is able to seamlessly switch to the grid-connected mode from the islanded mode, without inducing overcurrent and/or destabilizing the system [13].

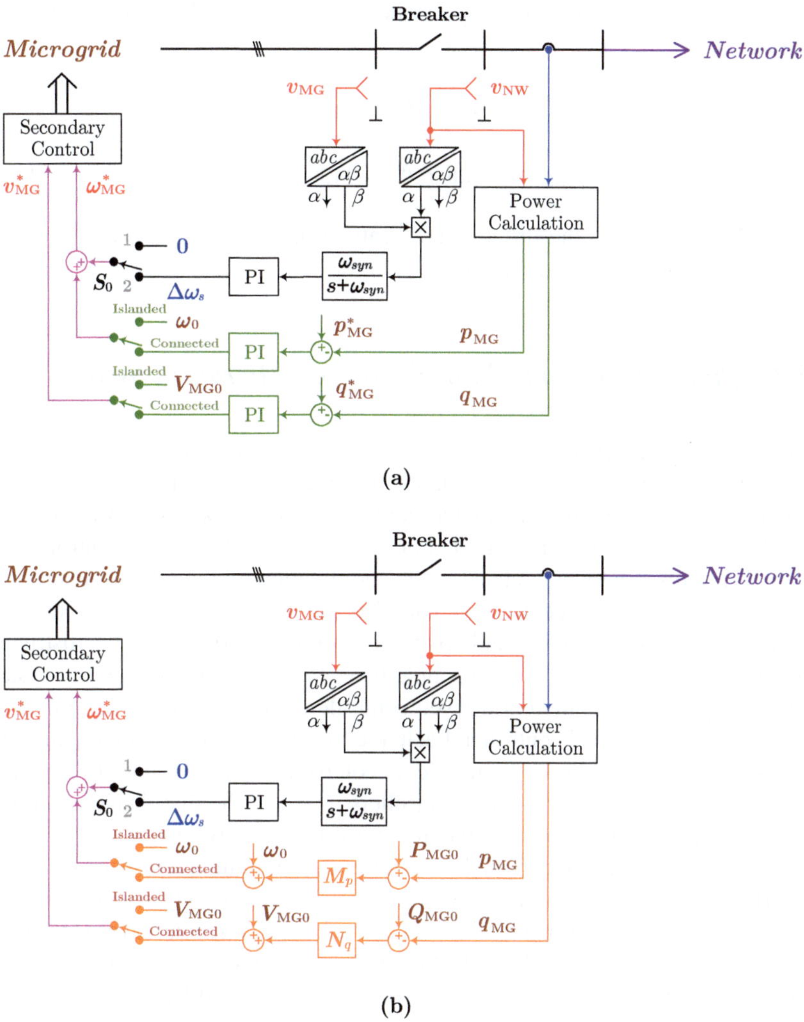

Fig. 1.21 Tertiary control of the microgrid. (**a**) Grid-following control and (**b**) grid-forming control

In the islanded mode, $\omega_{MG}^* = \omega_0$, and $v_{MG}^* = V_{MG0}$, where ω_0 and V_{MG0} are the desired values of the microgrid frequency and voltage, respectively. However, in the grid-connected mode, ω_{MG}^* and v_{MG}^* are related to the power import/export of the microgrid. As shown in Fig. 1.21, there are two different schemes to import/export the active and reactive powers [13, 37], including the grid-following control shown in Fig. 1.21a, and the grid-forming control shown in Fig. 1.21b. Here, p_{MG} and q_{MG} are the output active and reactive powers of the microgrid. In Fig. 1.21a, the microgrid will output constant active and reactive powers, i.e., p_{MG} and q_{MG} will follow their references p_{MG}^* and q_{MG}^*, respectively. In Fig. 1.21b, the droop control is applied, i.e.,

$$\omega_{\mathrm{MG}}^* = \omega_0 + M_p(P_{\mathrm{MG0}} - p_{\mathrm{MG}}) \tag{1.41}$$

$$v_{\mathrm{MG}}^* = V_{\mathrm{MG0}} + N_q(Q_{\mathrm{MG0}} - q_{\mathrm{MG}}) \tag{1.42}$$

where M_p and N_q are the droop coefficients; P_{MG0} and Q_{MG0} are the nominal values of p_{MG} and q_{MG}, respectively.

1.3.4 Quaternary Control for Networked-Microgrid Systems

For networked-microgrid systems (or multi-microgrid systems or microgrid cluster), frequency deviation, voltage deviation, and poor reactive power sharing may exist. To address these problems, additional control is needed, which is similar to the secondary control of the microgrid. For brevity, we call such a control as the *quaternary control,* which is related to the coordination of a number of microgrids. Thus, for the quaternary control, we can also use decentralized control structure, centralized control structure, and distributed control structure. The distributed quaternary control is taken as an example here. In a networked-microgrid system, each microgrid has a microgrid-level controller, and the networked-microgrid system has a system-level controller. Each microgrid-level controller communicates with its neighbors, while the system-level controller sends the control signals to a small fraction of microgrid-level controllers. Most of the distributed quaternary control algorithms are executed in the microgrid-level controllers.

Combining the aforementioned illustrations, the multilayer structure of the networked-microgrid system is shown in Fig. 1.22. The first layer is related to the power network and the primary control of converters. The second layer is related to the secondary control of GFMCs. The third layer is related to the tertiary control and quaternary control of microgrids. The power references of GFLCs are calculated in the third layer (precisely, the tertiary control) and dispatched to the first layer. This is the first connection manner between different layers. Moreover, from Fig. 1.22, the same controller may be at two different layers simultaneously, which is the second connection style between different layers.

1.4 Complex Behavior and Stability of Converter-Dominated Power Systems

1.4.1 Stability of Converter-Dominated Power Systems

A converter-dominated power system may suffer from various types of stability problems [1–5], which are categorized according to the physical cause of the

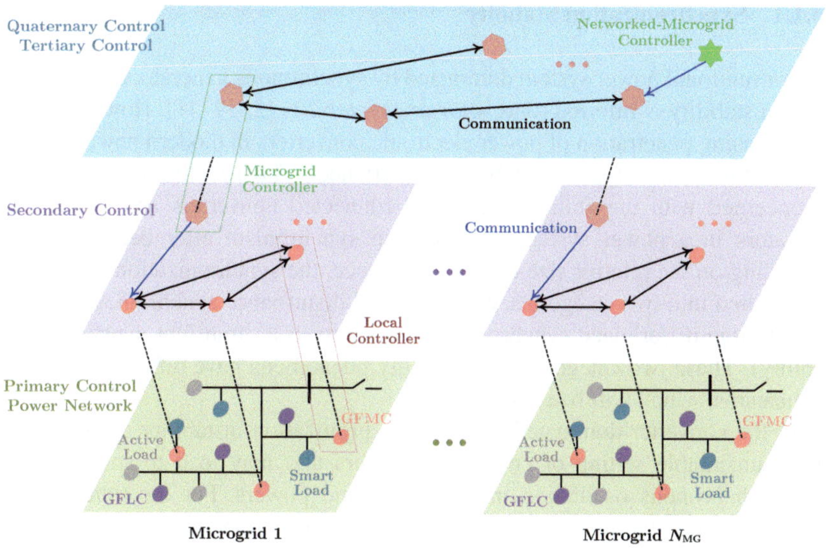

Fig. 1.22 Multilayer structure of networked-microgrid system, where N_{MG} is the number of the microgrids

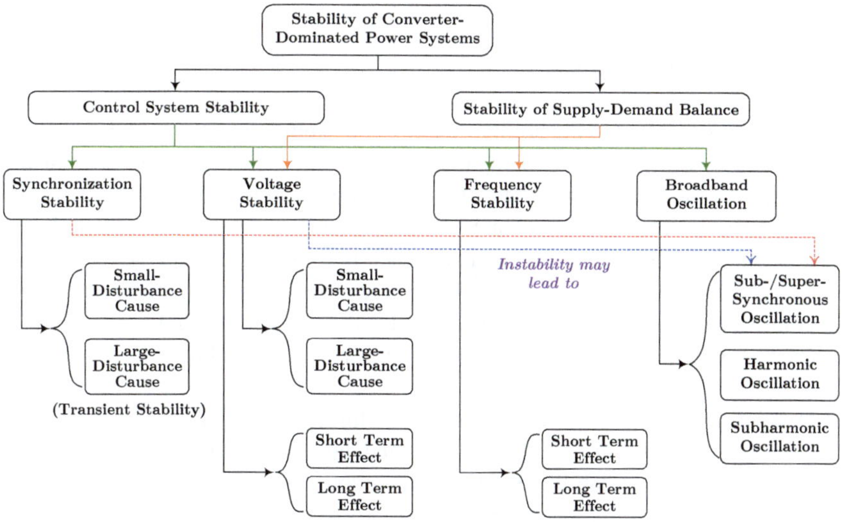

Fig. 1.23 Classification of the stability of converter-dominated power systems

instability, the physical nature of the resulting mode of instability, the relative size of the disturbance, and the relative timescale during which the instability occurs, as shown in Fig. 1.23.

1.4.1.1 Synchronization Stability

In a conventional power system dominated by synchronous generators, the synchronization stability is referred to as *rotor angle stability* [2, 39–41]. However, due to the increasing penetration of power electronic converters in modern power systems, the content of this type of stability is enriched. The synchronization stability is concerned with the ability of the grid-connected converters and synchronous generators in a power system to remain in synchronism after being disturbed. Depending on the relative size of the disturbance, the synchronization stability can be classified into two categories, namely, small-disturbance synchronization stability and large-disturbance synchronization stability (i.e., transient synchronization stability). These two categories of instability phenomena have different generation mechanisms, which will be explained later.

For the converter-dominated system, synchronization instability usually occurs due to unsuitable setting of control parameters and may lead to sustained sub-/super-synchronous oscillation in the angles and powers. For the timescale, the synchronization stability is about the synchronism of line-frequency waveforms; therefore, it is categorized as short-term stability. In other words, the slow system-level control (e.g., the secondary control) is usually not involved in this type of instability.

1.4.1.2 Voltage Stability

Voltage stability refers to the ability of a power system to maintain a steady voltage at all buses in the system after being disturbed. Usually, it is associated with insufficient supply of reactive power to meet the requirement of load demand [2, 3, 39, 40]. However, in the converter-dominated system, it may undergo a more complicated procedure to exhibit voltage instability, which is rooted at the interaction between converters and the grid (or among different converters) due to inappropriate choice of control parameters [30]. In other words, the supply-demand reactive power balance is not the only reason for inducing voltage instability in a converter-dominated system.

Depending on the relative size of the disturbance, voltage stability can also be categorized as small-disturbance type and large-disturbance type. For small-disturbance voltage instability, the bus voltage may exhibit a sustained sub-/super-synchronous oscillation, but drop (or rise) to a relatively low (or high) value when the circuit/control parameters slightly change. For large-disturbance voltage instability, the bus voltage may collapse quickly under a severe temporary fault. These two types of instability correspond to different generation mechanisms, which will be explained later.

From the viewpoint of the relative timescale, voltage stability can be classified as short-term type and long-term type. For the short-term type, it is concerned with the ability of a single bus to maintain its voltage in a normal range. In other words, when this type of instability occurs, the converter-based load/generator connected to such a bus will be disconnected from the grid. For the long-term type, it is also called voltage collapse, where cascaded events accompanying short-term voltage instability lead to a blackout or abnormally low voltage for a significant part of the power system. In this case, the system-level control (e.g., tap-changing transformers, reactive power support from some specific grid-connected converters, and current limiters of some converters) may be involved. Therefore, the short-term voltage instability is associated with the nonlinear dynamics (i.e., complex behavior) of converter-based loads/generators [30], whereas the long-term voltage instability is associated with the complex network features of converter-dominated systems [42].

1.4.1.3 Frequency Stability

Frequency stability refers to the ability of a power system to maintain a steady frequency following an incident where the supply-demand active power balance is significantly destroyed [2, 39, 40]. Frequency instability may result in a large excursion of the grid frequency, leading to tripping of a considerable number of converter-based loads/generators. Frequency instability can be interpreted from three distinct aspects:

1. Due to insufficient inertial provided by the grid-connected converters and synchronous generators, the initial time derivative of the frequency (also the rate of change of frequency, i.e., RoCoF) in some buses may become very high, which may increase the risk of pole slipping for the synchronous generators.
2. Under the aforementioned incident, the frequency will drop to a certain value before recovering. Such a value is denoted as the *frequency nadir.* If the nadir is too low, a load shedding protection may operate, leading to a blackout in a significant part of the power system. The specific value of the frequency nadir is determined by the primary frequency control of grid-connected converters and synchronous generators, and the damping of the loads. The RoCoF and the frequency nadir represent a short-term frequency stability.
3. The secondary frequency control or automatic generation control will bring the frequency back to its nominal value. However, in some specific cases, such control may lead to overspeed protection of steam turbines and/or overcurrent protection of grid-connected converters, inducing a long-term frequency stability problem.

1.4.1.4 Broadband Oscillation

Depending on the value of the oscillation frequency, broadband oscillation can be classified into three categories, as shown in Fig. 1.24.

1. Sub-/super-synchronous oscillation: The frequency of a sub-synchronous oscillation is lower than the line frequency, whereas the frequency of a super-synchronous oscillation varies from the line frequency to double-line frequency.

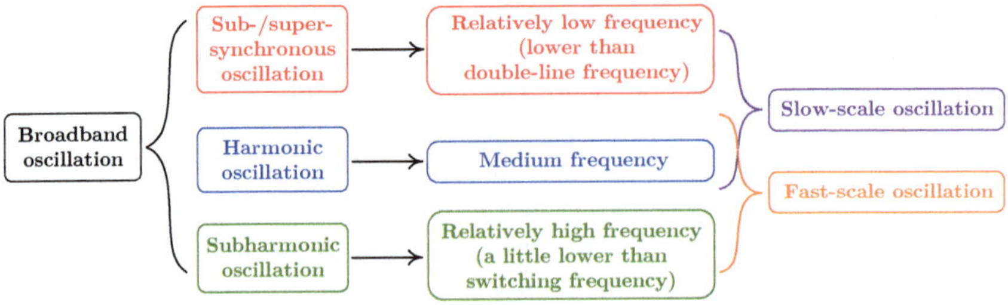

Fig. 1.24 Classification of the broadband oscillation

Usually, sub-synchronous and super-synchronous oscillations coexist in the current and voltage, and their oscillation frequencies are symmetric about the line frequency. However, only sub-synchronous oscillation may exist in the active/reactive power. Besides, as the aforementioned oscillation frequencies are much lower than the switching frequency, the switching behavior of grid-connected converters is usually not involved in generating such oscillations, which are therefore slow-scale oscillations [2, 43–45].

2. Harmonic oscillation: The frequency of harmonic oscillation is typically in the range from hundreds to a few thousands hertz [2, 43, 46], which is intermediate between the double-line frequency and the switching frequency. The current inner loops and time delays (including the modulator delay and the computation delay) of grid-connected converters are usually associated with the generation of harmonic oscillation. The switching behavior of grid-connected converters may also contribute to generating the oscillation. Depending on the involvement of switching behavior, harmonic oscillation can be regarded as a slow-scale oscillation if the switching behavior is not involved, but a fast-scale oscillation otherwise.
3. Subharmonic oscillation: The oscillation frequency is usually slightly lower than the switching frequency [5, 47–50]. Therefore, the switching behavior of grid-connected converters has a significant impact on the generation and evolution of subharmonic oscillation, which is therefore a fast-scale oscillation.

According to the characteristics of the aforementioned types of instability, synchronization instability, voltage instability, frequency instability, sub-/super-synchronous oscillation, and some harmonic oscillation are associated with slow-scale interaction, where the converters' switching behavior is not involved, while some harmonic oscillation and subharmonic oscillation are associated with fast-scale interaction, where the converters' switching behavior has remarkable impact. For the sake of analysis, a *continuous-time averaging approach* can be used to establish the converters' models when studying *slow-scale interaction*, while the *discrete-time iterative mapping approach* can be adopted to study *fast-scale interaction*. In this book, we focus on the slow-scale interaction which is more prevalent in practice, and the continuous-time state-space averaging model is adopted [5].

1.4.2 Continuous-Time Averaging Model of Grid-Connected Converters

The GFMC is taken as an example to illustrate the derivation of the averaged model of grid-connected converters. The circuit diagram of a GFMC is shown in Fig. 1.25. Here, O_1 is the midpoint of the DC-link, and O_2 is the common point of the three AC filter capacitors. It can be shown that $v_{O_1O_2} \approx 0$. In other words, O_1 is virtually connected to O_2.

For a specific phase shown in Fig. 1.26, the duty ratio of the top switch is d_ζ, whereas the duty ratio of the bottom switch is $(1 - d_\zeta)$, where $0 \leqslant d_\zeta \leqslant 1$ and $\zeta \in \{a, b, c\}$. When the top switch is closed, the bottom switch is open, as shown in Fig. 1.26a. In this case, we have

$$\begin{cases} \dfrac{\mathrm{d}i_{f\zeta}}{\mathrm{d}t} = \dfrac{1}{L_f}(\dfrac{v_{\mathrm{dc}}}{2} - v_{f\zeta} - r_f i_{f\zeta}) \\ \dfrac{\mathrm{d}v_{f\zeta}}{\mathrm{d}t} = \dfrac{1}{C_f}(i_{f\zeta} - i_{l\zeta}) \end{cases} \tag{1.43}$$

When the top switch is open, the bottom switch is closed, as shown in Fig. 1.26b. In this case, we have

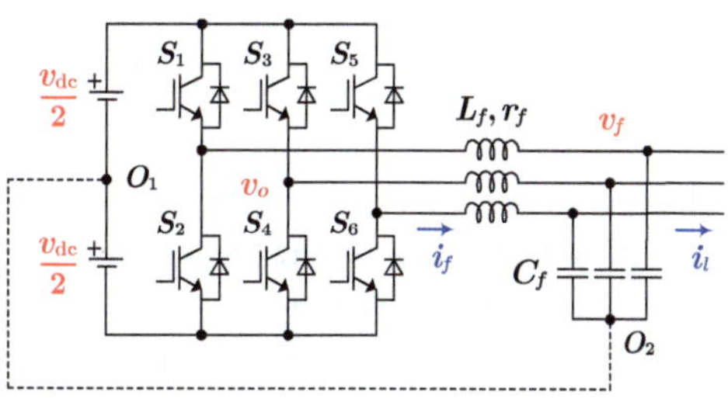

Fig. 1.25 Circuit diagram of a GFMC

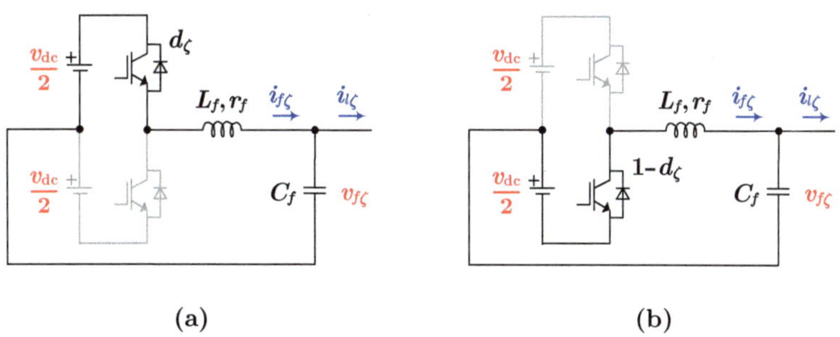

Fig. 1.26 Circuit topology (**a**) when the top switch is closed and the bottom switch is open and (**b**) when the top switch is open and the bottom switch is closed

$$\begin{cases} \frac{\mathrm{d}i_{f\zeta}}{\mathrm{d}t} = \frac{1}{L_f}(-\frac{v_{\mathrm{dc}}}{2} - v_{f\zeta} - r_f i_{f\zeta}) \\ \frac{\mathrm{d}v_{f\zeta}}{\mathrm{d}t} = \frac{1}{C_f}(i_{f\zeta} - i_{l\zeta}) \end{cases} \tag{1.44}$$

Combining (1.43) and (1.44), we get

$$\begin{cases} \frac{\mathrm{d}i_{f\zeta}}{\mathrm{d}t} = d_\zeta \frac{1}{L_f}(\frac{v_{\mathrm{dc}}}{2} - v_{f\zeta} - r_f i_{f\zeta}) + (1-d_\zeta)\frac{1}{L_f}(-\frac{v_{\mathrm{dc}}}{2} - v_{f\zeta} - r_f i_{f\zeta}) \\ \quad = (2d_\zeta - 1)\frac{1}{L_f}\frac{v_{\mathrm{dc}}}{2} + \frac{1}{L_f}(-v_{f\zeta} - r_f i_{f\zeta}) \\ \frac{\mathrm{d}v_{f\zeta}}{\mathrm{d}t} = d_\zeta \frac{1}{C_f}(i_{f\zeta} - i_{l\zeta}) + (1-d_\zeta)\frac{1}{C_f}(i_{f\zeta} - i_{l\zeta}) \\ \quad = \frac{1}{C_f}(i_{f\zeta} - i_{l\zeta}) \end{cases} \tag{1.45}$$

where $v_{\mathrm{ctr}\zeta} \triangleq 2d_\zeta - 1 \in [-1, 1]$. Here, $v_{\mathrm{ctr}\zeta}$ is the Phase-ζ modulation waveform participating in the PWM. Besides, $v_{o\zeta} = v_{\mathrm{ctr}\zeta} \times 0.5 v_{\mathrm{dc}}$, which is the Phase-$\zeta$ PWM voltage. Therefore,

$$\begin{cases} \frac{\mathrm{d}i_{f\zeta}}{\mathrm{d}t} = \frac{1}{L_f}(v_{o\zeta} - v_{f\zeta} - r_f i_{f\zeta}) \\ \frac{\mathrm{d}v_{f\zeta}}{\mathrm{d}t} = \frac{1}{C_f}(i_{f\zeta} - i_{l\zeta}) \end{cases} \tag{1.46}$$

which is the averaged model of GFMC in the *abc*-frame. In this model, the low-frequency properties of the system are persevered, while the detailed dynamics within a switching cycle are ignored. Therefore, the validity of averaged models is usually restricted to the low-frequency range up to an order of magnitude below the switching frequency.

On this basis, using the Clark transformation, the model in the $\alpha\beta$-frame is derived, which is similar to (1.46) but with $\zeta \in \{\alpha, \beta\}$. Here, the Clark transformation matrix is

$$T_{abc2\alpha\beta} = \frac{2}{3}\begin{bmatrix} 1 & -\frac{1}{2} & -\frac{1}{2} \\ 0 & -\frac{\sqrt{3}}{2} & -\frac{\sqrt{3}}{2} \end{bmatrix}. \tag{1.47}$$

Hence, we have

$$\begin{cases} \frac{\mathrm{d}}{\mathrm{d}t}\begin{bmatrix} i_{f\alpha} \\ i_{f\beta} \end{bmatrix} = \frac{1}{L_f}\left(\begin{bmatrix} v_{o\alpha} \\ v_{o\beta} \end{bmatrix} - \begin{bmatrix} v_{f\alpha} \\ v_{f\beta} \end{bmatrix} - r_f \begin{bmatrix} i_{f\alpha} \\ i_{f\beta} \end{bmatrix}\right) \\ \frac{\mathrm{d}}{\mathrm{d}t}\begin{bmatrix} v_{f\alpha} \\ v_{f\beta} \end{bmatrix} = \frac{1}{C_f}\left(\begin{bmatrix} i_{f\alpha} \\ i_{f\beta} \end{bmatrix} - \begin{bmatrix} i_{l\alpha} \\ i_{l\beta} \end{bmatrix}\right) \end{cases} \tag{1.48}$$

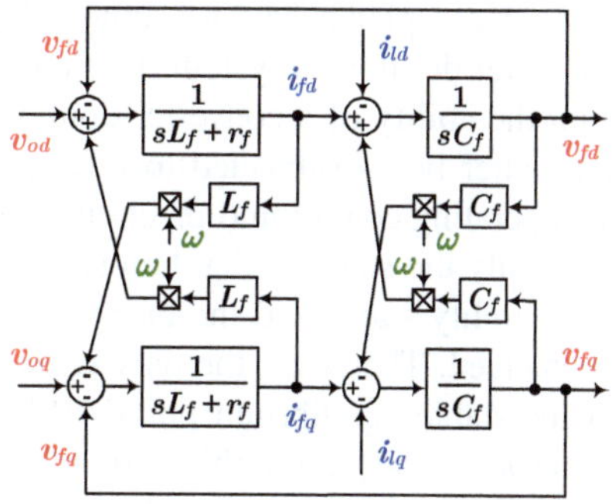

Fig. 1.27 Averaged model of GFMC in the dq-frame

Using the Park transformation, we obtain

$$T_{\alpha\beta 2dq} = \begin{bmatrix} \cos(\omega t) & \sin(\omega t) \\ -\sin(\omega t) & \cos(\omega t) \end{bmatrix}, \tag{1.49}$$

and the model in the dq-frame is derived, i.e.,

$$\begin{cases} \dfrac{\mathrm{d}}{\mathrm{d}t}\begin{bmatrix} i_{fd} \\ i_{fq} \end{bmatrix} = \dfrac{1}{L_f}\left(\begin{bmatrix} v_{od} \\ v_{oq} \end{bmatrix} - \begin{bmatrix} v_{fd} \\ v_{fq} \end{bmatrix} - r_f \begin{bmatrix} i_{fd} \\ i_{fq} \end{bmatrix}\right) \\ \dfrac{\mathrm{d}}{\mathrm{d}t}\begin{bmatrix} v_{fd} \\ v_{fq} \end{bmatrix} = \dfrac{1}{C_f}\left(\begin{bmatrix} i_{fd} \\ i_{fq} \end{bmatrix} - \begin{bmatrix} i_{ld} \\ i_{lq} \end{bmatrix}\right) \end{cases} \tag{1.50}$$

Using the Laplacian transformation, the corresponding s-domain representation can be derived, as shown in Fig. 1.27.

1.4.3 Complex Behavior and System Stability

The converter-dominated system is a nonlinear autonomous system, as reflected by (1.46). The nonlinearity is generated from some typical terms such as $\omega \times i_{fd}$, as shown in Fig. 1.27. The word "autonomous" means that the time dependence is absent in the state function [5].

Due to the rich nonlinearity, the converter-dominated systems may exhibit a variety of complex behavior such as stability changes of specific equilibrium points [EQPs] (which is associated with the local bifurcation) [51, 52], (intermittent) chaotic operation [5, 53], changes of basins of attraction (which is related to the global bifurcation), emergence/vanishment of hidden attractors [53], etc., which cannot be well described or explored by linear analytical techniques. In general, bifurcation can be regarded as a sudden change of qualitative behavior of a system (i.e., the appearance of phase portrait becomes topologically nonequivalent) under variation of one or more parameters [54].

Such complex behavior is crucial for understanding the instability mechanism of converter-dominated systems [1, 5]. For example, the transient stability is concerned with the ability of a converter-dominated system to return to its normal operation state after being subjected to a temporary severe disturbance [2]. It is noted that the operating condition (i.e., circuit and parameters) after removing the disturbance is usually the same as that before the occurrence of the disturbance, implying that the stability status and the basin of attraction of the usual operating point remain unchanged. Therefore, the only route to instability is that the trajectory moves out of the stable equilibrium point's (SEP's) basin of attraction upon removal of the disturbance [55]. From this point of view, the transient stability can be enhanced by increasing damping to keep the trajectory near the SEP during the disturbance [56] or by expanding the SEP's basin of attraction significantly [55]. However, the former method may not always be effective for all types of disturbances [55]. For the latter method, we need to identify the attractors/repellers and basins of attraction of interests and explore the mechanism behind the shrinking or expanding of these basins of attraction. Such change of the basins of attraction is associated with a specific type of bifurcation. Thus, knowing when (under what conditions) and how (in what way) such a bifurcation occurs should be of fundamental importance to understand the transient stability problem and develop the subsequent mitigation methods [5].

1.4.3.1 Origin of Small-/Large-Disturbance Instability

For a given set of parameters, the converter-dominated system may have multiple equilibrium solutions, in the sense that if the trajectory (i.e., system dynamics) starts at a point on a specific equilibrium solution, it stays permanently on that solution. Depending on the behavior of the trajectory starting at a point near the equilibrium solution, this solution is called an *attractor* if the trajectory converges to that solution, or a *repeller* if the trajectory diverges to other places. A converter-dominated system may have multiple attractors coexisting at the same time. Each attractor has its own basin of attraction, in the sense that the trajectory converges to a specific attractor if the initial condition is located in that attractor's basin of attraction.

Among these attractors, there is a usual operating point which is an SEP normally. In other words, the system usually operates at this SEP. From the engineering point of view, operating at attractors other than an SEP is not allowed, as these attractors always manifest themselves as some sustained oscillations, which may worsen the power quality, trigger overcurrent or overvoltage protection, and activate a torsional stress relay. Therefore, the SEP and its basin of attraction are of primary interest. As shown in Fig. 1.28, the phase space is divided into two areas, one being the SEP's basin of attraction and the other one being the unstable region. The separatrix between these two areas is called a *stability boundary*. Depending on the route to instability, the disturbances applied to the system can be classified into two categories:

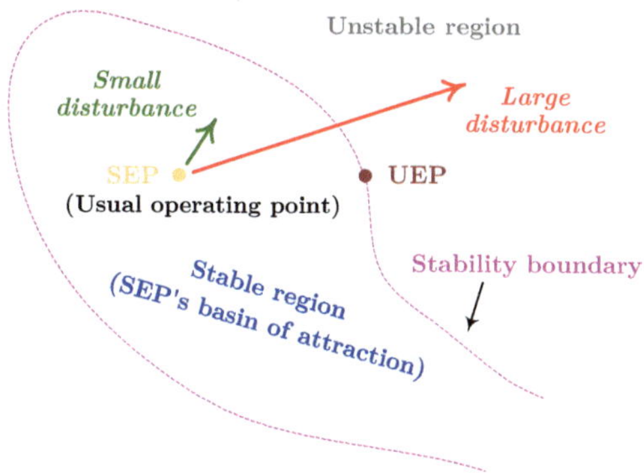

Fig. 1.28 SEP's basin of attraction (i.e., stable region), unstable region, and stability boundary separating the aforementioned two regions

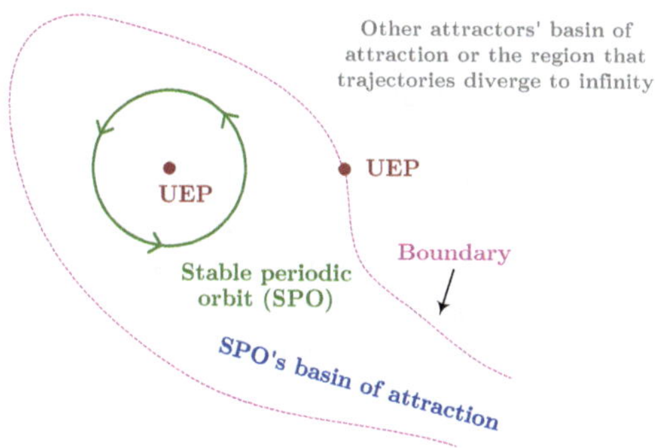

Fig. 1.29 Under a small disturbance, the usual operating point becomes a UEP and an SPO

1. Small disturbance: As shown in Fig. 1.28, the trajectory is perturbed from the SEP, but still in the SEP's basin of attraction under a small disturbance (e.g., the change of the grid impedance). In this case, the trajectory will return to the SEP, and the system remains stable. However, the disturbance may also change the stability of the usual operating point and/or significantly shrink the stability boundary. As shown in Fig. 1.29, the usual operating point becomes an unstable equilibrium point (UEP), and a stable periodic orbit (SPO) emerges nearby. This SPO takes the role of the original SEP and dominates the original SEP's basin of attraction. In this case, the system will operate at this SPO, which is regarded as an unstable operating state from the engineering point of view. Therefore, the instability induced by small disturbances is also referred to as *small-signal instability*. This type of instability is usually associated with a local bifurcation. The local bifurcations are referred to as bifurcations of EQPs, which are detected by examining small neighborhoods of the EQPs. Typical local bifurcations include saddle-node bifurcation and supercritical/subcritical Hopf bifurcation [5, 54]. Exploring such local bifurcation behavior is beneficial to understanding the essential generation mechanism of small-signal instability events [57] and unveil the relation between these events and typical power system instability problems (e.g., voltage instability) [30]. An example can be found in Chap. 2. Thus, the bifurcation analysis shows some advantages over

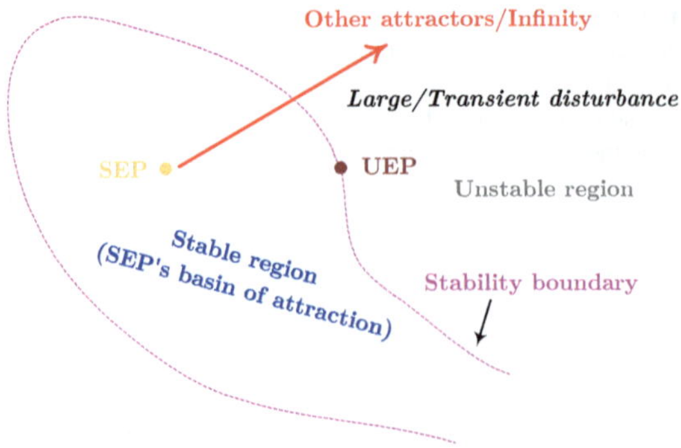

Fig. 1.30 Under a large disturbance, the stability of the usual operating point remains unchanged, but the trajectory is outside the SEP's basin of attraction

the impedance-based method [58–60] and the eigenvalue-based approach (i.e., Lyapunov's indirect method) [40, 61], regarding the complex instability behavior.

2. Large disturbance: As shown in Fig. 1.30, the usual operating point remains an SEP under a large disturbance (e.g., transient short-circuit fault, phase jump, malfunction of specific converters). However, after removing this disturbance, the trajectory is outside the SEP's basin of attraction. Therefore, it will not return to the usual operating point, but converges to other attractors or diverges to infinity. The instability induced by large disturbances is also referred to as *large-signal or transient instability*. In this type of instability, normally, there is no bifurcation involved. However, for such instability, the SEP's basin of attraction is of great interest. Large SEP's basin of attraction means high robustness against large disturbances. The sudden change of the appearance of SEP's basin of attraction (e.g., between Figs. 1.34 and 1.35) is usually associated with global bifurcations. The global bifurcations cannot be detected by looking at the small neighborhoods of the EQPs. Typical global bifurcations include homoclinic bifurcation and heteroclinic bifurcation [54]. For the transient instability problem, the widely used impedance-based method and eigenvalue-based approach are no longer effective, because these methods just expose the stability status of the usual operating point. On the other hand, some energy-based methods, such as the equal-area criterion and the Lyapunov's direct method, are adopted to estimate the SEP's basin of attraction and show effectiveness to derive the critical clearing time/angle [62–64]. However, these methods may suffer from the conservativeness problem and usually cannot capture the sudden change of the appearance of SEP's basin of attraction. Such deficiencies can be remedied by exploring the global bifurcation behavior.

In brief, the stability status of the usual operating point can be studied in terms of local bifurcation behavior when studying the small-signal stability, while the size of the SEP's basin of attraction is emphasized, and its sudden change can be exposed by the global bifurcation behavior for the transient stability, as shown in Fig. 1.31.

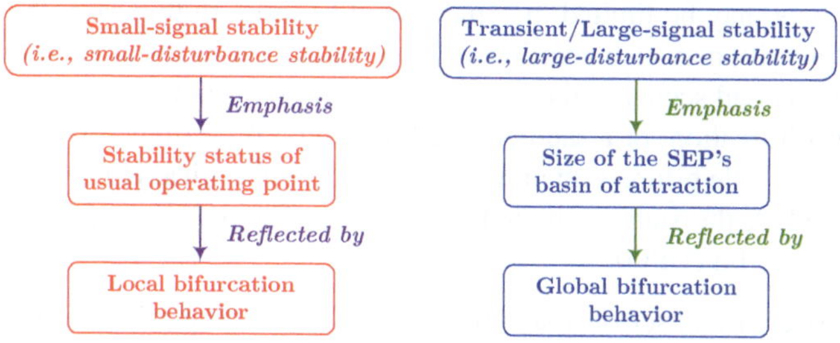

Fig. 1.31 Relation between the system stability and the bifurcation behavior

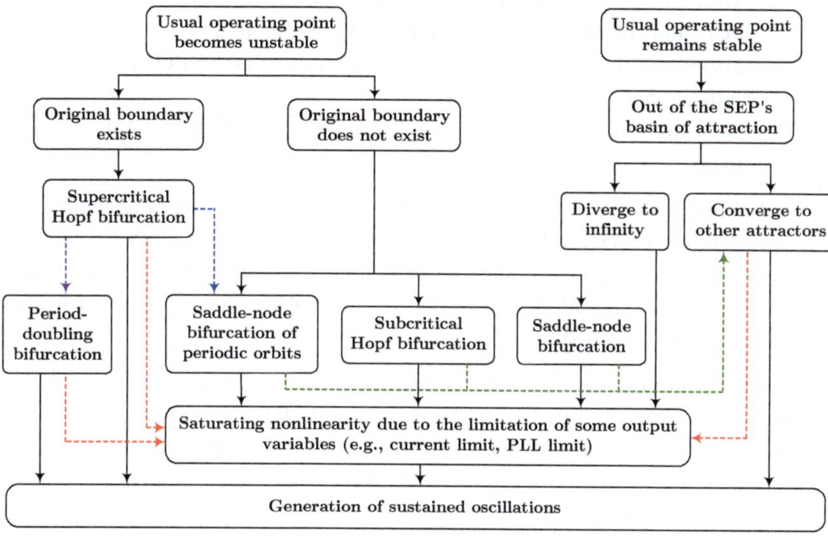

Fig. 1.32 Routes to generating the sustained oscillations

1.4.3.2 Origin of Sustained Oscillations

Sustained oscillations in grid-connected converters may be generated from various routes, as shown in Fig. 1.32. Here, we outline the key paths leading to sustained oscillations.

1. The usual operating point becomes unstable under two possible scenarios.

   a. In the first scenario, the original boundary exists, and a supercritical Hopf bifurcation (and period-doubling bifurcation) occurs, as outlined in Fig. 1.29. After the onset of a supercritical Hopf bifurcation, the SEP becomes a UEP and an SPO. The period of this SPO doubles upon the onset of a period-doubling bifurcation. This SPO manifests itself as a sustained oscillation. Furthermore, due to the inherent imposition of limits on some output variables (e.g., current limit and PLL output limit), the amplitudes of the oscillatory variables may be limited in some period of time. In this case, the saturating nonlinearity may also contribute to the generation of sustained oscillations.

b. In the second scenario, the original boundary does not exist, and a subcritical Hopf bifurcation, or a saddle-node bifurcation of periodic orbits (SNPO), or a saddle-node bifurcation may occur. For the subcritical Hopf bifurcation and the SNPO, the stability boundary shown in Fig. 1.28 is usually an unstable periodic orbit (UPO).

   i. After the onset of a subcritical Hopf bifurcation, the UPO shrinks and eventually swallows the SEP to become a UEP, and no SPO can be generated.
   ii. After the onset of a supercritical Hopf bifurcation, the SEP becomes a UEP and an SPO. Notably, a UPO (i.e., the boundary) surrounds the aforementioned SPO. Hence, after the onset of an SNPO, a UPO merges with the SPO and disappears.
   iii. After the onset of a saddle-node bifurcation, the SEP and a nearby UEP merge and again subside.

   For the aforementioned three bifurcations, no attractor exists near the usual operating point. Here, there are two possibilities. The trajectory may diverge to infinity or converge to a specific attractor in the outside space. In particular, for the former possibility, a sustained oscillation may generate from the saturating nonlinearity.

2. The usual operating point remains stable. In this case, the trajectory may be out of the SEP's basin of attraction under a large disturbance, as shown in Fig. 1.30. Then, the trajectory may diverge to infinity and converge to other attractors. The saturating nonlinearity plays a negligible, or partial, or full role in the generation of sustained oscillations.

1.4.3.3 Multiple/Coexisting Attractors, Invisible Attractors, and Hidden Attractors

Multiple attractors refer to the coexistence of two or more competing attractors within the phase space, which therefore are also named as coexisting attractors. Several examples of multiple attractors have been reported in the buck converter [5, 65], dual-output boost converter [66], islanded microgrids [53], and power systems [40, 67, 68]. Multiple attractors include the typical operating point, represented by the SEP, as well as invisible attractors, as illustrated in Fig. 1.33. Unlike the SEP, the invisible attractors cannot be identified by examining a small neighborhood of the usual operating point. They arise from non-local bifurcations and hold considerable practical significance, as highlighted in Fig. 1.30.

In addition, invisible attractors can be further classified into two categories: self-excited attractors and hidden attractors. If the generation process of a particular invisible attractor is independent upon any UEPs, it is considered a hidden attractor.

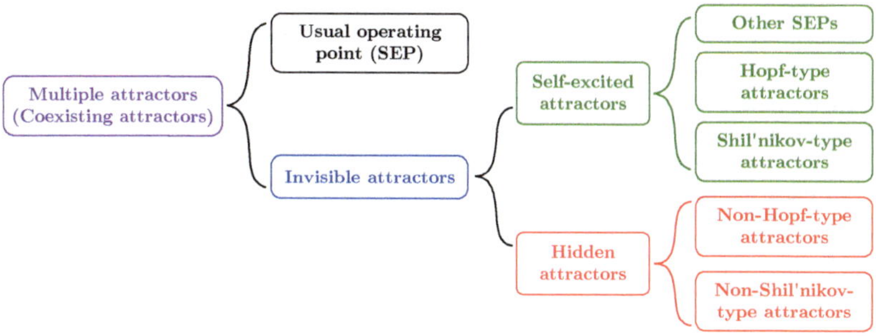

Fig. 1.33 Classification of multiple attractors

On the other hand, if the attractor is generated through a relationship with UEPs, it is classified as a self-excited attractor. In simpler terms, a hidden attractor is not created through a local bifurcation or a homoclinic/heteroclinic bifurcation.

Such classification is motivated by discovering the fundamental generating mechanism of the attractors, especially for chaotic attractors. The generation of conventional chaotic attractors (such as Lorenz system, Rössler system, and Chen system) can be explained by the Smale horseshoe map and Shil'nikov theorem [69]. In brief, one or more UEPs in the phase space are related to the generation process of these chaotic attractors. Therefore, such chaotic attractors are classified as self-excited attractors [70]. However, in the early 2010s, it was reported that distinct from the well-known Chua's attractor, a new type of chaotic attractors may exist in the Chua's circuit and is unavailable from the neighbors of any UEPs [71]. Moreover, it was found that in a simple three-dimensional autonomous quadratic system, there are an SEP, an SPO, and a chaotic attractor coexisting at the same time. Notably, no UEP exists in that system [72, 73]. Therefore, such chaotic attractors and periodic orbits are classified as hidden attractors.

For low-dimensional systems, we can simply use the unstable manifolds of UEPs to determine whether an invisible attractor is self-excited (if one or more unstable manifolds of UEPs converge to the attractor) or hidden (no unstable manifolds of UEPs converge to the attractor). However, high-dimensional systems (especially for converter-dominated systems) may exhibit more complicated behavior. For instance, the same attractor can be detected by the unstable manifolds of UEPs; however, when a specific parameter is varied, it cannot be detected by any unstable manifolds of UEPs. Therefore, for high-dimensional systems, the classification of an invisible attractor relies on examining the generation process of that attractor. More detailed discussions can be found in Chaps. 7 and 8.

1.4.3.4 Resynchronization and Multi-swing Stability

In the context of a grid-connected converter, the synchronization stability is typically indicated by the phase angle, denoted as δ. This angle represents the difference

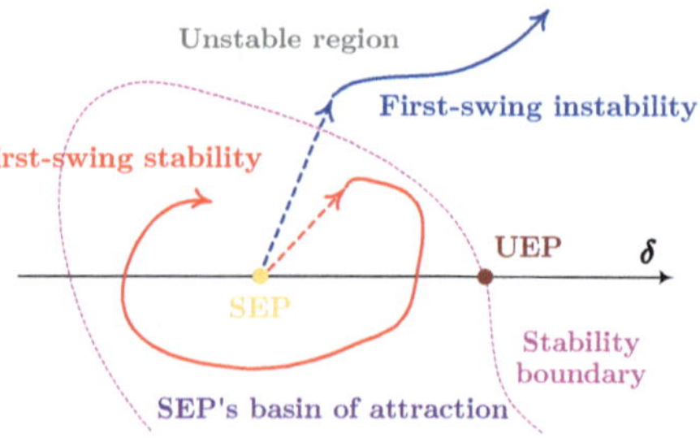

Fig. 1.34 First-swing stability and first-swing instability. Dashed curves represent the transient disturbances, and solid curves represent the trajectories after removing the disturbances

between the converter's output voltage and the grid voltage. Initially, when the system operates at the SEP, the phase angle δ remains constant, indicating that the converter is synchronized with the grid. However, when a transient disturbance is introduced, the system's behavior may vary depending on the operating point. Figure 1.34 illustrates the different outcomes. If the operating point remains within the SEP's basin of attraction, the trajectory of the system will eventually return to the SEP, ensuring that the converter remains synchronized with the grid. In the time domain, the phase angle δ will initially oscillate near the SEP value and gradually converge to it. This type of synchronization stability is referred to as first-swing stability, using terminology borrowed from synchronous generators [39]. On the other hand, if the operating point lies outside the SEP's basin of attraction, the system's trajectory will not return to the SEP but instead diverge towards other regions. In such a case, the converter loses its synchronization with the grid, and the phase angle δ rapidly increases to a significantly large value. This loss of synchronization is known as first-swing instability, i.e., δ cannot return to the SEP value within a cycle. Thus, for the first-swing stability, the converter's synchronization behavior within a cycle of δ is examined.

As depicted in Fig. 1.34, it is apparent that the UEP lies within the stability boundary. Consequently, if the phase angle δ exceeds the value of the UEP during a transient disturbance, the converter will undoubtedly lose synchronization with the grid. In simpler terms, to uphold first-swing stability, the trajectory must avoid surpassing the nearest UEP [74].

Nevertheless, the basin of attraction for the SEP can be significantly expanded following the occurrence of a specific global bifurcation, as depicted in Fig. 1.35. In this scenario, the UEP is no longer situated on the stability boundary but instead lies within the basin of attraction of the SEP. Moreover, the phase angle δ exhibits a periodic behavior with a period of 2π. Consequently, when a transient disturbance is introduced, the trajectory of the system may indeed pass through the UEP and converge towards the right-hand-side SEP. This convergence causes the phase angle δ to increase by at least 2π. It may initially appear that the converter loses synchronization with the grid, but eventually, it resynchronizes. This type of synchronization stability is commonly referred to as *multi-swing stability*. The resynchronization mechanism highlights the importance of studying global bifurcations in maintaining transient stability of the system.

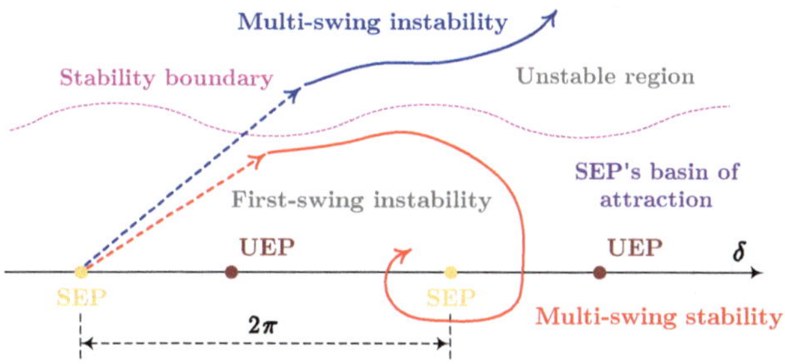

Fig. 1.35 Multi-swing stability and resynchronization. Dashed curve represents the transient disturbance, and solid curve represents the trajectory after removing the disturbance

1.5 Structure of This Book

This book is organized in two parts and ten chapters, as summarized in the organizational chart shown in Fig. 1.36. Part I focuses on the dynamics of single grid-connected converters, and Part II examines the entire connected system containing multiple converters.

In Part I, there are four chapters. Chapter 2 focuses on the small-disturbance stability. Chapters 3, 4, and 5 focus on the large-disturbance stability (i.e., transient stability).

- In Chap. 2, the local bifurcation behavior of the grid-following rectifier is studied, and several new routes to voltage instability in rectifier-based systems are identified, which are different from the conventional route via saddle-node bifurcation.
- In Chap. 3, a new path to generating the sustained oscillation is discovered. Conventionally, the sustained oscillation emerges via destabilizing the usual operating point. However, in the new path, the stability status of the usual operating point remains unchanged, but the trajectory moves to a faraway SPO under transient disturbances, leading to sustained oscillations.
- In Chap. 4, the similarity and differences between the homoclinic bifurcations of GFLCs and GFMCs are exposed, and a general set of criteria is established for studying the homoclinic bifurcation behavior in these converters. Notably, the homoclinic bifurcation dominates the change of the SEP's basin of attraction and therefore is highly relevant to the multi-swing stability and the transient stability improvement.
- In Chap. 5, the impact of the current control loop and the DC-link voltage control loop of GFLCs on the transient synchronization stability is studied, which extends the research work in Chap. 4. By studying the GFLC's complex behavior, we can simplify the analytical model without losing the accuracy to predict the stability, in some specific control parameters.

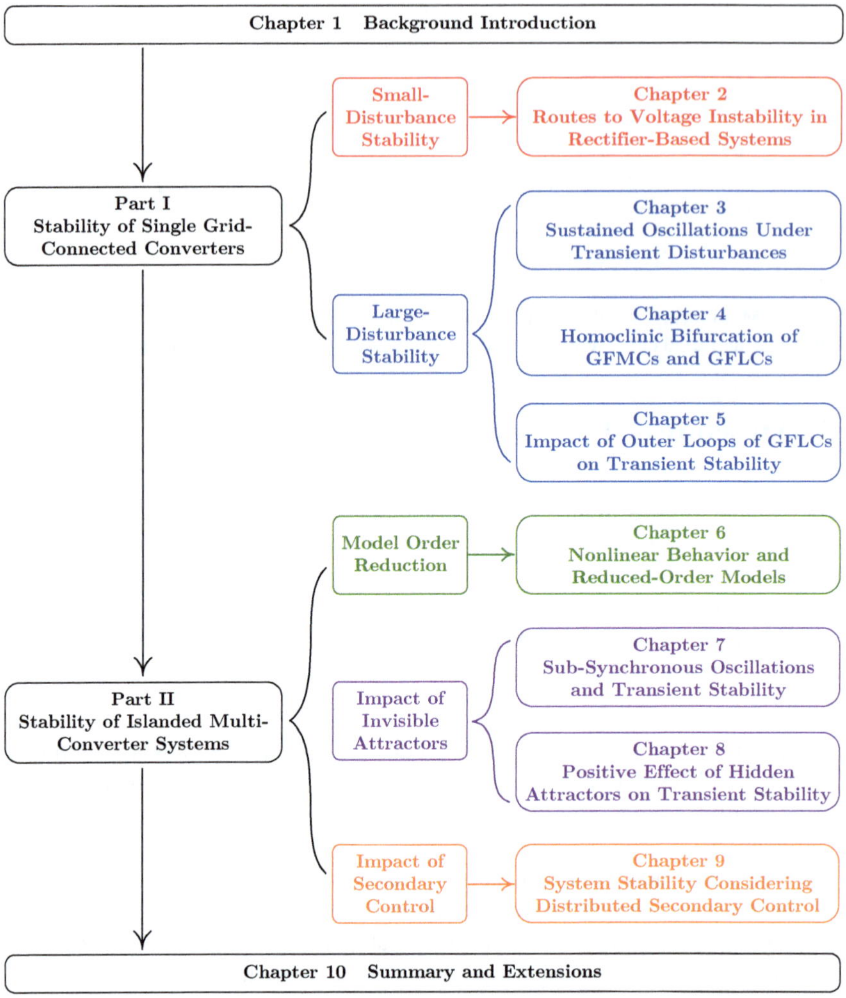

Fig. 1.36 Structure of this book

In Part II, there are four chapters. Chapter 6 studies the technique to reduce the order of microgrid models, Chapters 7 and 8 examine the impact of invisible attractors on the microgrids' transient stability, and Chap. 9 focuses on the impact of secondary control on the system stability.

- In Chap. 6, three typical types of slow-scale smooth bifurcations (i.e., Hopf, saddle-node, and homoclinic bifurcations) are considered, and adequate reduced-order models are derived to describe these nonlinear behaviors. On this basis, the procedure to establish the microgrid model is proposed.

- In Chap. 7, various types of hidden attractors are identified for an islanded microgrid containing a number of GFMCs. In contrary to the SPOs in Chaps. 3, 4, and 6, the aforementioned hidden attractors do not generate or subside from a homoclinic bifurcation. These hidden attractors enrich the form of sub-synchronous oscillations. Furthermore, the characteristics, bifurcation behaviors, and basins of attraction of these hidden attractors are studied, and transient stability enhancement methods are proposed.
- In Chap. 8, the transient synchronization stability of the islanded microgrid containing multiple GFMCs and GFLCs is studied. By analyzing the hidden attractors' characteristics and the related bifurcation behaviors, appropriate design guidelines are developed to enhance the microgrid transient stability. Besides, some specific hidden attractors are found to show benefits to maintain the transient synchronization stability, which is a departure from previous studies that primarily focus on the detrimental effects of hidden attractors.
- In Chap. 9, the saddle-node and Hopf bifurcation behaviors, and hidden attractors of the microgrid using the secondary control are studied. Besides, the impact of selecting communication network topology and driver controller location on the microgrid stability is investigated. Combining these results, a procedure to design the secondary control system is proposed.

Finally, in Chap. 10, we give a brief summary of our investigation into grid-connected converters, along with key extensions in this critical area of research.

Part I
Stability of Single Grid-Connected Converters

Chapter 2
Routes to Voltage Instability in Rectifier-Based Systems

Maintaining voltage stability within a given range at the point of common coupling (PCC) is crucial in a power system [2, 39]. Typically, a simple radial system is used to illustrate voltage stability, as depicted in Fig. 2.1a. The load at the PCC typically consists of a dynamic impedance load or a dynamic active/reactive power load, considering voltage dependence or exponential recovery dynamics. In the case of a dynamic active/reactive power load, the absorption of active and/or reactive power remains constant in the steady state. Power-voltage curves are derived from this radial system to investigate voltage stability, revealing that voltage instability is caused by a saddle-node bifurcation [40, 75–77], as shown in Fig. 2.1b. However, it is important to note that converter-based loads differ from the aforementioned types of loads, as they exhibit more nonlinear interactions with the grid, resulting in various pathways to voltage instability in converter-dominated systems.

In recent years, there has been growing interest in understanding the interaction between GFLCs and the power grid, as various instability phenomena related to GFLCs have been observed [1, 78–84]. However, the relationship between voltage instability in a power grid and the complex behavior of grid-following rectifiers has not been thoroughly investigated. It is important to examine the bifurcations caused by grid-following rectifiers, as they can lead to internal instability with significant consequences for the overall stability of the power system [2]. Additionally, certain complex behaviors of grid-following rectifiers, such as their heightened sensitivity to catastrophic bifurcations accompanied by supercritical Hopf bifurcations, the possibility of catastrophic bifurcations following subcritical Hopf bifurcations, and the potential interchange of supercritical and subcritical Hopf bifurcations, have not been adequately analyzed. To facilitate comprehension, this chapter focuses on studying the voltage instability of an LCL-type grid-following rectifier connected to a weak AC grid. This research serves as a foundation for future investigations into voltage stability in large-scale converter-dominated power grids.

J. Yang, C. K. Tse, *Complex Behavior of Grid-Connected Power Electronics Systems*, https://doi.org/10.1007/978-3-031-69340-3_2

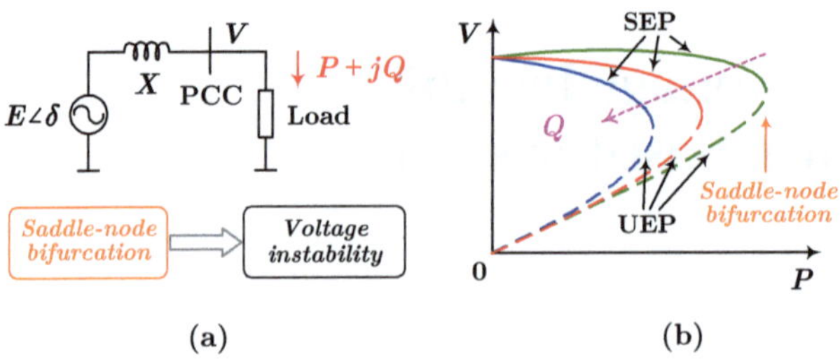

Fig. 2.1 (**a**) Schematic diagram of a simple radial system and route to voltage instability. (**b**) Power-voltage curves and saddle-node bifurcation of radial system

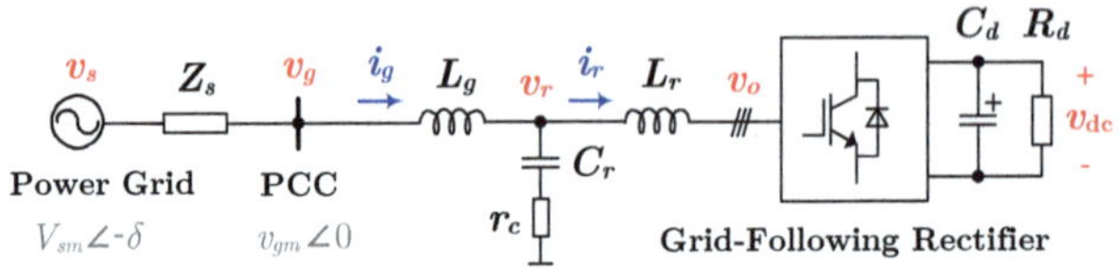

Fig. 2.2 Single-phase circuit model of an LCL-type grid-following rectifier connected to a weak AC grid. Here, the stray resistances of the electric reactors L_g and L_r are r_g and r_r, respectively

2.1 Instability Phenomena at a Glance

2.1.1 Topology and Control

The single-phase circuit model of an LCL-type grid-following rectifier connected to a weak AC grid is shown in Fig. 2.2, where L_g, C_r, and L_r are the grid-side inductance, capacitance, and converter-side inductance of the LCL filter, respectively. The weak AC grid is represented by a stiff voltage v_s with a large grid impedance Z_s connected in series. The impedance Z_s is modeled as a series connection of L_s and r_s. The stray resistances of the electric reactors L_g and L_r are r_g and r_r, respectively. Besides, r_c is the damping resistance of C_r [85]. Furthermore, C_d and R_d are the DC-link capacitor and load resistance of the grid-following rectifier, respectively. The magnitude and frequency of v_s are V_{sm} and ω_0, respectively. Also, v_g is the voltage of PCC, v_r is the voltage of C_r and r_c, v_o is the output voltage of the LCL filter, and v_{dc} is the DC-link voltage of the grid-following rectifier. Besides, the phase difference between v_g and v_s is defined as δ, as shown in Fig. 2.2. Here, v_{gm} is the magnitude of v_g. Moreover, i_g is the current of L_g, and i_r is the current of L_r. The above variables are expressed in the *abc*-frame. For example, i_g denotes i_{ga}, i_{gb}, and i_{gc}. Likewise, we represent the corresponding variables in the dq-frame as i_{gd} and i_{gq}.

The control structure in the dq-frame for the grid-following rectifier is shown in Fig. 2.3. The synchronous-reference-frame PLL is adopted here, where κ_p and κ_i are the proportional and integral gains of the PLL controller, respectively, and ω is the output frequency of the PLL. For the voltage controller, v_{dc}^* is the reference,

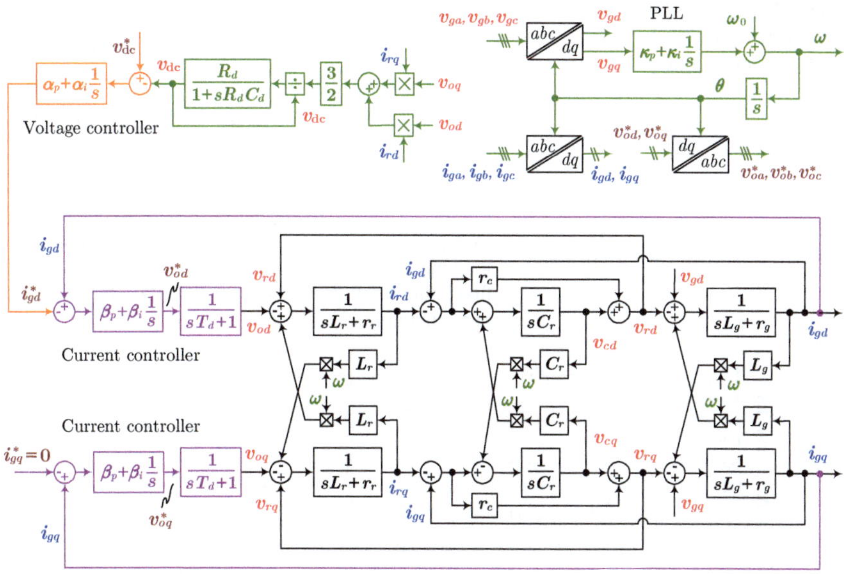

Fig. 2.3 Control block diagram in the dq-frame for the LCL-type grid-following rectifier

Table 2.1 Typical circuit parameters of the grid-following rectifier

Parameter	Value	Parameter	Value	Parameter	Value
V_{sm}	311 V	ω_0	100π rad/s	r_s	0.1 Ω
L_g	0.6 mH	C_r	60 μF	L_r	2 mH
r_g	0.05 Ω	r_c	1 Ω	r_r	0.1 Ω
C_d	3 mF	R_d	80 Ω		

and α_p and α_i are the proportional and integral gains, respectively. For the current controller, i_{gd}^* and i_{gq}^* are the references, and β_p and β_i are the proportional and integral gains, respectively. Focusing on the slow-scale dynamics, the time delay can be simplified as $\exp(-sT_d) \approx 1/(sT_d + 1)$, where $T_d = 1.5T_{\text{ctr}}$, and T_{ctr} is the control period of the grid-following rectifier. Finally, v_{od}^* and v_{oq}^* are the inputs of the time delay. Typical parameters and their values are summarized in Tables 2.1 and 2.2. It is noted that L_s "weakens" the AC power grid, playing an important role in the stability of the rectifier-based system. Therefore, L_s is selected as the bifurcation parameter in the following study.

2.1.2 Instability Phenomena

By utilizing MATLAB/Simulink simulations, we can generate the waveforms of the rectifier-based system, as depicted in Fig. 2.4. We adopt the default parameters

Table 2.2 Typical control parameters of the grid-following rectifier

Parameter	Value	p.u.	Parameter	Value	p.u.
v_{dc}^*	1 kV	/	f_{sw}	2 kHz	/
κ_p	0.413	1	κ_i	7.786	1
α_p	1.491	1	α_i	28.106	1
β_p	1.648	1	β_i	51.779	1

Note: f_{sw} is the switching frequency of the grid-following rectifier. Besides, the control frequency is set as $f_{\text{ctr}} = 2f_{\text{sw}}$, where $f_{\text{ctr}} = 1/T_{\text{ctr}}$. For the current-loop loop gain, the first phase margin is 78.5° for 102 Hz, the first gain margin is 10.7 dB for 662 Hz, and the second phase margin is 37.5° for 902 Hz

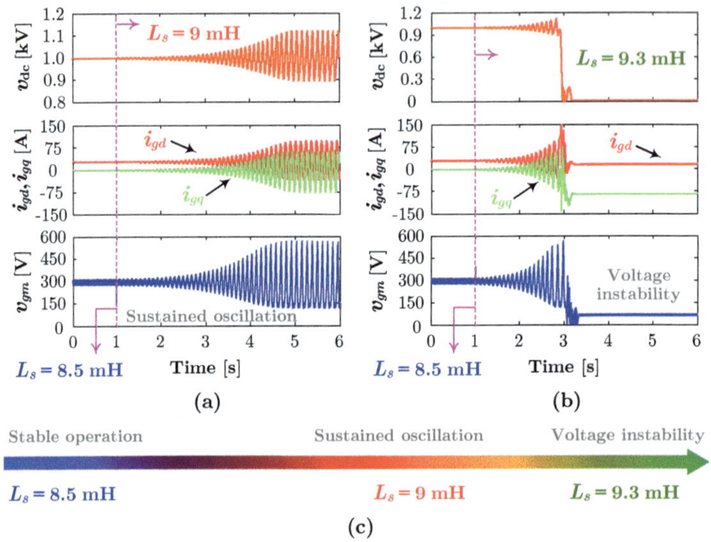

Fig. 2.4 Simulation waveforms showing dynamical behavior of a grid-following rectifier connected to a weak AC grid. All parameters follow Tables 2.1 and 2.2 with (**a**) $L_s = 8.5$ mH when $t < 1$ s, and $L_s = 9$ mH when $t > 1$ s. (**b**) $L_s = 8.5$ mH when $t < 1$ s, and $L_s = 9.3$ mH when $t > 1$ s. (**c**) Behavior of rectifier-based system under varying L_s

provided in Tables 2.1 and 2.2. Initially, the value of L_s is set to 8.5 mH, and the system operates normally. In Fig. 2.4a, at $t = 1$ s, we simulate the disconnection of a transmission line by setting L_s to 9 mH. Consequently, sustained low-frequency oscillation emerges, which can be observed in the waveforms of v_{dc}, v_{gm}, i_{gd}, and i_{gq}. In Fig. 2.4b, at $t = 1$ s, we further increase L_s to 9.3 mH. This leads to low-frequency oscillation, as evidenced by the waveforms of v_{dc}, v_{gm}, i_{gd}, and i_{gq}. After approximately 2 seconds, the system collapses, with v_{dc} and v_{gm} dropping almost to zero, i_{gd} reducing to 18 A, and i_{gq} decreasing to -82 A. It is worth mentioning that by slightly increasing L_s, the system can exhibit instability, sustained oscillation, or collapse, as demonstrated in Fig. 2.4c. Behind these phenomena, various types of bifurcation may occur.

2.2 System Model and Analysis in Weak AC Grids

2.2.1 Model of Rectifier-Based Systems

Referring to Fig. 2.3, the PLL can be modeled as

$$\frac{\mathrm{d}y_\omega}{\mathrm{d}t} = v_{gq} \tag{2.1}$$

$$\omega = \kappa_p v_{gq} + \kappa_i y_\omega + \omega_0 \tag{2.2}$$

where y_ω is the state variable of the PLL controller, and

$$\frac{\mathrm{d}\delta}{\mathrm{d}t} = \omega - \omega_0 \tag{2.3}$$

The dynamics of the voltage loop can be represented by

$$\frac{\mathrm{d}v_{\mathrm{dc}}}{\mathrm{d}t} = \frac{1}{C_d}\left(\frac{3}{2v_{\mathrm{dc}}}(v_{od}i_{rd} + v_{oq}i_{rq}) - \frac{v_{\mathrm{dc}}}{R_d}\right) \tag{2.4}$$

$$\frac{\mathrm{d}y_v}{\mathrm{d}t} = v_{\mathrm{dc}}^* - v_{\mathrm{dc}} \tag{2.5}$$

$$i_{gd}^* = \alpha_p(v_{\mathrm{dc}}^* - v_{\mathrm{dc}}) + \alpha_i y_v \tag{2.6}$$

where y_v is the state variable of the voltage controller. Moreover, the dynamics of the current loop can be described by

$$\frac{\mathrm{d}}{\mathrm{d}t}\begin{bmatrix} y_{id} \\ y_{iq} \end{bmatrix} = -\begin{bmatrix} i_{gd}^* \\ i_{gq}^* \end{bmatrix} + \begin{bmatrix} i_{gd} \\ i_{gq} \end{bmatrix} \tag{2.7}$$

$$\begin{bmatrix} v_{od}^* \\ v_{oq}^* \end{bmatrix} = \beta_p\left(-\begin{bmatrix} i_{gd}^* \\ i_{gq}^* \end{bmatrix} + \begin{bmatrix} i_{gd} \\ i_{gq} \end{bmatrix}\right) + \beta_i\begin{bmatrix} y_{id} \\ y_{iq} \end{bmatrix} \tag{2.8}$$

where y_{id} and y_{iq} are the state variables of the current controllers. The dynamics of the time delay block, LCL filter, and the grid impedance are described by

$$\frac{\mathrm{d}}{\mathrm{d}t}\begin{bmatrix} v_{od} \\ v_{oq} \end{bmatrix} = \frac{1}{T_d}\left(\begin{bmatrix} v_{od}^* \\ v_{oq}^* \end{bmatrix} - \begin{bmatrix} v_{od} \\ v_{oq} \end{bmatrix}\right) \tag{2.9}$$

$$\frac{\mathrm{d}}{\mathrm{d}t}\begin{bmatrix} i_{rd} \\ i_{rq} \end{bmatrix} = \frac{1}{L_r}\left(\begin{bmatrix} v_{rd} \\ v_{rq} \end{bmatrix} - \begin{bmatrix} v_{od} \\ v_{oq} \end{bmatrix} - \begin{bmatrix} r_r & -\omega L_r \\ \omega L_r & r_r \end{bmatrix}\begin{bmatrix} i_{rd} \\ i_{rq} \end{bmatrix}\right) \tag{2.10}$$

$$\frac{\mathrm{d}}{\mathrm{d}t}\begin{bmatrix} v_{cd} \\ v_{cq} \end{bmatrix} = \frac{1}{C_r}\left(\begin{bmatrix} i_{gd} \\ i_{gq} \end{bmatrix} - \begin{bmatrix} i_{rd} \\ i_{rq} \end{bmatrix} - \omega C_r \begin{bmatrix} 0 & -1 \\ 1 & 0 \end{bmatrix}\begin{bmatrix} v_{cd} \\ v_{cq} \end{bmatrix}\right) \tag{2.11}$$

$$\frac{\mathrm{d}}{\mathrm{d}t}\begin{bmatrix} i_{gd} \\ i_{gq} \end{bmatrix} = \frac{1}{L_s + L_g}\left(\begin{bmatrix} v_{sd} \\ v_{sq} \end{bmatrix} - \begin{bmatrix} v_{rd} \\ v_{rq} \end{bmatrix} - \begin{bmatrix} r_s + r_g & -\omega(L_s + L_g) \\ \omega(L_s + L_g) & r_s + r_g \end{bmatrix}\begin{bmatrix} i_{gd} \\ i_{gq} \end{bmatrix}\right) \tag{2.12}$$

and

$$\begin{bmatrix} v_{rd} \\ v_{rq} \end{bmatrix} = \begin{bmatrix} v_{cd} \\ v_{cq} \end{bmatrix} + r_c\left(\begin{bmatrix} i_{gd} \\ i_{gq} \end{bmatrix} - \begin{bmatrix} i_{rd} \\ i_{rq} \end{bmatrix}\right) \tag{2.13}$$

$$\begin{bmatrix} v_{gd} \\ v_{gq} \end{bmatrix} = \frac{L_s L_g}{L_s + L_g}\left(\frac{1}{L_s}\begin{bmatrix} v_{sd} \\ v_{sq} \end{bmatrix} + \frac{1}{L_g}\begin{bmatrix} v_{rd} \\ v_{rq} \end{bmatrix} + \left(\frac{r_g}{L_g} - \frac{r_s}{L_s}\right)\begin{bmatrix} i_{gd} \\ i_{gq} \end{bmatrix}\right) \tag{2.14}$$

where $v_{sd} = V_{sm}\cos\delta$, $v_{sq} = -V_{sm}\sin\delta$, and v_{cd} and v_{cq} are the voltages of C_r in the dq-frame.

Equations (2.1) to (2.14) represent the model of a rectifier-based system. The order of the system is 14, and the state vector is

$$\mathbf{u_o} = \begin{bmatrix} y_v & v_{\mathrm{dc}} & y_{id} & y_{iq} & v_{od} & v_{oq} & y_\omega & \delta & i_{rd} & i_{rq} & v_{cd} & v_{cq} & i_{gd} & i_{gq} \end{bmatrix}^T$$

where the element of $\mathbf{u_o}$ is denoted as $u_{o\ell}$. Here, $\ell = 1, 2, \cdots, 14$. Then, the system can be represented as

$$\frac{\mathrm{d}\mathbf{u_o}}{\mathrm{d}t} = \mathbf{F}(\mathbf{u_o}) \tag{2.15}$$

where the element of $\mathbf{F}(\mathbf{u_o})$ is denoted as $f_\ell(\mathbf{u_o}) \triangleq f_\ell(u_{o1}, u_{o2}, \cdots, u_{o14})$.

2.2.2 *Saddle-Node Bifurcation and Catastrophic Bifurcation*

Using the derived model, the bifurcation diagram showing the stability status of the EQPs is plotted in Fig. 2.5. Here, the solid curve represents the SEP, whereas the dashed curve represents the UEP. From Fig. 2.5, a Hopf bifurcation occurs when $L_s = 8.7$ mH, which converts the SEP to a UEP. For clarity, this EQP is denoted as EQP1 here. Furthermore, a generalized saddle-node bifurcation occurs when $L_s = 17.27$ mH, where two UEPs merge and vanish. It should be noted that the generalized saddle-node bifurcation differs from the saddle-node bifurcation where an SEP and a UEP merge and disappear. Due to the Hopf bifurcation, there is no saddle-node bifurcation in this case. Besides, in Fig. 2.4b, the system collapses

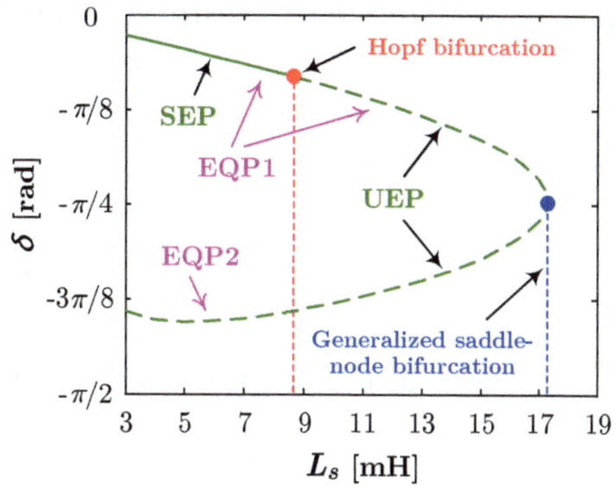

Fig. 2.5 Codimension-1 bifurcation diagram showing the EQPs. Solid curve denotes SEP, and dashed curve denotes UEP. System parameters are given in Tables 2.1 and 2.2 with varying L_s

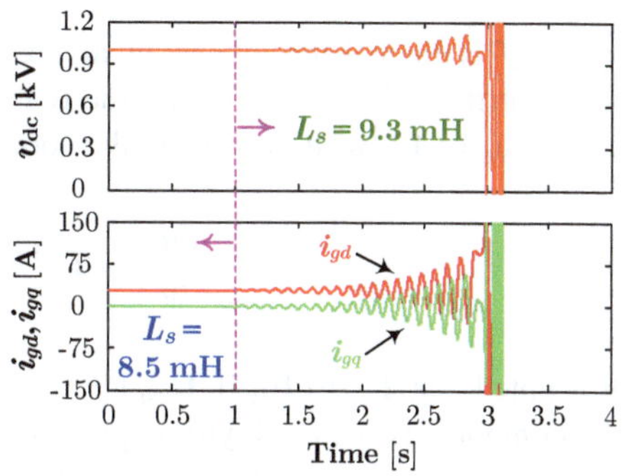

Fig. 2.6 Waveforms of v_{dc}, i_{gd}, and i_{gq} derived using the Runge-Kutta numerical method. All parameters follow Tables 2.1 and 2.2 with $L_s = 8.5$ mH when $t < 1$ s, and $L_s = 9.3$ mH when $t > 1$ s

when $L_s = 9.3$ mH, which is far from the generalized saddle-node bifurcation point. Therefore, the Hopf bifurcation induces a sustained oscillation as shown in Fig. 2.4a, but the (generalized) saddle-node bifurcation is not the cause of system collapse shown in Fig. 2.4b.

As studied previously [81, 83], the catastrophic bifurcation may occur when i_{gd} reaches a critical value. In that case, the current loop is driven into an overmodulation region, which significantly changes the system structure. The final behavior of the rectifier-based system shown in Fig. 2.4b is the result of a catastrophic bifurcation, as $v_{dc} \approx 0$, and an abundant amount of reactive current is injected. Therefore, the catastrophic bifurcation directly causes the system to collapse. However, there may be other factors contributing to the enlargement of the oscillation amplitudes of i_{gd} and i_{gq}. To probe further, the waveforms of v_{dc}, i_{gd}, and i_{gq} derived using the Runge-Kutta numerical method are provided, as shown in Fig. 2.6. It is noted that there is no limit in the PLL, voltage loop, and current loop in the numerical algorithm. From Fig. 2.6, the oscillation amplitudes of v_{dc}, i_{gd}, and i_{gq} enlarge fiercely after L_s switches to 9.3 mH, and finally these variables diverge to infinity, implying the existence of another bifurcation.

2.2.3 Existence of Unstable Periodic Orbit

First, we change the coordinates of (2.15) and move the EQP to the origin. The value of $\mathbf{u_o}$ in EQP1 is denoted as $\mathbf{U}_{\mathrm{oss}}$. Then, we can set $\mathbf{u_c} = \mathbf{u_o} - \mathbf{U}_{\mathrm{oss}}$. Therefore,

$$\frac{\mathrm{d}\mathbf{u_c}}{\mathrm{d}t} = \mathbf{F}(\mathbf{u_c} + \mathbf{U}_{\mathrm{oss}}) \tag{2.16}$$

The Jacobian, denoted as $\mathbf{A}$, can be derived as

$$\mathbf{A} = \left[\left. \frac{\partial f_\ell}{\partial u_{ck}} \right|_{\mathbf{u_c}=\mathbf{0}} \right]_{\ell,k=1,2,\cdots,14} = \mathbf{T}\mathcal{J}\mathbf{T}^{-1}$$

where $\mathcal{J}$ is the real Jordan canonical matrix and $\mathbf{T}$ is the corresponding transformation matrix, which is formed by the real generalized eigenvectors of $\mathbf{A}$. Then, we can set $\mathbf{u_e} = \mathbf{T}^{-1}\mathbf{u_c}$ to transform the coordinate to the eigenbasis, i.e.,

$$\frac{\mathrm{d}\mathbf{u_e}}{\mathrm{d}t} = \mathbf{T}^{-1}\mathbf{F}(\mathbf{T}\mathbf{u_e} + \mathbf{U}_{\mathrm{oss}}) \tag{2.17}$$

It is noted that $\mathbf{T}^{-1}\mathbf{F}(\mathbf{T}\mathbf{u_e} + \mathbf{U}_{\mathrm{oss}}) = \mathcal{J}\mathbf{u_e} + \mathcal{O}(\|\mathbf{u_e}\|^2)$. Here, $\mathcal{O}(\|\mathbf{u_e}\|^2)$ represents the components whose Taylor expansion in $\mathbf{u_e}$ starts with at least quadratic terms, e.g., $u_{e\ell}u_{ek}$, or $u_{e\ell}^2$, where $\ell, k = 1, 2, \cdots, 14$. Here, the element of $\mathbf{u_e}$ is denoted as $u_{e\ell}$.

It is observed that when $L_s \approx 8.7$ mH, a pair of complex eigenvalues of $\mathbf{A}$ will cross the imaginary axis and enter the right-half plane. These eigenvalues correspond to u_{e9} and u_{e10}. Meanwhile, the real part of the other eigenvalues remains negative. Hence, any trajectory in $u_{e\ell}$-direction where $\ell \neq 9, 10$ will converge to the nearby area of the origin, while such a trajectory in $u_{e\ell}$-direction where $\ell = 9, 10$ will either converge to the origin, exhibit sustained oscillation, or diverge to infinity. Some examples are also provided in Fig. 2.7. For Fig. 2.7a where $L_s = 8.5$ mH, the trajectory spins around in the vicinity of the (u_{e9}, u_{e10})-plane and finally converges to the origin. Besides, the trajectory in u_{e12}-direction quickly converges to the plane $u_{e12} \approx 0$. For Fig. 2.7b where $L_s = 9$ mH, the trajectory converges to the plane $u_{e12} \approx 0$ first, then spins arounds until it reaches an SPO. For Fig. 2.7c where $L_s = 9.3$ mH, the trajectory also converges to the plane $u_{e12} \approx 0$ first, and then after a few spins near the (u_{e9}, u_{e10})-plane, it finally diverges to infinity.

From Fig. 2.7, the system's nonlinear behavior is exhibited clearly in the u_{e9}-coordinate and the u_{e10}-coordinate. We thus choose different initial conditions: $u_{e\ell} \neq 0$ for $\ell = 9, 10$, and $u_{e\ell} = 0$ for $\ell \neq 9, 10$. We find that some trajectories

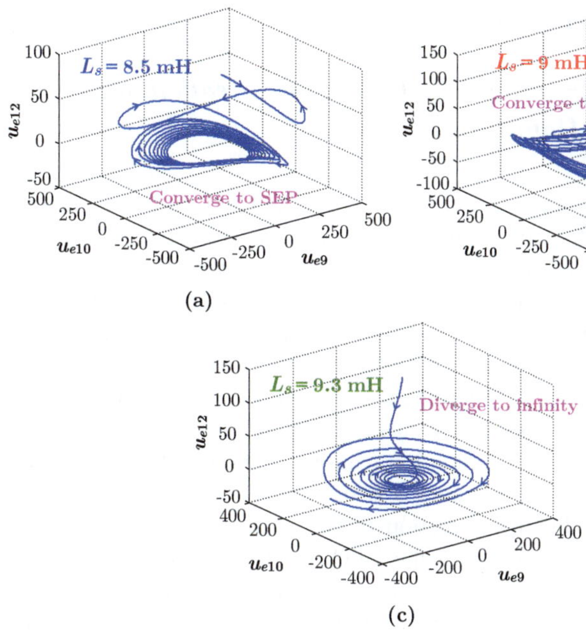

Fig. 2.7 Typical trajectories demonstrated in (u_{e9}, u_{e10}, u_{e12})-space, which are derived using the Runge-Kutta numerical method with running time of 1 s. Parameter values are given in Tables 2.1 and 2.2 with (**a**) $L_s = 8.5$ mH, (**b**) $L_s = 9$ mH, and (**c**) $L_s = 9.3$ mH

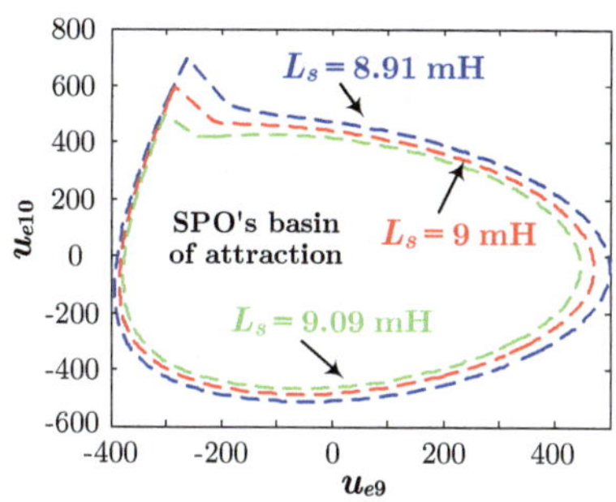

Fig. 2.8 SPO's basin of attraction in (u_{e9}, u_{e10})-plane, which is inside the dashed curves. Parameter values are given in Tables 2.1 and 2.2 with varying L_s

converge to the SPO, while others diverge to infinity. Then, the SPO's basin of attraction in the (u_{e9}, u_{e10})-plane for different values of L_s can be identified, as shown in Fig. 2.8. The SPO's basin of attraction is located inside the dashed curves, which implies the existence of a UPO. It is also observed from Fig. 2.8 that the SPO's basin of attraction shrinks as L_s increases, implying the contraction of the UPO.

2.3 Bifurcation Analysis

2.3.1 *Bautin Bifurcation, Saddle-Node Bifurcation of Periodic Orbits, and Hopf Bifurcation*

As shown in the foregoing section, an SPO and a UPO may coexist. In other words, four types of bifurcation may occur, namely, supercritical Hopf bifurcation, SNPO, subcritical Hopf bifurcation, and Bautin bifurcation, as illustrated in Fig. 2.9. According to Fig. 2.9a, the SPO is generated from a supercritical Hopf bifurcation. Then, its magnitude increases continuously until it merges with the UPO, where an SNPO occurs. After the onset of the SNPO, there is no periodic orbit, or more precisely, there is no attractor. According to Fig. 2.9b, the SEP absorbs the UPO and becomes a UEP after the onset of the subcritical Hopf bifurcation. Between these two cases shown in Fig. 2.9, a Bautin bifurcation occurs, which changes the Hopf bifurcation from supercritical to subcritical or vice versa. As the UPO is located outside the SPO, this Bautin bifurcation occurs in a subcritical mode [54, 86]. Furthermore, due to the presence of the UPO, the phase space is separated into two regions, one being the SEP/SPO's basin of attraction. Thus, if the initial state is in this region, the trajectory will be attracted to the SEP/SPO. On the contrary, if the initial state is outside this region, the trajectory will diverge to infinity.

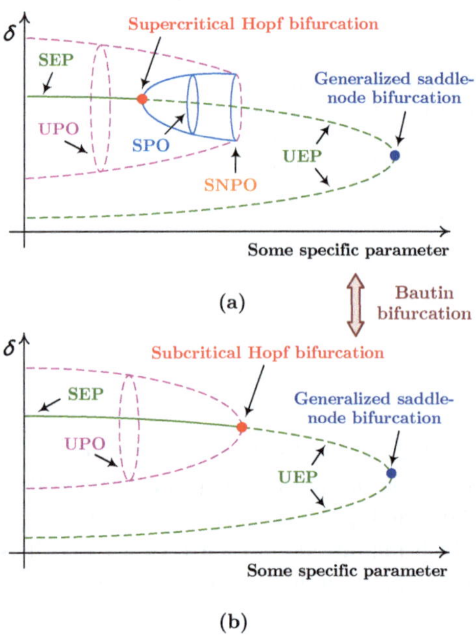

Fig. 2.9 Codimension-1 bifurcation diagrams corresponding to (**a**) supercritical Hopf bifurcation and SNPO and (**b**) subcritical Hopf bifurcation in the rectifier-based system. Between (**a**) and (**b**), a Bautin bifurcation occurs

2.3.2 Routes to Voltage Instability

For the rectifier-based power system, the trajectory cannot diverge to infinity in practice but be bounded, e.g., limited by the modulation of rectifiers. Hence, for Fig. 2.9a, after the onset of an SNPO, the grid current will increase dramatically and eventually reaches a critical value that triggers a catastrophic bifurcation [81, 83]. Then, the DC-link voltage and the PCC voltage collapses, leading to the voltage instability. As shown in Fig. 2.9b, a similar consequence exists after the onset of the subcritical Hopf bifurcation, where the catastrophic bifurcation and the voltage instability will occur eventually. It is also worth noting that for some specific ranges of parameters (e.g., small L_s and R_d), supercritical or subcritical Hopf bifurcation may not occur. The system will operate normally until the onset of a saddle-node bifurcation. As no attractor exists after the onset of the saddle-node bifurcation, the system collapses finally. Furthermore, under a transient disturbance (e.g., the grid voltage dip), the trajectory may be pushed out of the SEP/SPO's basin of attraction, resulting in the voltage instability. The routes to the voltage instability are summarized in Fig. 2.10.

2.3.3 Center Manifolds and First Lyapunov Coefficient

The center manifold is applied to convert the original high-order system into a reduced-order system preserving the salient nonlinear behavior. Based on the reduced-order system, the first Lyapunov coefficient can be derived, which can be used to predict the occurrence mode of Hopf bifurcation.

Define $\mathbf{y} = [u_{e1} \ \cdots \ u_{e8} \ u_{e11} \ \cdots \ u_{e14}]^T$ and $\mathbf{x} = [u_{e9} \ u_{e10}]^T$. For clarity, the elements of $\mathbf{x}$ and $\mathbf{y}$ are denoted as x_ℓ and y_k, respectively, where $\ell = 1, 2$, and $k = 1, 2, \cdots, 12$. Then, (2.17) can be divided as two subsystems, i.e.,

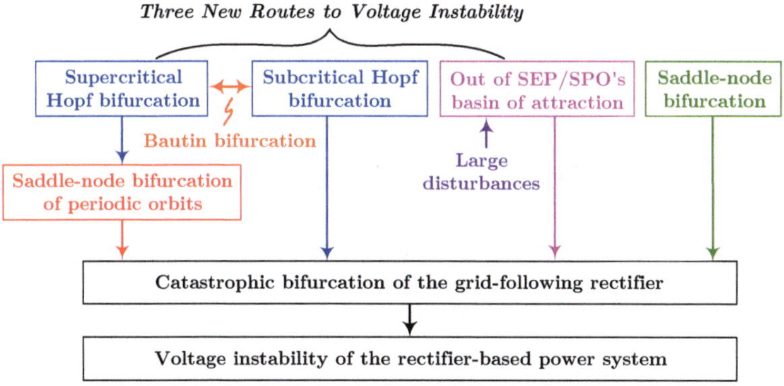

Fig. 2.10 Routes to the voltage instability in a weak AC grid

$$\frac{\mathrm{d}\mathbf{x}}{\mathrm{d}t} = \mathcal{J}_{\mathbf{b}}\mathbf{x} + \mathbf{P}(\mathbf{x}, \mathbf{y}) \tag{2.18}$$

$$\frac{\mathrm{d}\mathbf{y}}{\mathrm{d}t} = \mathcal{J}_{\mathbf{c}}\mathbf{y} + \mathbf{Q}(\mathbf{x}, \mathbf{y}) \tag{2.19}$$

where $\mathbf{x} \in \mathbb{R}^2$, $\mathbf{y} \in \mathbb{R}^{12}$, $\mathcal{J}_{\mathbf{b}} \in \mathbb{R}^2 \times \mathbb{R}^2$, and $\mathcal{J}_{\mathbf{c}} \in \mathbb{R}^{12} \times \mathbb{R}^{12}$. Here, $\mathbb{R}^\ell$ represents the ℓ-dim real-value vector, while $\mathbb{R}^\ell \times \mathbb{R}^k$ represents the $\ell \times k$ real-value matrix. Besides, $\mathbf{P}(\mathbf{x}, \mathbf{y})$ and $\mathbf{Q}(\mathbf{x}, \mathbf{y})$ are real-value functions, which have Taylor expansions in $[\mathbf{x}\ \mathbf{y}]^{\mathrm{T}}$ starting with at least quadratic terms, e.g., $x_\ell y_k$, x_ℓ^2 or y_k^2. The elements of $\mathbf{P}(\mathbf{x}, \mathbf{y})$ and $\mathbf{Q}(\mathbf{x}, \mathbf{y})$ are denoted as $p_\ell(\mathbf{x}, \mathbf{y})$ and $q_k(\mathbf{x}, \mathbf{y})$, respectively. Furthermore,

$$\mathcal{J}_{\mathbf{b}} = \begin{bmatrix} \alpha_b & \omega_b \\ -\omega_b & \alpha_b \end{bmatrix}, \quad \mathcal{J}_{\mathbf{c}} = \begin{bmatrix} \mathcal{J}_{\mathbf{c1}} & 0 \\ 0 & \mathcal{J}_{\mathbf{c2}} \end{bmatrix}, \quad \mathcal{J} = \begin{bmatrix} \mathcal{J}_{\mathbf{c1}} & 0 & 0 \\ 0 & \mathcal{J}_{\mathbf{b}} & 0 \\ 0 & 0 & \mathcal{J}_{\mathbf{c2}} \end{bmatrix}.$$

Here, $\lambda_b = \alpha_b + j\omega_b$ and $\overline{\lambda}_b = \alpha_b - j\omega_b$ are the corresponding complex eigenvalues of $\mathbf{A}$, where α_b may become positive.

Defining $z = x_1 + jx_2$, the differential equations can be rewritten as

$$\frac{\mathrm{d}z}{\mathrm{d}t} = \overline{\lambda}_b z + g(z, \overline{z}, \mathbf{y}) \tag{2.20}$$

$$\frac{\mathrm{d}\mathbf{y}}{\mathrm{d}t} = \mathcal{J}_{\mathbf{c}}\mathbf{y} + \mathbf{H}^{\mathbf{f}}(z, \overline{z}, \mathbf{y}) \tag{2.21}$$

where $z, \overline{z} \in \mathbb{C}^1$. Here, $\mathbb{C}^\ell$ represents the ℓ-dim complex-value vector. Besides, $g(z, \overline{z}, \mathbf{y})$ and $\mathbf{H}^{\mathbf{f}}(z, \overline{z}, \mathbf{y})$ are complex-value functions. The elements of $\mathbf{H}^{\mathbf{f}}(z, \overline{z}, \mathbf{y})$ are denoted as $h_k^f(z, \overline{z}, \mathbf{y})$. Note that $g(z, \overline{z}, \mathbf{y}) = p_1(\mathbf{x}, \mathbf{y}) + jp_2(\mathbf{x}, \mathbf{y})$, and $\mathbf{H}^{\mathbf{f}}(z, \overline{z}, \mathbf{y}) = \mathbf{Q}(\mathbf{x}, \mathbf{y})$, where $x_1 = 0.5(z + \overline{z})$ and $x_2 = 0.5j(\overline{z} - z)$.

Using the Taylor expansions in z and $\overline{z}$, we get

$$\begin{aligned} \frac{\mathrm{d}z}{\mathrm{d}t} = \overline{\lambda}_b z + \frac{1}{2}G_{20}z^2 + G_{11}z\overline{z} + \frac{1}{2}G_{02}\overline{z}^2 + \frac{1}{2}G_{21}z^2\overline{z} \\ + \langle \mathbf{M_{10}}, \mathbf{y}\rangle z + \langle \mathbf{M_{01}}, \mathbf{y}\rangle \overline{z} + \cdots \end{aligned} \tag{2.22}$$

$$\frac{\mathrm{d}\mathbf{y}}{\mathrm{d}t} = \mathcal{J}_{\mathbf{c}}\mathbf{y} + \frac{1}{2}\mathbf{H_{20}}z^2 + \mathbf{H_{11}}z\overline{z} + \frac{1}{2}\mathbf{H_{02}}\overline{z}^2 + \cdots \tag{2.23}$$

where

$$G_{\ell_1\ell_2} = \left. \frac{\partial^\ell}{\partial z^{\ell_1} \partial \overline{z}^{\ell_2}} g(z, \overline{z}, \mathbf{0}) \right|_{z=0},$$

$$\overline{m}_{\ell_1\ell_2,k} = \left.\frac{\partial^{\ell+1}}{\partial y_k \partial z^{\ell_1} \partial \overline{z}^{\ell_2}} g(z, \overline{z}, \mathbf{y})\right|_{z=0,\mathbf{y}=\mathbf{0}},$$

$$\langle \mathbf{M}_{\ell_1\ell_2}, \mathbf{y} \rangle = \sum_{k=1}^{12} \overline{m}_{\ell_1\ell_2,k} \times y_k,$$

$$\mathbf{H}_{\ell_1\ell_2} = \left.\frac{\partial^{\ell}}{\partial z^{\ell_1} \partial \overline{z}^{\ell_2}} \mathbf{H}^{\mathbf{f}}(z, \overline{z}, \mathbf{0})\right|_{z=0}$$

It is noted that $\ell = \ell_1 + \ell_2$, $G_{\ell_1\ell_2}, G_{21}, m_{\ell_1\ell_2,k} \in \mathbb{C}^1$, and $\mathbf{M}_{\ell_1\ell_2}, \mathbf{H}_{\ell_1\ell_2} \in \mathbb{C}^{12}$, where $\ell = 1, 2$, and $k = 1, 2, \cdots, 12$. As shown in Fig. 2.7, the trajectories are attracted to a specific surface near the origin, and such a surface is called the *center manifold* [54, 87], which can be represented as

$$\mathbf{y} = \mathbf{V}(z, \overline{z}) = \frac{1}{2}\mathbf{W}_{20} z^2 + \mathbf{W}_{11} z\overline{z} + \frac{1}{2}\mathbf{W}_{02} \overline{z}^2 + \cdots \tag{2.24}$$

where $\mathbf{W}_{20}, \mathbf{W}_{02} \in \mathbb{C}^{12}$, and $\mathbf{W}_{11} \in \mathbb{R}^{12}$. Since

$$\frac{\mathrm{d}\mathbf{y}}{\mathrm{d}t} = \frac{\mathrm{d}\mathbf{y}}{\mathrm{d}z}\frac{\mathrm{d}z}{\mathrm{d}t} + \frac{\mathrm{d}\mathbf{y}}{\mathrm{d}\overline{z}}\overline{\left(\frac{\mathrm{d}z}{\mathrm{d}t}\right)},$$

we have

$$\mathbf{W}_{20} = \frac{1}{2}\left(\overline{\lambda}_b \mathbf{I} - \frac{1}{2}\mathcal{J}_{\mathbf{c}}\right)^{-1} \mathbf{H}_{20},$$

$$\mathbf{W}_{11} = (2\alpha_b \mathbf{I} - \mathcal{J}_{\mathbf{c}})^{-1} \mathbf{H}_{11}, \quad ,$$

$$\mathbf{W}_{02} = \frac{1}{2}\left(\lambda_b \mathbf{I} - \frac{1}{2}\mathcal{J}_{\mathbf{c}}\right)^{-1} \mathbf{H}_{02},$$

where $\mathbf{I}$ is the identity matrix. As an example, the center manifold in (u_{e9}, u_{e10}, u_{e12})-space when $L_s = 9$ mH is exhibited in Fig. 2.11. As shown in Figs. 2.7b and 2.11, the trajectory wanders around the center manifold when it gets close to the origin. In other words, the dynamics of the system represented by (2.17) are essentially determined by the center manifold.

Combining (2.22), (2.23), and (2.24), a reduced system describing the dynamics of the center manifold can be derived as

$$\begin{aligned}\frac{\mathrm{d}z}{\mathrm{d}t} &= \overline{\lambda}_b z + \frac{1}{2}G_{20} z^2 + G_{11} z\overline{z} + \frac{1}{2}G_{02}\overline{z}^2 \\ &+ \left(\frac{1}{2}G_{21} + \langle \mathbf{M}_{10}, \mathbf{W}_{11} \rangle + \frac{1}{2}\langle \mathbf{M}_{01}, \mathbf{W}_{20} \rangle\right) z^2\overline{z} + \cdots\end{aligned} \tag{2.25}$$

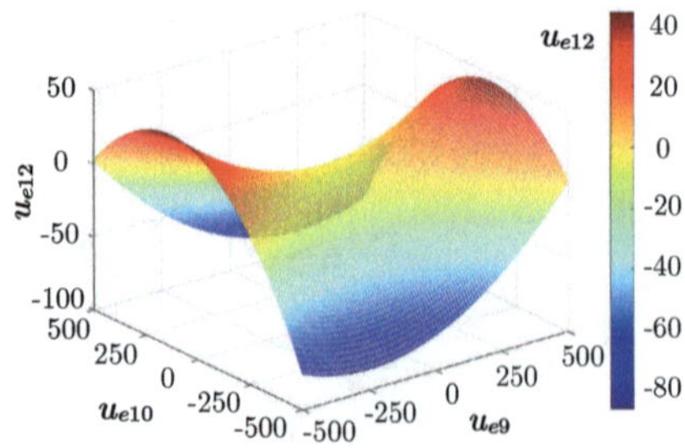

Fig. 2.11 Center manifold demonstrated in (u_{e9}, u_{e10}, u_{e12})-space. Parameter values are given in Tables 2.1 and 2.2 with $L_s = 9$ mH. It is also noted that $u_{e9} = x_1 = \mathrm{Re}(z)$, $u_{e10} = x_2 = \mathrm{Im}(z)$, and $u_{e12} = y_{10}$

Basically, this reduced order system can be regarded as a projection of the original 14th-order system on the center manifold.

Using the Taylor series expansion, the coordinates can be changed [54, 87], as given by

$$\frac{\mathrm{d}\vartheta}{\mathrm{d}t} = \overline{\lambda}_b \vartheta + c_1 \vartheta^2 \overline{\vartheta} + \cdots \tag{2.26}$$

where $\vartheta, c_1 \in \mathbb{C}^1$, and

$$c_1 = \frac{G_{20}G_{11}(2\overline{\lambda}_b + \lambda_b)}{2|\lambda_b|^2} + \frac{|G_{11}|^2}{\overline{\lambda}_b} + \frac{|G_{02}|^2}{2(2\overline{\lambda}_b - \lambda_b)}$$
$$+ \frac{1}{2}G_{21} + \langle \mathbf{M_{10}}, \mathbf{W_{11}} \rangle + \frac{1}{2}\langle \mathbf{M_{01}}, \mathbf{W_{20}} \rangle$$

Furthermore,

$$l_1(0) \triangleq \mathrm{Re}(c_1)\Big|_{\alpha_b \to 0} \tag{2.27}$$

which is the first Lyapunov coefficient. Specifically, $l_1(0) < 0$ implies a supercritical mode, while $l_1(0) > 0$ implies a subcritical mode.

2.4 Design-Oriented Analysis

2.4.1 Saddle-Node Bifurcation of Periodic Orbits and Supercritical Hopf Bifurcation

As discussed above, when a supercritical Hopf bifurcation occurs, the real part of the complex eigenvalue λ_b is positive, and the first Lyapunov coefficient is negative, i.e., $\alpha_b < 0$ and $l_1(0) < 0$. In this case, an SPO can be observed numerically. On this

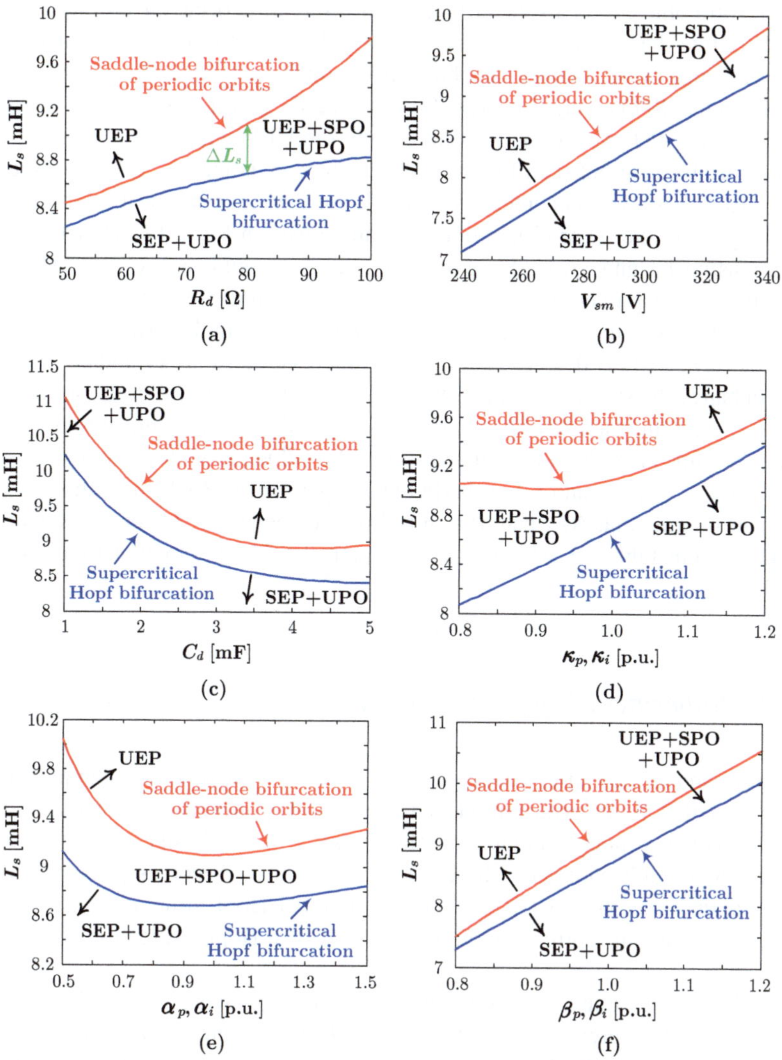

Fig. 2.12 Codimension-2 bifurcation diagrams corresponding to supercritical Hopf bifurcation and SNPO on (**a**) R_d-L_s plane; (**b**) V_{sm}-L_s plane; (**c**) C_d-L_s plane; (**d**) κ_p-L_s plane ($\kappa_p = \kappa_i$ in p.u. value); (**e**) α_p-L_s plane ($\alpha_p = \alpha_i$ in p.u. value); and (**f**) β_p-L_s plane ($\beta_p = \beta_i$ in p.u. value). Other parameters follow the default set provided in Tables 2.1 and 2.2. Above the red curve, only UEP exists. Between the red and blue curves, UEP, SPO, and UPO exist. Below the blue curve, SEP and UPO exist

basis, if the SPO vanishes and the trajectory diverges to infinity, an SNPO occurs. The codimension-2 bifurcation diagrams corresponding to SNPO and supercritical Hopf bifurcation can be derived, as shown in Fig. 2.12. In Fig. 2.12a, the bifurcation diagrams are displayed in the R_d-L_s plane. On the solid red curve, an SNPO occurs.

On the solid blue curve, a supercritical Hopf bifurcation occurs. Here, we focus on EQP1. Below the blue curve, the EQP1 is an SEP; moreover, a UPO, surrounding the EQP1, also exists. Between the blue curve and the red curve, the EQP1 becomes a UEP, and an SPO emerges nearby. Besides, the UPO still exists, which surrounds the UEP and SPO. Above the red curve, only the UEP exists. For a specific value of R_d, the distance between the two solid curves is defined as ΔL_s. We see that as R_d decreases, ΔL_s becomes small, implying the reduction of the SPO regime. Similarly, from the other bifurcation diagrams given in Fig. 2.12, decreasing V_{sm}, or increasing κ_p and κ_i, or decreasing β_p and β_i, would reduce ΔL_s. However, adjusting other parameters (e.g., C_d, α_p and α_i) has little effect on ΔL_s.

In practice, we may try to keep the system operating in the SEP area. From this point of view, we may decrease L_s, or C_d, or increase R_d, V_{sm}, κ_p and κ_i, or β_p and β_i to enhance the system stability. Due to the non-monotonicity of the solid curves in Fig. 2.12e, it may not be effective to adjust α_p and α_i to maintain system stability. Using the same procedure, we can investigate the influence of other parameters (e.g., L_g, L_r, C_r, r_s, r_g, r_r, and r_c) on SNPO and supercritical Hopf bifurcation as well. It is worth noting that ΔL_s can be reduced by increasing L_g, L_r, or C_r. Results are summarized in Table 2.3, which provides practical guidelines to improve the system stability.

2.4.2 *Subcritical Hopf Bifurcation*

As shown above, the Hopf bifurcation always occurs in the supercritical mode. Here, we also see that the subcritical mode may be caused by increasing L_g and L_r (which will be explained later). As an example, we set $L_g = 2.7$ mH and $L_r = 9$ mH as the new default parameters here. In this case, if α_b becomes positive and the first Lyapunov coefficient $l_1(0) > 0$, a subcritical Hopf bifurcation occurs. The bifurcation diagrams corresponding to the subcritical Hopf bifurcation are given in Fig. 2.13.

The bifurcation diagrams shown in Fig. 2.13a are presented in the R_d-L_s plane. Solid curves of different colors correspond to different values of L_g and L_r. On the solid curves, a subcritical Hopf bifurcation occurs. Here, we focus on the EQP1. Below the solid curves, the EQP1 is an SEP; moreover, a UPO, surrounding the EQP1, also exists. Above the solid curves, the EQP1 becomes a UEP, and no periodic orbit exists anymore. From Fig. 2.13a, we may reduce L_s or increase R_d to avoid triggering a subcritical Hopf bifurcation. Besides, we may decrease L_g and L_r to enlarge the SEP regime to maintain the stability of the system. The bifurcation diagrams shown in Fig. 2.13b and c are also presented in the R_d-L_s plane. For Fig. 2.13d, e, and f, the bifurcation diagrams are presented in the V_{sm}-L_s plane. We see that increasing C_r, or decreasing C_d, or increasing κ_p and κ_i, or increasing α_p and α_i, or increasing β_p and β_i, would convert the UEP region into an SEP

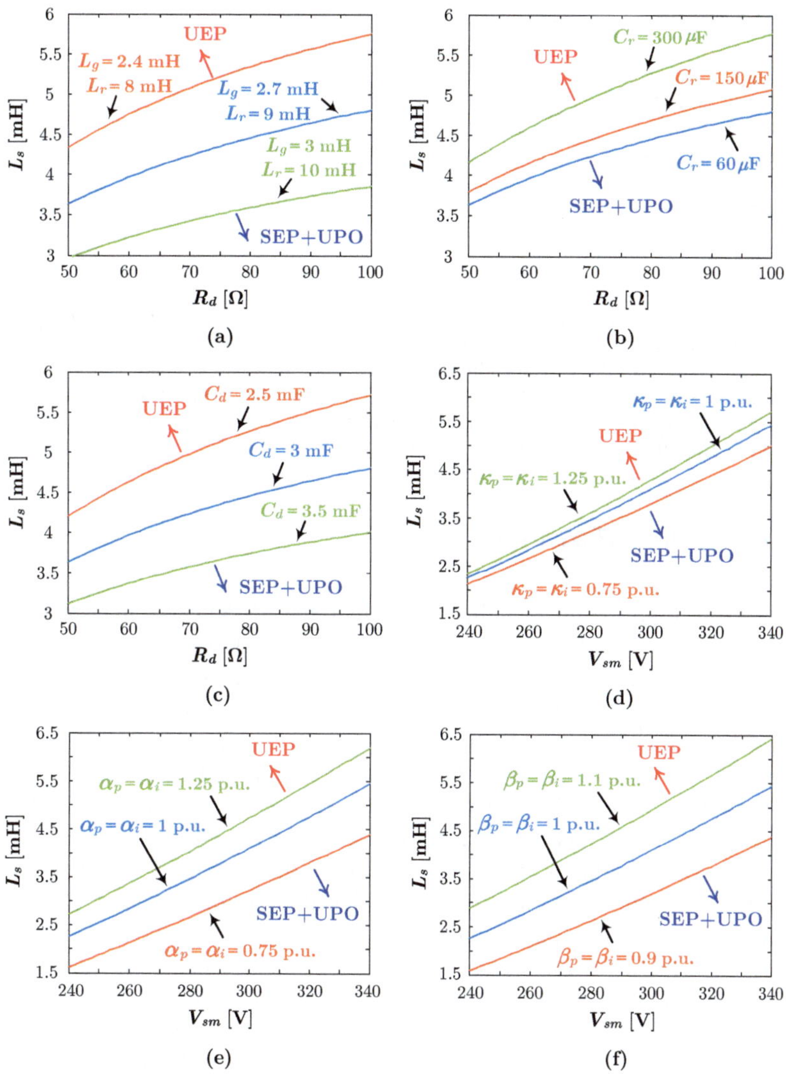

Fig. 2.13 Codimension-2 bifurcation diagrams corresponding to subcritical Hopf bifurcation on (**a**) R_d-L_s plane for varying L_g and L_r; (**b**) R_d-L_s plane for varying C_r; (**c**) R_d-L_s plane for varying C_d; (**d**) V_{sm}-L_s plane for varying κ_p and κ_i; (**e**) V_{sm}-L_s plane for varying α_p and α_i; and (**f**) V_{sm}-L_s plane for varying β_p and β_i. Other parameters adopt the default values provided in Tables 2.1 and 2.2 with $L_g = 2.7$ mH and $L_r = 9$ mH. Above the solid curves, only UEP exists. Below the solid curves, SEP and UPO exist

region. Moreover, we may increase V_{sm} to enhance the system stability. Likewise, the influence of other parameters (such as r_s, r_g, r_r, and r_c) on the subcritical Hopf bifurcation can be derived. Results are summarized in Table 2.3.

Table 2.3 Stability enhancement trends

	L_s	r_s	V_{sm}	L_g,L_r	r_g	r_r	C_r	r_c	C_d	R_d	κ_p,κ_i	α_p,α_i	β_p,β_i
SNPO and supercritical Hopf bifurcation	↓	↑	↑	↓	↑	↑	↑	↑	↓	↑	↑	↓↑	↑
Increasing SPO regime	/	↓	↑	↓	↓	↓	↓	↓	–	↑	↓	–	↑
Subcritical Hopf bifurcation	↓	↓	↑	↓	↑	↑	↑	↑	↓	↑	↑	↑	↑
(Generalized) Saddle-node bifurcation	↓	↓	↑	–	↓	↓	–	–	–	↑	–	–	–

Note: "↑" ("↓") denotes improvement in system stability, or enlargement in SPO regime, by increasing (decreasing) the corresponding parameter. "↓↑" denotes improvement in system stability by decreasing the corresponding parameter when its value is small, and by increasing the parameter when its value is large. "–" denotes weak or negligible effect on system stability or SPO regime by increasing the corresponding parameter. Note that increasing SPO regime means being away from the onset of Bautin bifurcation

2.4.3 Bautin Bifurcation

As shown in Fig. 2.12, as ΔL_s becomes small, the SNPO curve may merge with the supercritical Hopf bifurcation curve. Then, Hopf bifurcation may occur in the subcritical mode. As analyzed above, we can reduce ΔL_s by decreasing R_d, V_{sm}, or β_p and β_i, or increasing κ_p and κ_i, L_g and L_r, or C_r. However, a (generalized) saddle-node bifurcation may be triggered when R_d or V_{sm} is decreased, as shown in Table 2.3. The mode of Hopf bifurcation may alter only when κ_p and κ_i are significantly increased. A similar problem exists when changing C_r. Therefore, to probe further, we vary L_g, L_r, β_p, and β_i. Relevant bifurcation diagrams are presented in Fig. 2.14a and b, connecting four different types of bifurcation. For Fig. 2.14a, the bifurcation diagrams are displayed in the L_g-L_s plane. Here, $L_r = (10/3) \times L_g$, and other parameters follow the default set provided in Tables 2.1 and 2.2. On the solid red curve, an SNPO occurs. On the solid blue curve, a supercritical Hopf bifurcation occurs. On the dashed curve, a subcritical Hopf bifurcation occurs. These three curves gather at one point, i.e., Bautin bifurcation point. We also find that $l_1(0)$ may fail to indicate the type of Hopf bifurcation near the Bautin bifurcation point. For example, although $l_1(0) < 0$ on the purple dashed curve, a Hopf bifurcation occurs in the subcritical mode. The limited terms of Taylor expansions may cause a discrepancy in $l_1(0)$. Thus, the numerical method is applied here to eliminate the discrepancy. The bifurcation diagrams given in Fig. 2.14b are displayed in the β_p-L_s plane. Here, $\beta_p = \beta_i$ in p.u. value, and other parameters follow the default set provided in Tables 2.1 and 2.2. According to Fig. 2.3, the current-loop loop gain can be derived as

$$L_i(s) = \frac{G_{ci}(s)G_d(s)Z_c(s)Y_r(s)Y_g(s)}{1 + Z_c(s)\left(Y_r(s) + Y_g(s)\right)}, \tag{2.28}$$

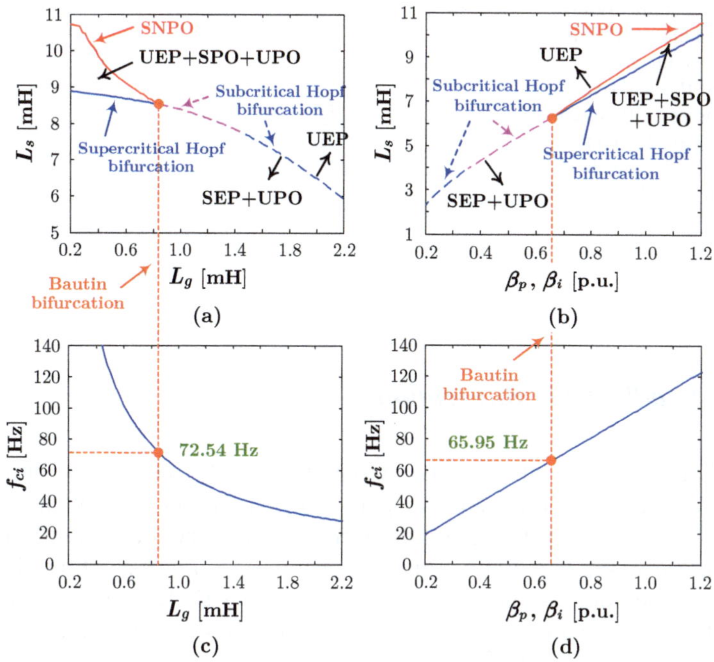

Fig. 2.14 Codimension-2 bifurcation diagrams corresponding to super- and subcritical Hopf bifurcations, SNPO, and Bautin bifurcation on (**a**) L_g-L_s plane where $L_r = (10/3) \times L_g$ and (**b**) β_p-L_s plane where $\beta_p = \beta_i$ in p.u. value. Other parameters adopt the default values provided in Tables 2.1 and 2.2. The gain crossover frequency of the current-loop loop gain when the grid-following rectifier connecting to an ideal grid (i.e., $L_s = 0$, and $r_s = 0$) is given for (**c**) varying L_g and L_r and (**d**) varying β_p and β_i

where $G_d(s) = \exp(-sT_d)$, $G_{ci}(s) = \beta_p + \beta_i/s$, $Y_r(s) = 1/(sL_r + r_r)$, $Z_c(s) = r_c + 1/(sC_r)$, $Y_g(s) = 1/(sL_g + r_g)$.

Then, the first gain crossover frequency of $L_i(s)$ (i.e., f_{ci}) can be derived, as shown in Fig. 2.14c and d. From Fig. 2.14, a large f_{ci} implies supercritical Hopf bifurcation and SNPO, while a small f_{ci} implies subcritical Hopf bifurcation. Besides, $f_{ci} \approx 70$ Hz for Bautin bifurcation. In other words, we may increase the bandwidth of the current loop to change the subcritical mode of Hopf bifurcation to supercritical mode. It is also noteworthy that another Hopf bifurcation may occur if the bandwidth of the current loop is significantly increased.

Remarks In practice, subcritical Hopf bifurcation is much more dangerous than supercritical Hopf bifurcation. After the onset of a subcritical Hopf bifurcation, no attractor exists, and the trajectory will diverge to infinity. In this case, the current amplitudes may be enlarged drastically, driving the current loop into an overmodulation region and causing the system to collapse. Such instability is irreversible. Therefore, f_{ci} should be increased to avoid subcritical Hopf bifurcation, and a practical way to achieve this is to increase β_p and β_i or decrease L_g and L_r.

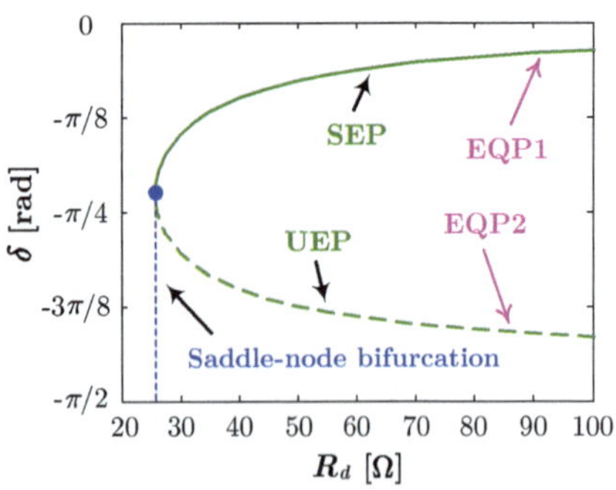

Fig. 2.15 Codimension-1 bifurcation diagram showing the EQPs. Solid curve denotes SEP, and dashed curve denotes UEP. System parameters are given in Tables 2.1 and 2.2 with varying R_d and $L_s = 5$ mH

2.4.4 *Saddle-Node Bifurcation*

The voltage instability may also be induced through the saddle-node bifurcation. A sample saddle-node bifurcation diagram with varying R_d is shown in Fig. 2.15, where $L_s = 5$ mH. In this case, the supercritical or subcritical Hopf bifurcation does not occur, and thus EQP1 remains stable when varying R_d. From Fig. 2.15, the saddle-node bifurcation occurs when $R_d \approx 25.8\ \Omega$. Besides, decreasing R_d may induce the saddle-node bifurcation. In a likewise fashion, the influence of other parameters can be derived, as summarized in Table 2.3.

2.5 Verification

In this section, exact cycle-by-cycle simulations based on MATLAB/Simulink with a step size of 2 μs and relative and absolute tolerances of 10^{-5} are presented to verify the above analytical results. In the simulations, the primitive switching behavior of the power electronic converter is fully considered. All the controllers are discretized through the Tustin discretization method and implemented in MATLAB functional blocks for every control cycle. Besides, the computational delay and the modulator delay are also included in the simulations. Default parameters given in Tables 2.1 and 2.2 are adopted, unless otherwise specified.

2.5.1 *Saddle-Node Bifurcation of Periodic Orbits and Supercritical Hopf Bifurcation*

The influence of the load R_d is demonstrated in Fig. 2.16, where $L_s = 8.6$ mH. Initially, we set $R_d = 100\ \Omega$, then $R_d = 65\ \Omega$ for $t = 5 \sim 15$ s, $R_d = 100\ \Omega$ for $t = 15 \sim 25$ s, and $R_d = 55\ \Omega$ for $t = 25 \sim 30$ s. The system operates normally before $t = 5$ s, exhibits sustained low-frequency oscillation at $t = 5 \sim 15$ s,

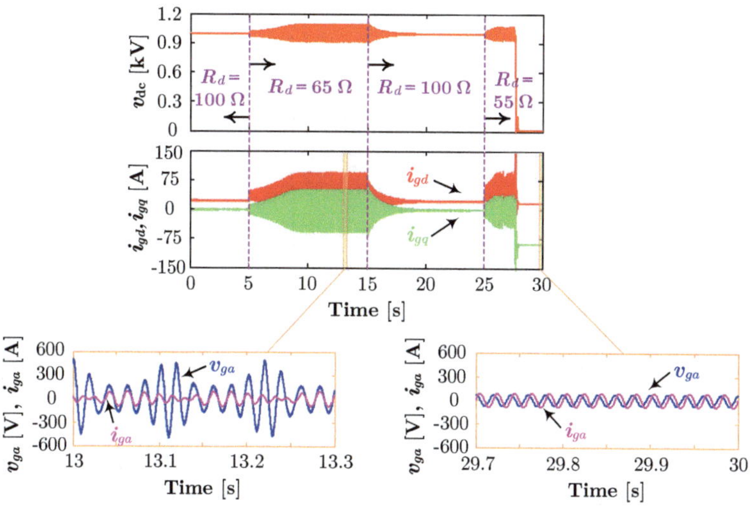

Fig. 2.16 Simulated waveforms of v_{dc}, i_{gd}, i_{gq}, v_{ga}, and i_{ga} when R_d changes. Other parameters follow their default values, and $L_s = 8.6$ mH. Only phase A of v_g and i_g (i.e., v_{ga}, i_{ga}) are shown here

recovers to normal working state at $t = 15 \sim 25$ s, and eventually collapses after $t = 25$ s. After the system collapses, v_g exhibits a significant dip. It is also noted that the oscillation frequency of the aforementioned sustained oscillation is about 8.803 Hz, which is far below the grid frequency. These results verify the above theoretical analysis and confirm the bifurcation diagrams shown in Fig. 2.12a.

2.5.2 *Basin of Attraction*

The simulation waveforms corresponding to a voltage dip are presented in Fig. 2.17. During the voltage dip, i_{gd} and i_{gq} stay inside their limits, and the modulation waveforms are far from the overmodulation region. In Fig. 2.17a, V_{sm} drops from 340 V to 240 V, lasting for 60 ms. At $t = 0.16$ s when v_s recovers, $v_{dc} = 944$ V, $v_{gm} = 66$ V, $i_{gd} = 96.5$ A, and $i_{gq} = -24.4$ A. However, after the grid voltage v_s recovers, v_{dc} and v_{gm} keep decreasing, and i_{gd} and $|i_{gq}|$ keep increasing. Eventually, v_{dc} and v_{gm} almost drop to zero, and the system collapses. In Fig. 2.17b, V_{sm} drops from 340 V to 240 V, lasting for 50 ms. At $t = 0.15$ s when v_s recovers, $v_{dc} = 954$ V, $v_{gm} = 98$ V, $i_{gd} = 79$ A, and $i_{gq} = -24.6$ A. After $t = 0.15$ s, the grid voltage v_s recovers. Fierce but converging oscillation emerges in v_{dc}, v_{gm}, i_{gd}, and i_{gq}. It is noted that the system becomes normal after a few seconds. Thus, when the grid voltage recovers, the trajectory is outside the SEP's basin of attraction for Fig. 2.17a, while the trajectory stays inside the SEP's basin of attraction for Fig. 2.17b. This

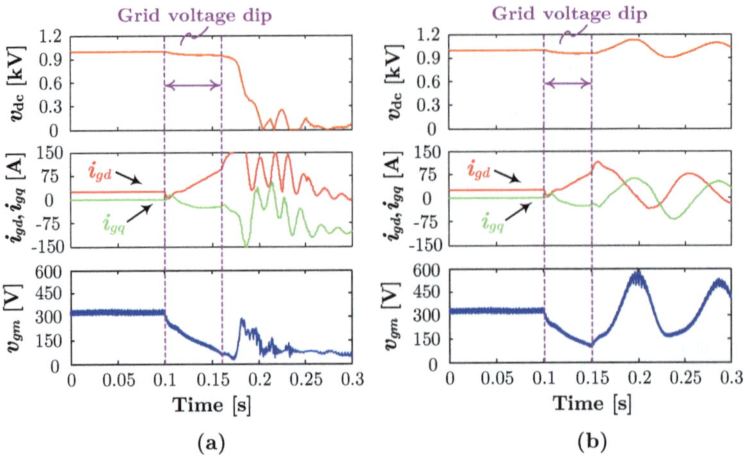

Fig. 2.17 Simulated waveforms of v_{dc}, v_{gm}, i_{gd}, and i_{gq} when the grid exhibits a voltage dip. (**a**) $V_{sm} = 240$ V when $t = 0.1 \sim 0.16$ s, and $V_{sm} = 340$ V at other times. (**b**) $V_{sm} = 240$ V when $t = 0.1 \sim 0.15$ s, and $V_{sm} = 340$ V at other times. Other parameters are set at their default values, and $L_s = 8.6$ mH

basin of attraction implies the existence of a UPO, which verifies the analysis given in Sects. 2.2 and 2.3.

2.5.3 Bautin Bifurcation

The influence of the current loop on the Bautin bifurcation is demonstrated in Fig. 2.18, where $\beta_p = \beta_i = 0.5$ p.u. in Fig. 2.18a, and $\beta_p = \beta_i = 0.8$ p.u. in Fig. 2.18b. For Fig. 2.18a, the system operates normally when L_s is 3.7 mH and 5.1 mH, but exhibits a diverging oscillation and eventually collapses when $L_s = 5.2$ mH, implying a subcritical Hopf bifurcation. For Fig. 2.18b, the system operates normally when $L_s = 7$ mH, exhibits a sustained oscillation when $L_s = 7.4$ mH, and finally collapses when $L_s = 7.5$ mH, implying a supercritical Hopf bifurcation and an SNPO. These results verify the existence of Bautin bifurcation and confirm the bifurcation diagrams shown in Fig. 2.14.

2.5.4 Routes to Voltage Instability

As indicated in Sect. 2.3, there are four routes to voltage instability. Figures 2.16 and 2.18b verify that the voltage instability can be induced by successive onsets of supercritical Hopf bifurcation and SNPO. Specifically, Fig. 2.18a shows voltage

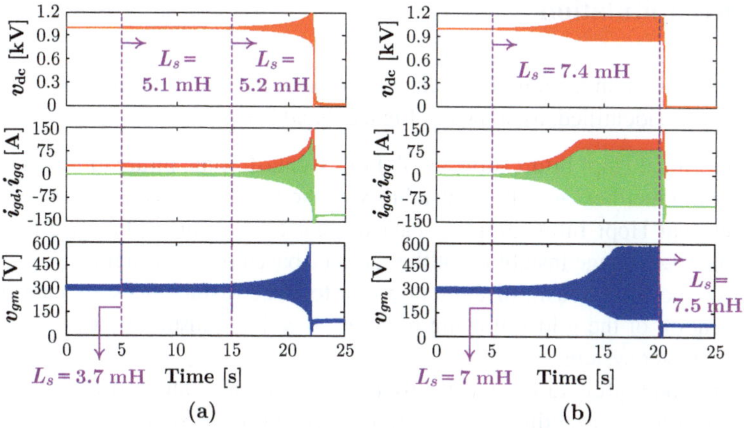

Fig. 2.18 Simulated waveforms of v_{dc}, v_{gm}, i_{gd} and i_{gq}. (**a**) $\beta_p = \beta_i = 0.5$ p.u. (**b**) $\beta_p = \beta_i = 0.8$ p.u. Other parameters are set at their default values

Fig. 2.19 Simulated waveforms of v_{dc}, v_{gm}, i_{gd}, and i_{gq} when R_d changes. Other parameters are set at their default values, and $L_s = 5$ mH

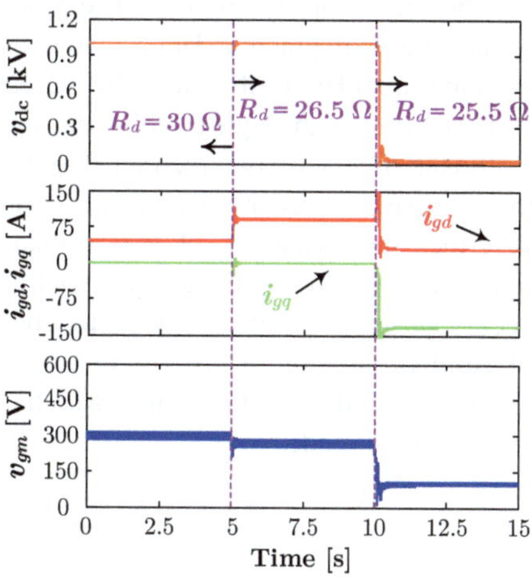

instability being induced by the onset of a subcritical Hopf bifurcation. Also, Fig. 2.17a verifies voltage instability being induced when the system trajectory is out of the SEP/SPO's basin of attraction. The route to voltage instability through the saddle-node bifurcation is demonstrated in Fig. 2.19, where $R_d = 30\ \Omega$ for $t = 0 \sim 5$ s, $R_d = 26.5\ \Omega$ for $t = 5 \sim 10$ s, and $R_d = 25.5\ \Omega$ for $t = 10 \sim 15$ s. The system operates normally before $t = 10$ s, but collapses rapidly after $t = 10$ s. Furthermore, v_{gm} drops to 100 V from 270 V after $t = 10$ s. These results confirm the saddle-node bifurcation depicted in Fig. 2.15 and verify the route to voltage instability through a saddle-node bifurcation.

2.6 Recapitulation

In a power system containing grid-following rectifiers, four routes to voltage instability are identified, as shown in Figs. 2.9 and 2.10, i.e.,

1. Supercritical Hopf bifurcation → SNPO → catastrophic bifurcation of the grid-following rectifier → voltage instability of the rectifier-based power system
2. Subcritical Hopf bifurcation → catastrophic bifurcation of the grid-following rectifier → voltage instability of the rectifier-based power system
3. Out of the SEP's basin of attraction due to large disturbances → catastrophic bifurcation of the grid-following rectifier → voltage instability of the rectifier-based power system
4. Saddle-node bifurcation → catastrophic bifurcation of the grid-following rectifier → voltage instability of the rectifier-based power system

It is also found that the occurrence modes of Hopf bifurcation may be transformed through a Bautin bifurcation. Notably, the system collapses suddenly after the onset of subcritical Hopf bifurcation. Therefore, the subcritical mode of Hopf bifurcation is undesirable in practice. Hence, we may trigger the Bautin bifurcation to eliminate the subcritical Hopf bifurcation by increasing the bandwidth of the current loop.

Furthermore, through establishing the center manifold, the original high-order nonlinear model of the rectifier-based system (i.e., 14th-order) can be reduced to a real second-order system, or a complex first-order system, which essentially describes the dynamics of the original system near the EQP. In the reduced first-order system, the linear-term coefficient determines whether Hopf bifurcation occurs or not, and the cubic-term coefficient determines the occurrence mode of Hopf bifurcation. Thus, the real part of the cubic-term coefficient is the first Lyapunov coefficient, the sign of which determines whether it is supercritical mode or subcritical mode.

Chapter 3
Sustained Oscillations Under Transient Disturbances for a Grid-Forming Converter

Sustained oscillation is usually generated via the onset of a supercritical Hopf bifurcation, as illustrated in Chap. 2. For this bifurcation, some specific parameters must vary to change the stability status of the usual EQP [1, 5].

However, in many recent publications, it has been reported that the sustained oscillation may emerge in a GFMC under a transient disturbance [41, 88]. This is counterintuitive as the operating condition after removing the disturbance is exactly the same as that before the occurrence of the disturbance. In other words, the location and the stability status of the EQPs remain unchanged before and after the disturbance. Therefore, the aforementioned oscillation is not generated from the supercritical Hopf bifurcation.

In this chapter, the physical origin of such sustained oscillation will be explored, and a new type of bifurcation will be exposed in the GFMC, which can explain the emergence and vanishment of the aforementioned oscillation.

3.1 Topology and Control of Grid-Forming Converters

The single-phase circuit model of a GFMC is shown in Fig. 3.1, where L_f and C_f are the inductance and capacitance of the output filter, respectively. The grid is represented as a stiff voltage v_g with an inductance L_l connected in series. The frequency of v_g is $\omega_g = 2\pi f_g$. In this chapter, $\omega_g = \omega_0$ unless otherwise specified. Also, i_f is the current of L_f, v_f is the voltage of C_f, i_l is the current of L_l, V_{gm} is the magnitude of v_g, and v_{fm} is the magnitude of v_f. Besides, p and q are the output instantaneous active power and reactive power, respectively.

The phase difference between v_f and v_g is defined as δ. Here, v_f is controlled to track its reference $v_{f,\mathrm{ref}}$ whose magnitude and phase are $v_{fm,\mathrm{ref}}$ and θ, respectively. Furthermore, θ is generated through the active power-frequency droop control, while $v_{fm,\mathrm{ref}}$ is generated through a reactive power-voltage magnitude droop

J. Yang, C. K. Tse, *Complex Behavior of Grid-Connected Power Electronics Systems*, https://doi.org/10.1007/978-3-031-69340-3_3

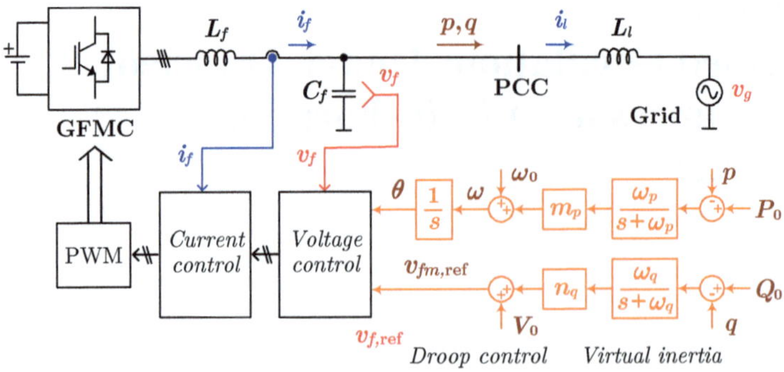

Fig. 3.1 Single-phase circuit model and control strategy of a GFMC

Table 3.1 Typical parameters of the GFMC

Parameter	Value	p.u.	Parameter	Value	p.u.
V_{gm}	311 V	1	L_f	1.8 mH	/
L_l	15 mH	/	C_f	300 μF	/
ω_0	100π	1	V_0	311 V	1
P_0	25 kW	1	Q_0	0	0
ω_p	$2\pi \times 0.1$	0.1	ω_q	$2\pi \times 0.1$	0.1
P_m	50 kW	2	Q_m	15 kVar	0.6
m_p	$0.04\, \omega_0/P_m$	0.04	n_q	$0.05\, V_0/Q_m$	0.05

control, as shown in Fig. 3.1. Moreover, ω_p is the cutoff frequency of the low-pass filter in the active power control, while ω_q is the cutoff frequency of the low-pass filter in the reactive power control. Notably, the former low-pass filter offers a virtual inertia, where small ω_p means large virtual inertia. Besides, m_p and n_q are the droop coefficients; P_0 is the nominal value of active power; Q_0 is the nominal value of reactive power; and V_0 is the magnitude of v_f when $q = Q_0$. Typical parameters and their values are summarized in Table 3.1.

3.2 A Glimpse at Transient Instability Phenomenon

Simulation waveforms generated from a full-circuit cycle-by-cycle switching model with typical parameters shown in Table 3.1 are given in Fig. 3.2.

A symmetrical voltage dip of v_g from $t = 2$ to 8 s is applied, with $V_{gm} = 240$ V. Before $t = 2$ s, δ, p, q, and i_l are all in the steady state, and in particular, $\delta = 0.318\pi$. During $t = 2$ to 8 s, oscillations occur in δ, p, q, and i_l. After $t = 8$ s, V_{gm} recovers to its nominal value of 311 V. However, δ does not return to its original value but continues to oscillate after $t = 8$ s. Furthermore, the oscillations in p, q and i_l

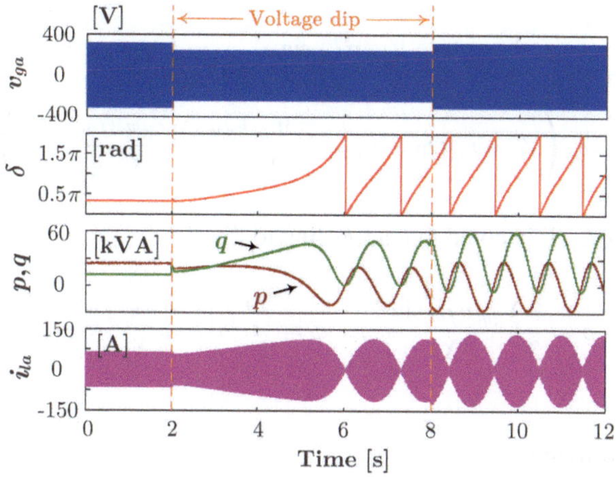

Fig. 3.2 Dynamical behavior of a GFMC during and after a symmetrical voltage dip. The values of typical parameters are shown in Table 3.1. Only phase A of v_g and i_l (i.e., v_{ga}, i_{la}) are shown here. At $t = 2$ s, V_{gm} drops from 311 V to 240 V. At $t = 8$ s, V_{gm} recovers to 311 V

become more fierce after $t = 8$ s. It should be noted that the magnitude of i_f (which participates in the current control, as shown in Fig. 3.1) is far from the inverter's current limit in the simulation, and no saturation is triggered.

It is interesting to note that the system exhibits a sustained oscillation, neither diverging nor converging to the original SEP after the perturbation is removed. Specifically, the system structure and parameters after the voltage recovers are the same as those before the voltage dip, implying that such oscillation is not induced by the Hopf bifurcation. This instability phenomenon, along with its generation mechanism and suppression method, will be analyzed in this chapter.

3.3 Modeling and Decomposition

As shown in Fig. 3.2, the frequency of the sustained oscillation is very low, i.e., 1 Hz approximately. On the other hand, for the GFMC, the dynamics of the power loop are much slower than the dynamics of the voltage loop and the current loop. In other words, the power loop may have much stronger impact on the aforementioned sustained oscillation. Therefore, in this chapter, the inner-loop dynamics are neglected, and we only consider the power-loop dynamics, i.e., $v_f \approx v_{f,\text{ref}}$, and $v_{fm} \approx v_{fm,\text{ref}}$. Then, according to Fig. 3.1, the GFMC can be modeled as

$$\begin{cases} \dot{\delta} &= y \\ \dot{y} &= \omega_p \left(-y - \dfrac{3m_p}{2} \dfrac{V_{gm} v_{fm} \sin\delta}{X_l} + m_p P_0 \right) \\ \dot{v}_{fm} &= \omega_q \left(V_0 + n_q Q_0 - v_{fm} - \dfrac{3n_q}{2} \dfrac{v_{fm}^2 - V_{gm} v_{fm} \cos\delta}{X_l} \right) \end{cases} \tag{3.1}$$

where $X_l = \omega_0 L_l$. It is noted that according to Fig. 3.1,

$$p = \frac{3}{2} \frac{V_{gm} v_{fm} \sin\delta}{X_l}, \quad q = \frac{3}{2} \frac{v_{fm}^2 - V_{gm} v_{fm} \cos\delta}{X_l}$$

3.3.1 Nullclines

The nullclines can be derived as follows. First, the δ-nullcline of (3.1) is $y = 0$, where $\dot{\delta} = 0$. This is a plane in the 3-dim (δ, y, v_{fm})-space. Then, the y-nullcline of (3.1) is

$$y \triangleq h_1(\delta, v_{fm}) = -\frac{3}{2} \frac{m_p V_{gm} v_{fm}}{X_l} \sin\delta + m_p P_0 \tag{3.2}$$

where $\dot{y} = 0$. It is noted that $y = h_1(\delta, v_{fm})$ is a periodic function in δ, which has a period of 2π. As shown in Fig. 3.3a, $y = h_1(\delta, v_{fm})$ is a surface in the (δ, y, v_{fm})-space. For clarity, two periods of δ are provided in Fig. 3.3. Furthermore, the v_{fm}-nullcline of (3.1) is

$$\begin{aligned} v_{fm} &\triangleq h_2(\delta) \\ &= \frac{1}{2} \left(V_{gm} \cos\delta - \frac{2X_l}{3n_q} + \sqrt{\left(V_{gm} \cos\delta - \frac{2X_l}{3n_q} \right)^2 + \frac{8X_l}{3n_q}(V_0 + n_q Q_0)} \right) \end{aligned} \tag{3.3}$$

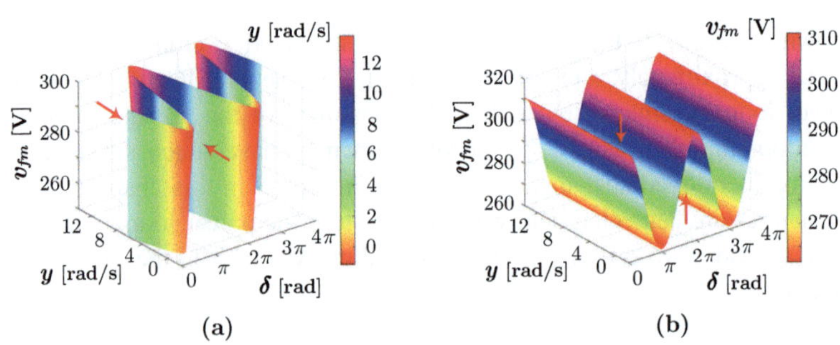

Fig. 3.3 Nullclines represented by (3.1). (**a**) y-nullcline $y = h_1(\delta, v_{fm})$, and red arrows represent the vector field in y-direction. (**b**) v_{fm}-nullcline $v_{fm} = h_2(\delta)$, and red arrows represent the vector field in v_{fm}-direction

where $\dot{v}_{fm} = 0$. As shown in Fig. 3.3b, $v_{fm} = h_2(\delta)$ is also a surface in the (δ, y, v_{fm})-space.

For the δ-nullcline, $\dot{\delta} > 0$ if $y > 0$, and $\dot{\delta} < 0$ if $y < 0$. For the y-nullcline, $\dot{y} > 0$ if $y < h_1(\delta, v_{fm})$, and $\dot{y} < 0$ if $y > h_1(\delta, v_{fm})$. Moreover, for the v_{fm}-nullcline, $\dot{v}_{fm} > 0$ if $v_{fm} < h_2(\delta)$, and $\dot{v}_{fm} < 0$ if $v_{fm} > h_2(\delta)$. Therefore, any trajectory in the phase space will converge to some place near the above nullclines as $t \to +\infty$.

3.3.2 Equilibrium Points

According to (3.1), the EQPs can be derived by setting $\dot{\delta} = 0$, $\dot{y} = 0$, and $\dot{v}_{fm} = 0$. Then, the Jacobian of (3.1) can be derived accordingly. From the eigenvalues of the Jacobian, the stability of the EQPs can be assessed. The EQPs may vary as some parameters change. Here, n_q and V_{gm} are taken as examples to study the bifurcation of the EQPs. Other parameters are set with default values, as given in Table 3.1. The corresponding bifurcation diagrams are derived, as shown in Fig. 3.4. For the EQPs, $y = 0$; the bifurcation diagrams in the (V_{gm},δ)-plane and in the (V_{gm},v_{fm})-plane are shown in Fig. 3.4a, b, respectively. Here, the solid curves represent the SEP, whereas the dashed curves represent the UEP.

For specific values of n_q, there are an SEP and a UEP when V_{gm} is large. Then, as V_{gm} decreases, the two EQPs will get close to each other and eventually merge. After that, there is no EQP as V_{gm} further decreases. Hence, a saddle-node bifurcation occurs here [5]. The value of V_{gm} when there is only one EQP is denoted as $V_{gm,\mathrm{b}}$, which is also the bifurcation point. From Fig. 3.4, as n_q decreases, $V_{gm,\mathrm{b}}$ also decreases, and the variation range of δ of the EQPs increases. Also, the value of v_{fm} of the EQPs increases within a small variation range.

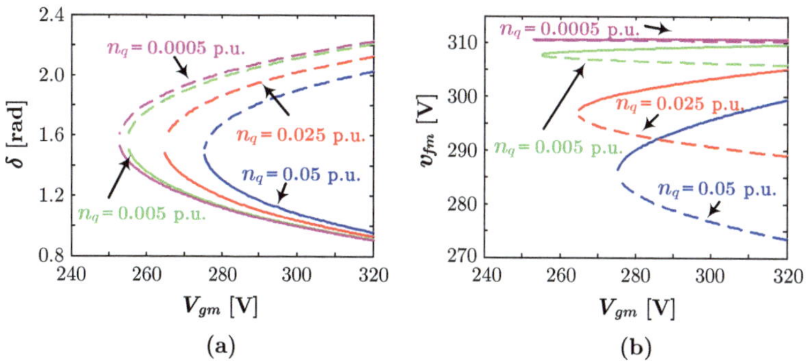

Fig. 3.4 Bifurcation diagrams showing EQPs with solid curves denoting the SEP and dashed curves denoting the UEP. (**a**) Bifurcation diagram in the (V_{gm},δ)-plane for different values of n_q. (**b**) Bifurcation diagram in the (V_{gm},v_{fm})-plane for different values of n_q

3.3.3 System Structure

The sustained oscillation is a behavior typically observed from a second-order nonlinear system [89]. Thus, in modeling a system that exhibits such sustained oscillations, a second-order structure can be identified as the dominant system, which is responsible for the emergence and vanishment of the sustained oscillation. In this subsection, we attempt to find such dominant structure for (3.1).

In general, as shown in Fig. 3.2, δ may converge to the SEP or suffer sustained oscillation. For the case where δ converges to the SEP, the variation of δ can be assumed to be small so that $\delta \approx \delta_{\text{SEP}}$ and $\cos\delta \approx \cos\delta_{\text{SEP}}$, where δ_{SEP} is the value of δ when the system operates at the SEP. Thus, (3.1) can be approximated as

$$\begin{cases} \dot{\delta} = y \\ \dot{y} = \omega_p \left(-y - \dfrac{3m_p}{2} \dfrac{V_{gm} v_{fm,\text{SEP}}}{X_l} \sin\delta + m_p P_0 \right) \end{cases} \tag{3.4}$$

and

$$\dot{v}_{fm} = \omega_q \left(V_0 + n_q Q_0 - v_{fm} - \frac{3n_q}{2} \frac{v_{fm}^2 - V_{gm} v_{fm} \cos\delta_{\text{SEP}}}{X_l} \right) \tag{3.5}$$

where $v_{fm,\text{SEP}} = h_2(\delta_{\text{SEP}})$. The original third-order system is decomposed to two non-interacting subsystems, one being second order and the other first order.

If δ exhibits sustained oscillation, so does v_{fm}. From Fig. 3.3b, we can judge from the vector field that the trajectory has to intersect the nullcline $v_{fm} = h_2(\delta)$ every half period of δ, i.e., π in phase angle. The projections of typical trajectories and the nullcline $v_{fm} = h_2(\delta)$ in the $(\delta,\ v_{fm})$-plane are provided in Fig. 3.5a. As seen from Fig. 3.5a, the oscillation in v_{fm} is restricted within

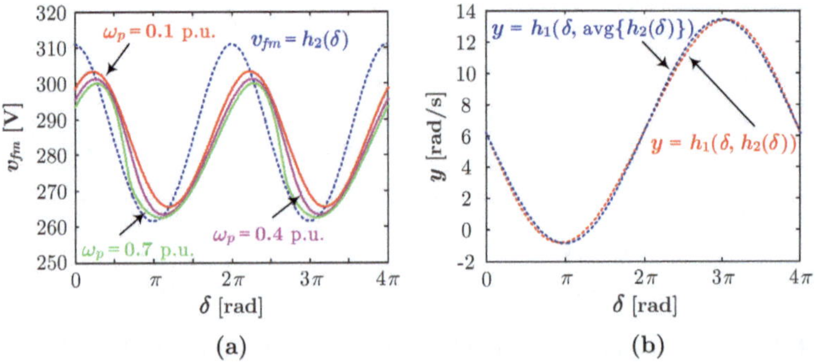

Fig. 3.5 (**a**) Projections of typical trajectories and nullcline $v_{fm} = h_2(\delta)$ in $(\delta,\ v_{fm})$-plane. Blue dashed curve is v_{fm}-nullcline, and solid curves are trajectories with $\omega_q = 1$ p.u. (**b**) Plots of y-nullcline $y = h_1(\delta, h_2(\delta))$ in red and y-nullcline $y = h_1(\delta, \text{avg}\{h_2(\delta)\})$ in blue

the area of $\min\{h_2(\delta)\} < v_{fm} < \max\{h_2(\delta)\}$. In the worst case, the oscillation in v_{fm} coincides with the nullcline $v_{fm} = h_2(\delta)$. In Fig. 3.5b, the y-nullcline $y = h_1(\delta, h_2(\delta))$ in the (δ, y)-plane is plotted as red dashed curve. In addition, the y-nullcline $y = h_1(\delta, \text{avg}\{h_2(\delta)\})$ is given as the blue dashed curve. Here, $\text{avg}\{h_2(\delta)\}$ represents the average value of $h_2(\delta)$. As shown in Fig. 3.5b, the red dashed curve almost coincides with the blue dashed curve, implying that the system with $v_{fm} = h_2(\delta)$ has nearly the same dynamics as the system with $v_{fm} = \text{avg}\{h_2(\delta)\}$. From this point of view, (3.1) can be simplified as

$$\begin{cases} \dot{\delta} = y \\ \dot{y} = \omega_p \left(-y - \frac{3}{2} \frac{m_p V_{gm} \text{avg}\{h_2(\delta)\}}{X_l} \sin \delta + m_p P_0 \right) \end{cases} \tag{3.6}$$

and

$$\dot{v}_{fm} = \omega_q \left(V_0 + n_q Q_0 - v_{fm} - \frac{3 n_q}{2} \frac{v_{fm}^2 - V_{gm} v_{fm} \cos \delta}{X_l} \right) \tag{3.7}$$

Once again, the original third-order system is decomposed to two subsystems of second order and first order. The dynamical behavior of δ is determined by (3.6) and unaffected by v_{fm}. With an oscillating δ, the dynamical behavior of v_{fm} is determined by (3.7).

In summary, the second-order system involving δ and y constitutes the main structure that determines the dynamics of the system represented by (3.1). Although the state of v_{fm} alone cannot cause sustained oscillation, it may change the values of $v_{fm,\text{SEP}}$ and $\text{avg}\{h_2(\delta)\}$ leading to oscillation. Comparing (3.4) with (3.6), we see that they share a similar structure. In the next section, this structure is utilized to study the complex behavior of the GFMC.

3.4 Emergence and Vanishment Mechanisms of Sustained Oscillation

3.4.1 *Existence of the Periodic Orbit*

For clarity and analytical convenience, we rewrite (3.4) or (3.6) as

$$\begin{cases} \dot{\delta} = y \\ \dot{y} = \omega_p \left(-y - m_p P_{\max} \sin \delta + m_p P_0 \right) \end{cases} \tag{3.8}$$

where $P_{\max} = 1.5 V_{gm} V_0 / X_l$. Here, we use the phase portrait to explore the complex behavior of (3.8). Specifically, the δ-nullcline of (3.8) is $y = 0$, while the y-nullcline of (3.8) is

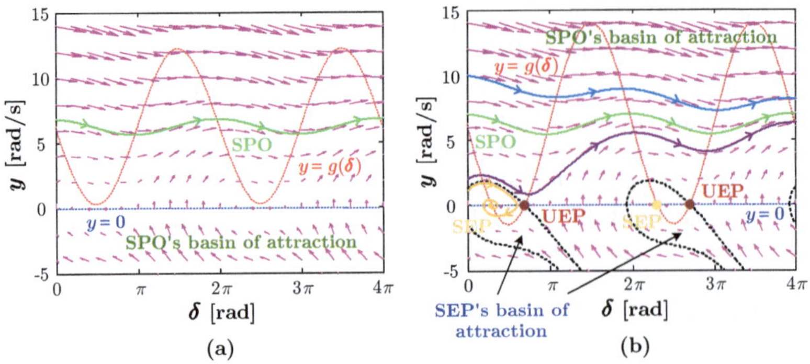

Fig. 3.6 Phase portraits of (3.8), where $\omega_p = 0.1$ p.u., $P_0 = 1$ p.u., and $m_p = 0.04$ p.u. The vector field according to (3.8) is drawn in purple. Red dashed curve is the y-nullcline, and blue dashed line is the δ-nullcline. Black dashed curve is the boundary of the SEP's basin of attraction and the SPO's basin of attraction. Green solid curve is the SPO. SPO, virtual boundary and all trajectories are derived using Runge-Kutta numerical method. (**a**) $V_{gm} = 240$ V and $P_{\max} = 0.9503$ p.u. $< P_0$; (**b**) $V_{gm} = 311$ V and $P_{\max} = 1.2315$ p.u. $> P_0$

$$y \triangleq g(\delta) = -m_p P_{\max} \sin\delta + m_p P_0 \tag{3.9}$$

where $y = g(\delta)$ is a periodic function whose period is 2π. These two nullclines are plotted in Fig. 3.6 for illustration.

From (3.9), there are two EQPs when $P_{\max} > P_0$, namely, an SEP and a UEP, where the UEP is also a saddle, and there is one EQP when $P_{\max} = P_0$. Moreover, there is no EQP when $P_{\max} < P_0$. The phase portraits of (3.8) for $P_{\max} < P_0$ and $P_{\max} > P_0$ are shown in Fig. 3.6a, b, respectively. The vector field, the solutions, some typical trajectories, and the boundary of basins of attraction are also shown in Fig. 3.6.

3.4.1.1 No Equilibrium Point

From Fig. 3.6a, we observe that the vectors near $y = 15$ all point to the bottom-right, while the vectors on $y = 0$ all point vertically upward. Moreover, there is no EQP in the area between $y = 0$ and $y = 15$. Then, according to Poincaré-Bendixson Theorem [89], there must be an SPO in the region $0 < y < 15$, and all trajectories in this region will be attracted to this SPO.

3.4.1.2 Two Equilibrium Points

For $P_{\max} > P_0$, there are an SEP and a UEP. As the UEP is a saddle, some nearby trajectories may be attracted by the UEP, while others are being repelled [89]. Therefore, a virtual boundary (related to the stable manifolds of the UEP, which will be discussed later) can be conceived, which divides the (δ, y)-plane into two parts. This virtual boundary is plotted as the black dashed curve in Fig. 3.6b. It is noted that the vector field is tangential to the virtual boundary, as will be explained later. Hence, no trajectory can cross the virtual boundary.

For the area surrounded by the virtual boundary and lines $y = 0$ and $y = 15$, we can see from Fig. 3.6b that all trajectories will be forced to go inside this area. As there is no EQP in this area, there must be an SPO according to Poincaré-Bendixson Theorem [89]. Applying Runge-Kutta numerical method, the SPO can be found as the green solid curve shown in Fig. 3.6b.

As two distinct stable solutions coexist, there must be two different basins of attraction in the (δ, y)-plane. The SEP's basin of attraction is located inside the virtual boundary, whereas the SPO's basin of attraction is located outside the virtual boundary. It is noteworthy that the virtual boundary is also derived by numerically setting different initial conditions and determining which solution the trajectory is finally attracted to.

In this case, for any trajectory, its final behavior depends on which basin of attraction it is initially located in. Here, three examples are given in Fig. 3.6b. For the yellow solid curve, its initial position is near the virtual boundary but still in the SEP's basin of attraction. Thus, the trajectory eventually converges to the SEP. For the purple solid curve, its initial position is near the virtual boundary and in the SPO's basin of attraction. Thus, the final behavior is an SPO. For the blue solid curve, the initial position is above the SPO, and it will therefore converge to the SPO. Furthermore, even when the system has operated at the SEP, a large perturbation may drive the system out of its original basin of attraction and bring it to the SPO's basin of attraction. Such perturbations may be caused by a voltage dip, a fault of a transmission line (i.e., disconnection, short circuiting to ground), phase jump, or frequency disturbance of the grid voltage.

3.4.1.3 Transient Instability and Sustained Oscillation

Now, we are ready to explain the mechanism behind the simulation waveforms shown in Fig. 3.2.

When $t < 2$ s, $V_{gm} = 311$ V. According to Fig. 3.4, there are an SEP and a UEP. The system operates at the SEP and can be described by (3.4) and (3.5), where $v_{fm,\text{SEP}} = 298$ V according to Fig. 3.4b. For (3.4), $P_{\max} = 1.5V_{gm}v_{fm,\text{SEP}}/X_l = 1.18$ p.u. $> P_0$. Therefore, the phase portrait of (3.4) is similar to Fig. 3.6b. In this case, the SEP and the SPO coexist.

When $t = 2$ to 8 s, $V_{gm} = 240$ V. According to Fig. 3.4, no EQP exists. Then, sustained oscillation may emerge. The system can be described by (3.6) and (3.7), where $\text{avg}\{h_2(\delta)\} = 285.5$ V according to (3.3). For (3.6), $P_{\max} = 1.5V_{gm}\text{avg}\{h_2(\delta)\}/X_l = 0.87$ p.u. $< P_0$. Therefore, the phase portrait of (3.6) is similar to Fig. 3.6a. In this case, the SPO is the only attractor in the phase space.

When $t > 8$ s, $V_{gm} = 311$ V. The system returns to its original working condition, and the phase portrait of the 2-dim subsystem is similar to that given in Fig. 3.6b.

However, as the system is operating outside the SEP's basin of attraction when v_g recovers, the trajectory cannot return to the SEP but will converge to the SPO. Then, another sustained oscillation emerges. More precisely, the system can be described by (3.6) and (3.7), where $\text{avg}\{h_2(\delta)\} = 286.3$ V according to (3.3). For (3.6), $P_{\max} = 1.5V_{gm}\text{avg}\{h_2(\delta)\}/X_l = 1.13\,\text{p.u.} > P_0$.

3.4.2 Variation of Basins of Attraction

We now vary ω_p and keep m_p, $P_{\max}$ and P_0 unchanged, and observe the variation of the basins of attraction. The phase portraits for $\omega_p = 0.1$ p.u., 0.5 p.u., and 0.68 p.u. are presented in Fig. 3.7a–c, respectively. We can see that the SPO is close to the UEP, and its magnitude increases as ω_p increases. In addition, the SEP's basin of attraction broadens, while the SPO's basin of attraction shrinks, as ω_p increases. At

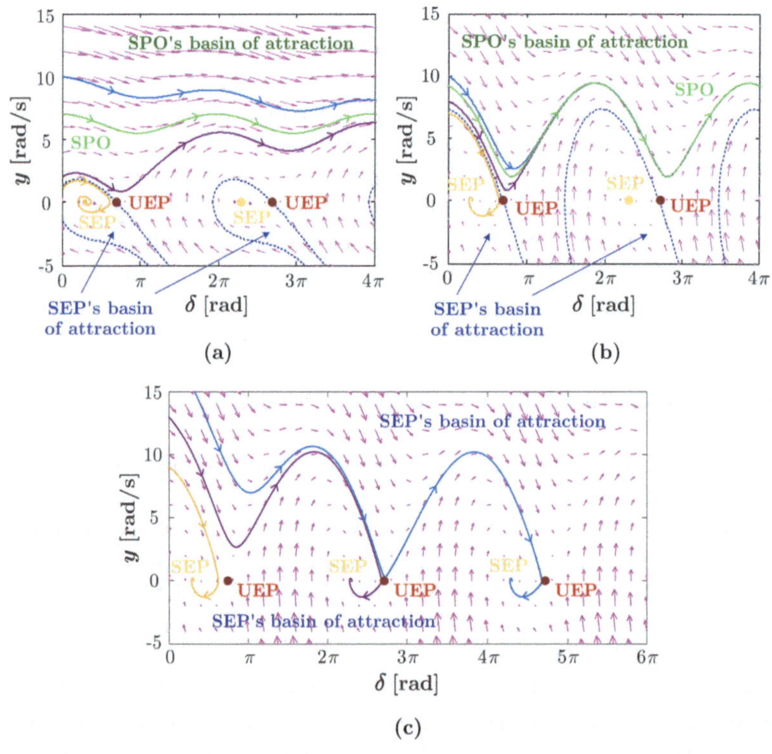

Fig. 3.7 Phase portraits of (3.8), where $P_{\max} = 1.2315$ p.u., $P_0 = 1$ p.u., and $m_p = 0.04$ p.u. The vector field according to (3.8) is drawn in purple. Blue dashed curve is the boundary of the SEP's basin of attraction and the SPO's basin of attraction. Green solid curve is the SPO. Other solid curves are the typical trajectories. (**a**) $\omega_p = 0.1$ p.u.; (**b**) $\omega_p = 0.5$ p.u.; and (**c**) $\omega_p = 0.68$ p.u.

last, when $\omega_p = 0.68$ p.u., the SPO's basin of attraction vanishes, and the whole phase plane becomes the SEP's basin of attraction. From this viewpoint, there must be a bifurcation dominating this dramatical change of the basins of attraction.

3.4.3 Homoclinic Bifurcation

As shown in Fig. 3.6b, when ω_p is small and the SPO is far away from the UEP, the SPO can be expressed as

$$y = a_0 + a_1 \sin\delta + a_2 \cos\delta \tag{3.10}$$

where a_0, a_1, and a_2 can be derived by putting (3.10) in (3.8). The coefficient of the second-order components is very small and can be neglected.

However, when the SPO is close to the UEP, a large amount of harmonics are present in the SPO, making the analytical expression rather complicated. From Fig. 3.7, we may conjecture that for a certain value of ω_p (denoted as $\omega_{p,\text{hb}}$), the SPO may pass the UEP. In this case, the SPO disappears, and a homoclinic loop emerges. In other words, a homoclinic bifurcation occurs at $\omega_p = \omega_{p,\text{hb}}$. It is noted that as ω_p is varied, the EQPs remain unchanged, but the SPO, the virtual boundary, the vector field, and the manifolds change significantly. Hence, the homoclinic bifurcation is a type of global bifurcation, where the qualitative behavior significantly changes in the entire phase space. Such a mechanism is quite different from the Hopf bifurcation, where a stable focus is divided into an unstable focus and an SPO [5].

We now probe into the mechanism of the homoclinic bifurcation by examining the UEP's manifolds. For the UEP, there are two stable manifolds and two unstable manifolds. For the stable manifolds, $\mathbf{x} \to$ UEP as $t \to +\infty$, where $\mathbf{x} = \left[\delta \ y\right]^{\mathrm{T}}$ for the 2-dim subsystem, or $\mathbf{x} = \left[\delta \ y \ v_{fm}\right]^{\mathrm{T}}$ for the 3-dim system. For the unstable manifolds, $\mathbf{x} \to$ UEP as $t \to -\infty$. The manifolds for different values of ω_p are shown in Fig. 3.8. These manifolds are found using Runge-Kutta numerical method. From Fig. 3.8, when $\omega_p \ll \omega_{p,\text{hb}}$, the unstable manifold with $y > 0$ is on the top of and far away from the stable manifold with $y > 0$. Then, as ω_p increases, the unstable manifold becomes much closer to the stable manifold. Finally, when $\omega_p = \omega_{p,\text{hb}}$, the unstable manifold overlaps with the stable manifold. In this case, the homoclinic loop connects the UEP's unstable manifold to its stable manifold. After that, if we keep increasing ω_p, the homoclinic loop will be broken, and the unstable manifold with $y > 0$ is on the bottom of the stable manifold with $y > 0$.

On this basis, we can more readily investigate the change of the attractors and their domains. For $\omega_p < \omega_{p,\text{hb}}$, the SEP is surrounded by the stable manifolds of the UEP. It is noted that no trajectory can cross other trajectories [89]. Therefore, the trajectories inside the stable manifolds of the UEP cannot escape, which forms the SEP's basin of attraction. Similarly, the area outside the stable manifolds of the UEP forms the SPO's basin of attraction.

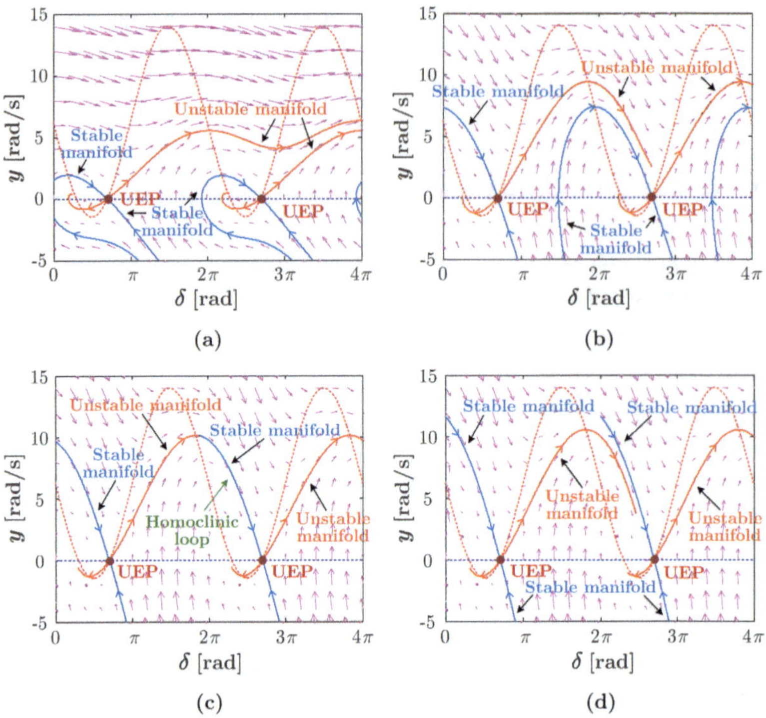

Fig. 3.8 Stable and unstable manifolds of the UEP, where $P_{\max} = 1.2315$ p.u., $P_0 = 1$ p.u., and $m_p = 0.04$ p.u. Solid curves are the manifolds of UEP. Vector field according to (3.8) is drawn in purple. Red dashed curve is the y-nullcline, and blue dashed line is the δ-nullcline. (**a**) $\omega_p = 0.1$ p.u.; (**b**) $\omega_p = 0.5$ p.u.; (**c**) $\omega_p = \omega_{p,\text{hb}}$; and (**d**) $\omega_p = 0.8$ p.u.

Moreover, if we compare the stable manifolds of the UEP with the virtual boundaries mentioned above (e.g., Figs. 3.7a and 3.8a, 3.7b and 3.8b), we can see that the virtual boundaries are the stable manifolds of the UEP in essence. For $\omega_p = \omega_{p,\text{hb}}$, the SPO vanishes, and the SEP's basin of attraction is below the homoclinic loop. For $\omega_p > \omega_{p,\text{hb}}$, the SEP is no longer surrounded by the stable manifolds of the UEP. According to the manifolds and the vector field, all trajectories except the stable manifolds of the UEP will converge to the SEP. No SPO exists in this case. Therefore, the entire phase space, except the stable manifolds of the UEP, becomes the SEP's basin of attraction.

In summary, initially, there are two attractors, namely, the SEP and the SPO, and each of them occupies a specific area in the phase space to form its basin of attraction. Moreover, once the homoclinic bifurcation is triggered, only one attractor exists, which is the SEP, and the entire phase space becomes the SEP's basin of attraction.

3.5 Stability Enhancement

As shown previously, the GFMC may become unstable due to the existence of the SPO. Therefore, in order to enhance stability of the system, we may attempt to keep the trajectory within the SEP's basin of attraction or eliminate the SPO through the homoclinic bifurcation.

3.5.1 Critical Disturbances to Adjacent Basin of Attraction

For the two-attractor case, a transient disturbance with sufficient intensity and sufficient duration will drive the trajectory from the SEP's basin of attraction to the SPO's basin of attraction. Here, the critical intensity and duration of such transient disturbances are found numerically. At first, the system operates at the SEP. Then, under a specific transient disturbance, the system may return to the SEP or exhibit sustained oscillation after a transient process. If the given disturbance is not enough to drive the trajectory to cross the boundary of basins of attraction, the trajectory converges to the SEP. Moreover, if the given disturbance is strong enough (i.e., with sufficiently high intensity or long duration), the trajectory may enter the SPO's basin of attraction. Here, three different kinds of disturbances are studied, namely, voltage dip, frequency disturbance, and phase jump.

3.5.1.1 Voltage Dip

Consider a voltage dip of duration Δt_{dip} and magnitude ΔV_{gm}. Using (3.1), the critical disturbances in the $(\Delta V_{gm}, \Delta t_{\text{dip}})$-plane are found and plotted as the solid curves in Fig. 3.9. Hence, for a given disturbance with specific ΔV_{gm} and Δt_{dip}, if $(\Delta V_{gm}, \Delta t_{\text{dip}})$ is below the respective solid curve, the system will stay in the SEP's basin of attraction. However, if $(\Delta V_{gm}, \Delta t_{\text{dip}})$ is above the respective solid curve, the system will enter the SPO's basin of attraction. Here, we examine the effects of ω_p, ω_q, m_p, and n_q on the critical disturbances. In Fig. 3.9a, the critical disturbances with different values of ω_p are given. Specifically, for small ΔV_{gm}, as ω_p increases, it requires a larger Δt_{dip} to push the trajectory into the SPO's basin of attraction. However, for large ΔV_{gm}, as ω_p increases, a smaller Δt_{dip} will be enough to push the system into the SPO's basin of attraction. The critical disturbances with varying ω_q, m_p, and n_q are given in Fig. 3.9b–d, respectively. Clearly, as ω_q, m_p or n_q is decreased, a larger Δt_{dip} or ΔV_{gm} is needed to push the trajectory to the other basin of attraction.

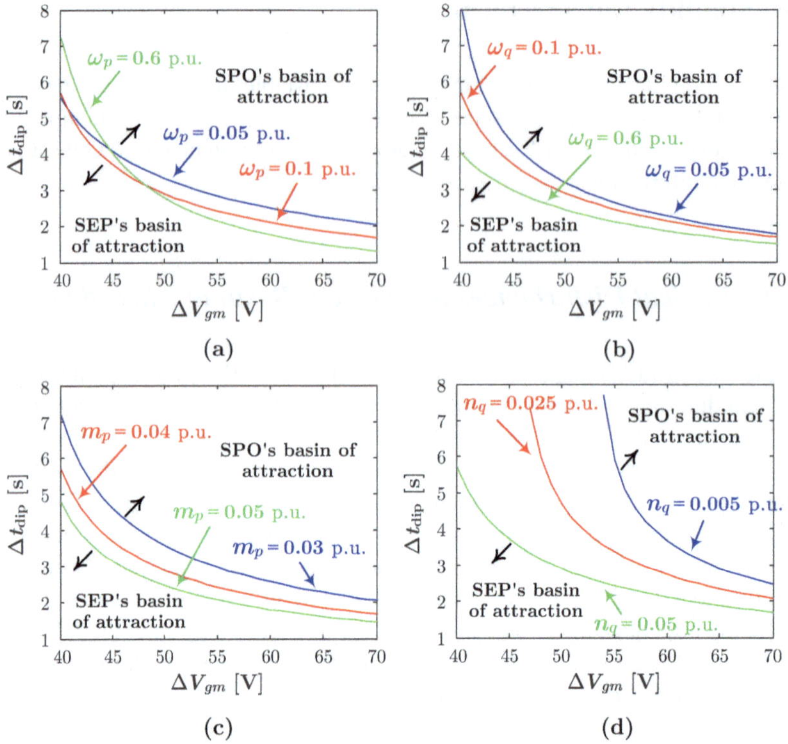

Fig. 3.9 Critical disturbances described in $\Delta t_{\rm dip}$ and ΔV_{gm} under voltage dip scenario are shown as solid curves. Below the curves, the system is still in the SEP's basin of attraction, while above the curves, it will fall in the SPO's basin of attraction. Parameter values are given in Table 3.1 with (**a**) varying ω_p, (**b**) varying ω_q, (**c**) varying m_p, and (**d**) varying n_q

3.5.1.2 Frequency Disturbance

For a frequency disturbance, suppose the duration is $\Delta t_{\rm fd}$ and the magnitude of the frequency drop is Δf_g. Thus, under frequency disturbance, (3.1) becomes

$$\begin{cases} \dot{\delta} &= y \\ \dot{y} &= \omega_p \left(-y - \dfrac{3m_p}{2} \dfrac{V_{gm} v_{fm} \sin\delta}{X_l} + m_p P_0 + 2\pi \Delta f_g \right) \\ \dot{v}_{fm} &= \omega_q \left(V_0 + n_q Q_0 - v_{fm} - \dfrac{3n_q}{2} \dfrac{v_{fm}^2 - V_{gm} v_{fm} \cos\delta}{X_l} \right) \end{cases} \tag{3.11}$$

Then, the critical disturbances in the $(\Delta f_g, \Delta t_{\rm fd})$-plane can be derived, as given in Fig. 3.10. The critical disturbances with varying ω_p, ω_q, m_p, and n_q are given in Fig. 3.10a–d, respectively. We see that if ω_p, ω_q, or n_q is decreased, or m_p is

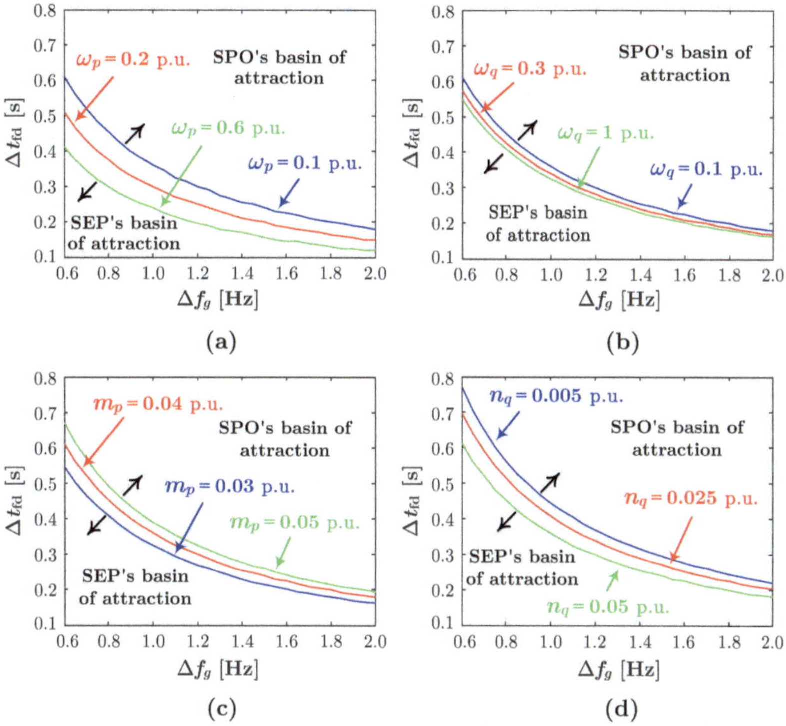

Fig. 3.10 Critical disturbances described in Δt_{fd} and Δf_g under frequency disturbance are shown as solid curves. Below the curves, the system is still in the SEP's basin of attraction, while above the curves, it will fall in the SPO's basin of attraction. Parameter values are given in Table 3.1 with (**a**) varying ω_p, (**b**) varying ω_q, (**c**) varying m_p, and (**d**) varying n_q

increased, a larger Δt_{fd} or Δf_g is needed to push the trajectory out of its original basin of attraction.

3.5.1.3 Phase-Jump Disturbance

For a phase-jump disturbance, suppose a sudden phase lag of $\Delta\theta_g$ is applied to v_g. The system described by (3.1) is now subject to a phase jump occurring at $t = t_\theta$, i.e.,

$$\delta\left(t_\theta^+\right) = \delta\left(t_\theta^-\right) + \Delta\theta_g \tag{3.12}$$

The critical disturbances are plotted in Fig. 3.11. The critical disturbances with varying ω_p, ω_q, m_p, and n_q are given in Fig. 3.11a–d, respectively. It can be observed that if ω_p or m_p is increased, or ω_q or n_q is decreased, a larger $\Delta\theta_g$ is needed to push the trajectory out of its original basin of attraction.

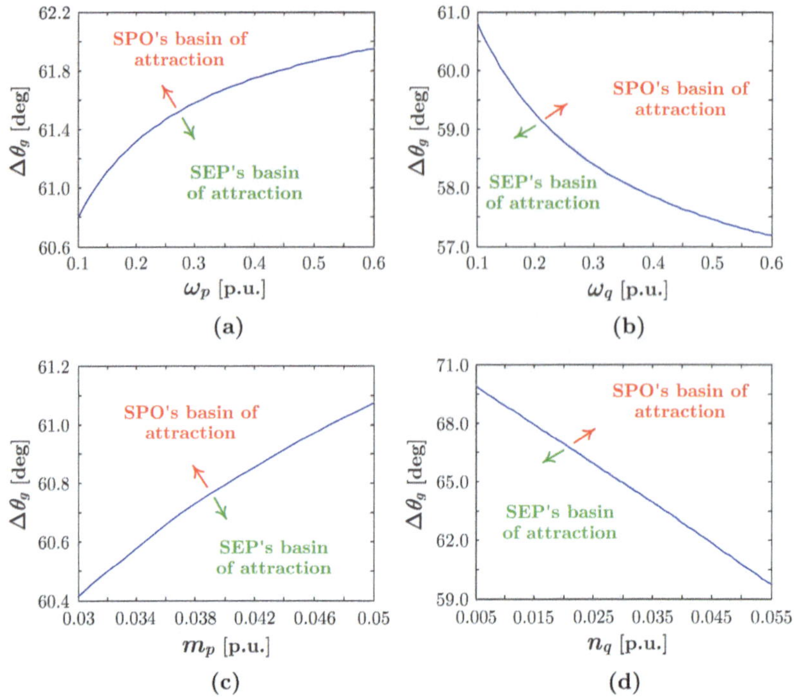

Fig. 3.11 Critical disturbances described in $\Delta\theta_g$ under phase jump scenario are shown as solid curves. Below the curves, the system is still in the SEP's basin of attraction, while above the curves, it will fall in the SPO's basin of attraction. Parameter values are given in Table 3.1 with (**a**) varying ω_p, (**b**) varying ω_q, (**c**) varying m_p, and (**d**) varying n_q

The above analytical results are summarized in Table 3.2. In general, reducing ω_q or n_q would mitigate the effects of disturbances. However, the SPO still exists, and a large enough disturbance can still drive the system into the state of sustained oscillation.

3.5.2 Bifurcation Diagrams and Design-Oriented Analysis

As discussed earlier in Sect. 3.4, there are different numbers of attractors before and after the onset of homoclinic bifurcation. Bifurcation diagrams showing the boundary of the two regimes can be obtained numerically. From Sect. 3.4, the maximum y-range of the SPO is smaller than $\max\{g(\delta)\} = m_p(P_m + P_0)$. As analyzed in Sect. 3.3, the v_{fm}-range of the orbit satisfies $\min\{h_2(\delta)\} < v_{fm} < \max\{h_2(\delta)\}$. To find the SPO, we can set the initial condition as $\delta(0) = 0$, $y(0) = m_p(P_m + P_0)$, and $v_{fm}(0) = V_{fm1}$, where $V_{fm1} > \max\{h_2(\delta)\}$. However, the v_{fm}-coordinate of the SPO may be below $v_{fm,\text{SEP}}$, where $v_{fm,\text{SEP}}$ is the value of v_{fm}

Table 3.2 Mitigation of the effects of disturbances

Type of disturbance		ω_p	ω_q	m_p	n_q
Voltage dip	Small ΔV_{gm}	↑	↓	↓	↓
	Large ΔV_{gm}	↓			
Frequency disturbance		↓	↓	↑	↓
Phase jump		↑	↓	↑	↓

Note: "↑" ("↓") denotes that by increasing (decreasing) the corresponding parameter, disturbance of larger magnitude or longer duration is needed to knock the trajectory off its original basin of attraction

for the SEP. In this case, the trajectory may converge to the SEP first, and the SPO cannot be found. In order to address this issue, two different initial conditions should be set. The second initial condition is set as $\delta(0) = 0$, $y(0) = m_p(P_m + P_0)$, and $v_{fm}(0) = V_{fm2}$, where $V_{fm2} < \min\{h_2(\delta)\}$. If the trajectories with these two initial conditions all converge to the SEP, then the SPO does not exist, confirming that there is only one attractor in the phase space. The bifurcation diagrams corresponding to this operating condition are shown in Fig. 3.12.

For Fig. 3.12a, the bifurcation diagrams are displayed in the (ω_q, m_p)-plane. Three solid curves with different colors are also provided in Fig. 3.12a, corresponding to different values of ω_p. On the solid curves, homoclinic bifurcation occurs. Above the solid curves, two attractors exist. Below the solid curves, one attractor exists. As low-frequency oscillation is undesirable for most practical applications, the two-attractor case should be avoided. From Fig. 3.12a, we may reduce m_p or increase ω_p to move the operating point from the two-attractor area to the single-attractor area, so that the attractor of SPO can be eliminated. However, due to the small variation and the non-monotonicity of the solid curves, it may not be effective to adjust ω_q to eliminate the SPO. Other bifurcation diagrams in Fig. 3.12 are displayed in the (ω_p, m_p)-plane. We can see that decreasing P_0, n_q, or L_l, or increasing Q_0 or V_0, can convert the two-attractor area into a one-attractor area. From the design point of view, in order to avoid sustained oscillation, the bifurcation diagrams given in Fig. 3.12 define the boundaries of stable operation in terms of parameter ranges, within which the system would operate with one stable attractor.

Table 3.3 summarizes the above analytical results. The physical implication of Table. 3.3 is that in order to eliminate the SPO, we may:

(i) Increase the cutoff frequency of the low-pass filters, e.g., increasing ω_p (also reducing the virtual inertia), and ω_q for small ω_q
(ii) Reduce the droop coefficients by decreasing m_p and n_q
(iii) Reduce the input active power to the grid by decreasing P_0
(iv) Increase the input reactive power to the grid by increasing Q_0 and V_0, or decreasing L_l

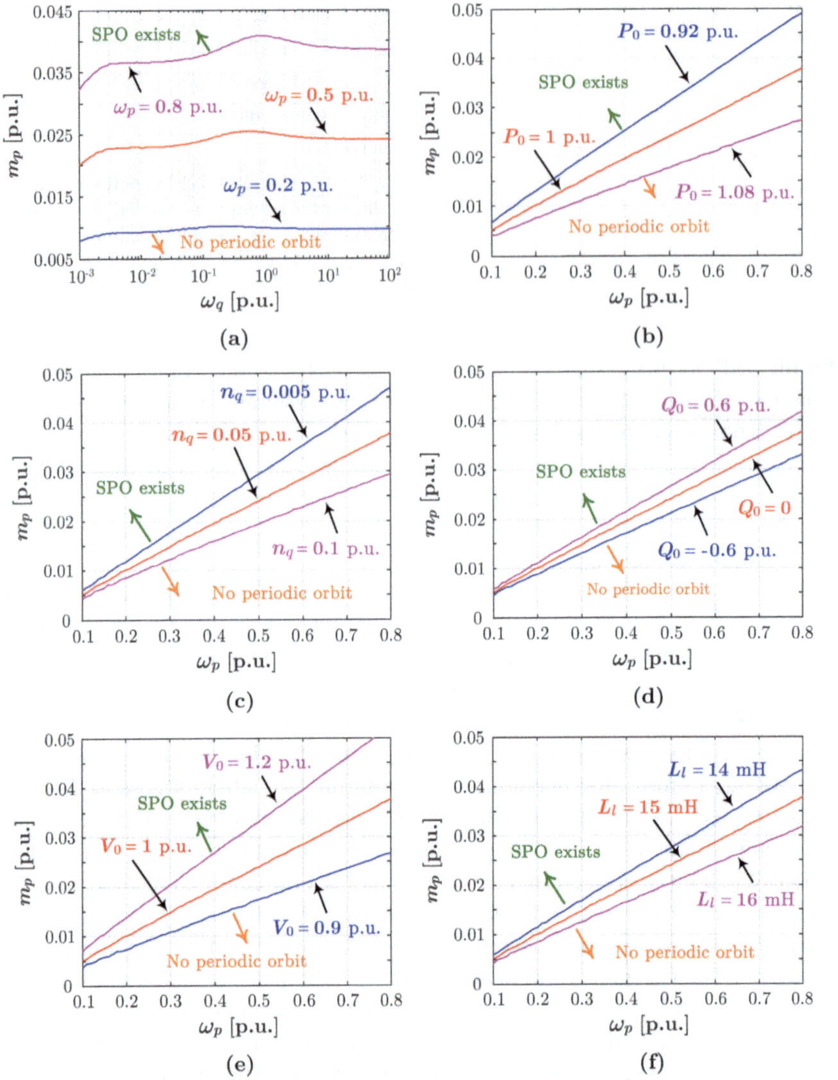

Fig. 3.12 Bifurcation diagrams (**a**) in the (ω_q,m_p)-plane for varying ω_p; (**b**) in the (ω_p,m_p)-plane for varying P_0; (**c**) in the (ω_p,m_p)-plane for varying n_q; (**d**) in the (ω_p,m_p)-plane for varying Q_0; (**e**) in the (ω_p,m_p)-plane for varying V_0; and (**f**) in the (ω_p,m_p)-plane for varying L_l. Above the solid curves, SPO and SEP exist. Below the solid curves, only SEP exists

3.5.3 Remarks

In practice, a sufficiently large virtual inertia can guarantee frequency stability [3]. Thus, it is not desirable to reduce the virtual inertia to suppress sustained oscillation. Besides, the input power of a GFMC depends on the actual power demand from and

Table 3.3 Elimination of the SPO

ω_p	ω_q		m_p	n_q	P_0	Q_0	V_0	L_l
↑	Small	Large	↓	↓	↓	↑	↑	↓
	↑	↓						

Note: "↑" ("↓") denotes that the SPO can be removed by increasing (decreasing) the corresponding parameter

supply to the grid. Hence, it may not be effective to change the input active power and reactive power either. As mentioned above, we may consider adjusting the droop coefficients m_p and n_q to eliminate the SPO. However, as indicated in Fig. 3.12c, when ω_p is small (which is usually the case), decreasing n_q may not widen the one-attractor area significantly. Thus, reducing m_p seems to offer a more effective way to exclude the SPO and hence prevent sustained oscillation in practice.

3.6 Simulation and Experimental Verification

In this section, cycle-by-cycle simulations and experimental measurements are provided to verify the above analytical findings. Default values given in Table 3.1 are adopted unless otherwise specified.

3.6.1 *Simulation Verification*

3.6.1.1 Disturbances to Induce Sustained Oscillations

The simulation waveforms corresponding to a voltage dip disturbance are found in Fig. 3.2. Here, the consequence of frequency disturbance and phase jump is verified. Under frequency disturbance with $\Delta f_g = 1\,\text{Hz}$ and $\Delta t_{\text{fd}} = 0.4\,\text{s}$, sustained low-frequency oscillations in δ, p, q, and i_l exist after the disturbance is removed, as shown in Fig. 3.13a. Under phase jump with $\Delta\theta_g = 62°$, sustained oscillations are observed in Fig. 3.13b for δ, p, q, and i_l. These results confirm the existence of SPO and verify the critical disturbances shown in Figs. 3.10a and 3.11a.

3.6.1.2 Homoclinic Bifurcation

Initially the system is set to operate in the SPO, i.e., sustained low-frequency oscillation, as shown in Fig. 3.14. Here, m_p is set to 0.03 p.u., while ω_p varies from the initial value of 0.1 p.u. As ω_p increases to 0.6 p.u., the oscillation period increases with more obvious distortion in $\delta(t)$, and the oscillation amplitude in p, q, and i_l reduces slightly. Moreover, when ω_p is increased to 0.7 p.u., the low-

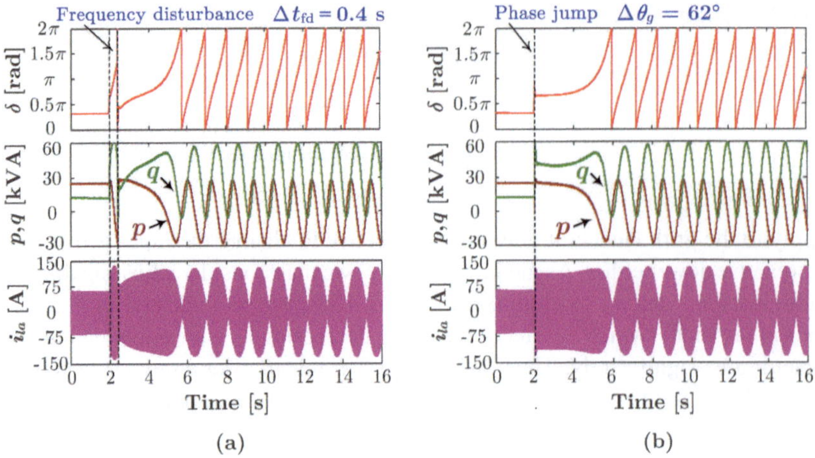

Fig. 3.13 (**a**) Simulation waveforms of δ, p, q, and i_{la} during and after a frequency disturbance with $f_g = 49$ Hz for $2 < t < 2.4$ s, and $f_g = 50$ Hz otherwise. (**b**) Simulation waveforms of δ, p, q and i_{la} before and after a phase jump applied to v_g at $t = 2$ s, with $\Delta\theta_g = 62°$. Default values given in Table 3.1 are used

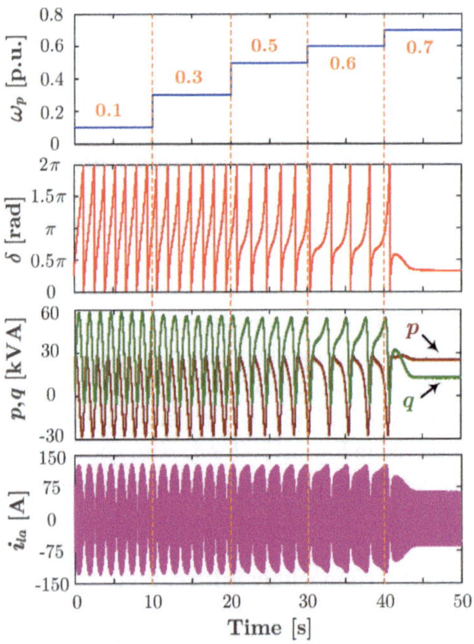

Fig. 3.14 Simulation waveforms of ω_p, δ, p, q, and i_{la}. $\omega_p = 0.1$ p.u. for $0 \leq t < 10$ s; $\omega_p = 0.3$ p.u. for $10 \leq t < 20$ s; $\omega_p = 0.5$ p.u. for $20 \leq t < 30$ s; $\omega_p = 0.6$ p.u. for $30 \leq t < 40$ s; and $\omega_p = 0.7$ p.u. for $40 \leq t < 50$ s. $m_p = 0.03$ p.u., and other parameter values follow Table 3.1

frequency oscillations subside, confirming the vanishment of the SPO after the onset of homoclinic bifurcation. This verifies the theoretical analysis in Sect. 3.4 and confirms the bifurcation diagrams shown in Fig. 3.12.

3.6.1.3 Two-Attractor Regime to Single-Attractor Regime (Elimination of Sustained Oscillation)

The transformation from a two-attractor regime to a single-attractor regime is demonstrated in Fig. 3.15. Here, $\omega_p = 0.8$ p.u., and other parameters are in their default values for $0 < t < 10$ s. At the beginning, the system is set to operate in

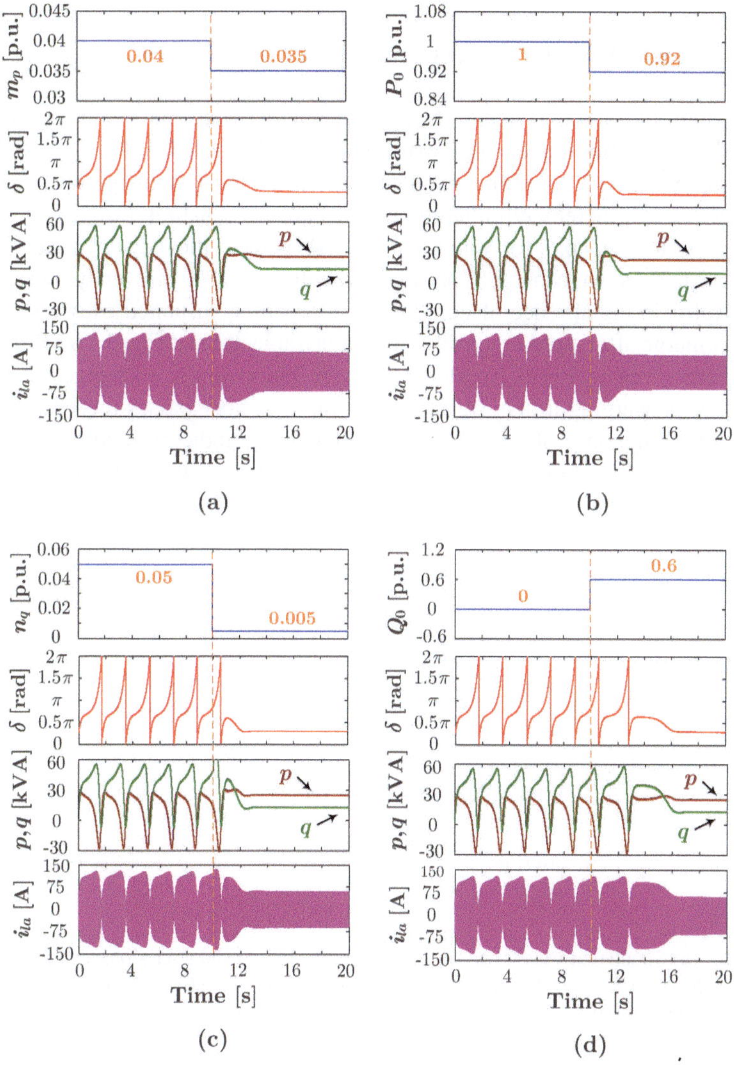

Fig. 3.15 Simulation waveforms of δ, p, q, i_{la}, and some variables. Parameters are set as $\omega_p = 0.8$ p.u. and all others in their default values for $0 \le t < 10$ s. Then, for $t \ge 10$ s, parameters are changed to **(a)** $m_p = 0.035$ p.u.; **(b)** $P_0 = 0.92$ p.u.; **(c)** $n_q = 0.005$ p.u.; **(d)** $Q_0 = 0.6$ p.u.

sustained low-frequency oscillation. Then, a parameter is varied aiming to eliminate the sustained oscillation, i.e., transforming the two-attractor regime to one-attractor regime. In Fig. 3.15a, m_p decreases to 0.035 p.u. from 0.04 p.u. at $t = 10$ s. In Fig. 3.15b, P_0 decreases to 0.92 p.u. from 1 p.u. at $t = 10$ s. In Fig. 3.15c, n_q decreases to 0.005 p.u. from 0.05 p.u. at $t = 10$ s. In Fig. 3.15d, Q_0 increases to 0.6 p.u. from 0 at $t = 10$ s. In each case, oscillation subsides after $t = 10$ s, verifying the regime change and the bifurcation diagrams shown in Fig. 3.12.

3.6.1.4 Single-Attractor System Under Large Disturbance

The system with default values shown in Table 3.1 may run into sustained low-frequency oscillation under large disturbance, as illustrated in Figs. 3.2 and 3.13. According to Table 3.3, we may reduce m_p to ensure the system operate in a single-attractor regime, thereby eliminating any sustained oscillation. From Fig. 3.12, we may set m_p to 0.004 p.u. for which $\Delta\theta_g \ll 90°$ is the critical disturbance (see also Fig. 3.11c). Thus, we can apply phase jump with $\Delta\theta_g = 90°$ to verify our analysis. Specifically, in the absence of homoclinic bifurcation, sustained oscillation would always emerge after a phase jump as shown in Table 3.2 and Fig. 3.11c. Simulation results are given in Fig. 3.16. Here, after the phase jump, δ changes from 0 to 2π before converging to its original value (the trajectory passes the UEP), and no sustained oscillation is observed, implying that the periodic-orbit attractor is absent in the phase space. Thus, a single-attractor regime is more desirable in practice.

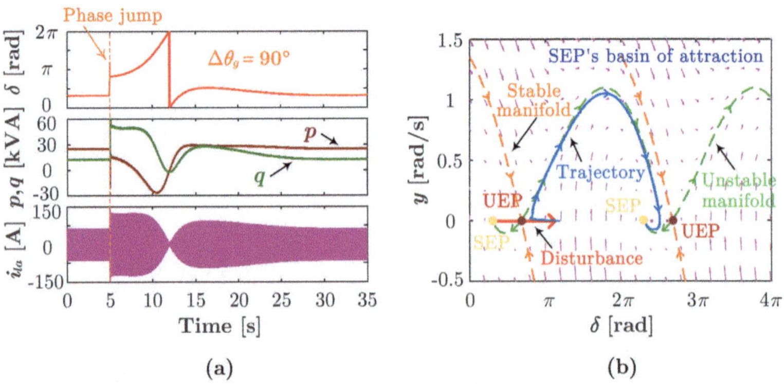

Fig. 3.16 (**a**) Simulation waveforms of δ, p, q, and i_{la} before and after a phase jump applied to v_g at $t = 5$ s, with $\Delta\theta_g = 90°$. $m_p = 0.004$ p.u., and other parameters follow Table 3.1. (**b**) The corresponding phase portrait in the (δ, y)-plane, where the manifolds are given using (3.4), and $v_{fm,\mathrm{SEP}} = 298$ V according to Fig. 3.4b

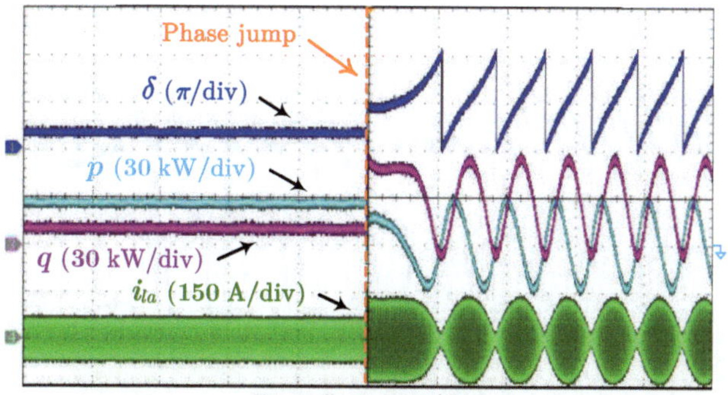

Fig. 3.17 Experimental waveforms of δ, p, q, and i_{la} before and after a phase jump is applied to v_g, with $\Delta\theta_g = 90°$. Besides, $m_p = 0.03$ p.u., and other parameters follow Table 3.1

3.6.2 Experimental Verification

In our experiments, all parameters follow their default values shown in Table 3.1, except that $m_p = 0.03$ p.u. and ω_p is varied.

3.6.2.1 Large Disturbance

Here, $\omega_p = 0.1$ p.u., and the system is initially set to operate at the SEP. Then, a sudden phase lag is applied to v_g, with $\Delta\theta_g = 90°$. The measured waveforms are shown in Fig. 3.17. We see that after the phase jump, sustained low-frequency oscillation in δ, p, q, and i_l emerges, verifying that the trajectory is pushed to the adjacent basin of attraction under transient disturbance. This also confirms the existence of the SPO, and verifies the bifurcation diagrams shown in Fig. 3.12.

3.6.2.2 Homoclinic Bifurcation

At the beginning, the system is set to operate with sustained low-frequency oscillation, which is triggered by a phase jump disturbance. As shown in Fig. 3.18, as ω_p increases, the sustained oscillation changes its form. Specifically, the oscillation period increases and the waveform of δ becomes distorted. Finally, when ω_p increases to 0.7 p.u., the oscillation subsides, revealing that the SPO vanishes and the SEP becomes the only attractor once triggering the homoclinic bifurcation. This agrees with the analytical findings and simulation results provided in Fig. 3.14.

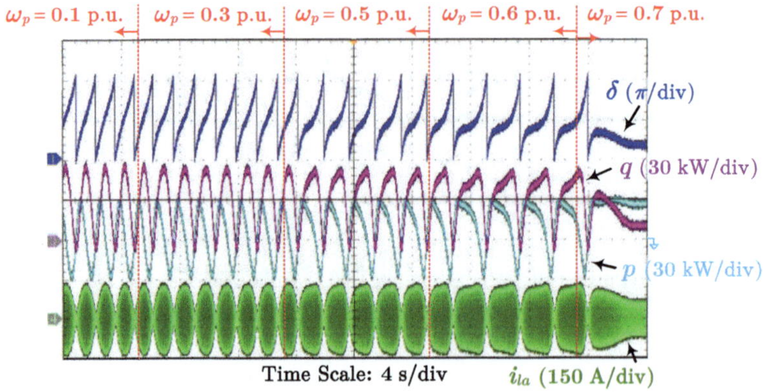

Fig. 3.18 Experimental waveforms of δ, p, q, and i_{la}. Initially, $\omega_p = 0.1$ p.u., and the system is set to operate with sustained oscillation. After that, ω_p is set as 0.3 p.u., 0.5 p.u., 0.6 p.u., and 0.7 p.u. Besides, $m_p = 0.03$ p.u., and other parameter values follow Table 3.1

3.7 Recapitulation

The loss of stability of a power converter leading to emergence of sustained oscillation has been known to be associated with supercritical Hopf bifurcation in the nonlinear system's viewpoint. In a GFMC, moreover, sustained oscillation or exhibition of an SPO has been found to emerge as well as to vanish as a result of homoclinic bifurcation whose mechanism is significantly different from the standard Hopf bifurcation.

1. The SEP and an SPO may coexist in the phase space. In this case, the trajectory may be driven to escape the SEP's basin of attraction and enter the SPO's basin of attraction under a transient disturbance. As a result, the system operates at the SPO, which manifests itself as a sustained oscillation. This is a new path leading to sustained oscillation.
2. Once the homoclinic bifurcation is triggered, only the SEP exists. In this case, the trajectory can always return to the SEP under any given transient disturbance. Thus, no sustained oscillation may appear when the system being subjected to a disturbance.

Therefore, the defining boundary of homoclinic bifurcation can thus be used to indicate the existence or non-existence of sustained oscillation. The control parameters should be chosen according to the bifurcation boundary so as to avoid losing stability to a sustained oscillation and to allow the system to resynchronize with the grid under whatever transient disturbances.

Chapter 4
Homoclinic Bifurcation Behaviors of Grid-Following and Grid-Forming Converters

As illustrated in Chap. 3, sustained oscillations may emerge in a GFMC under a transient disturbance. The physical origin of this type of sustained oscillations has been found to be an SPO, which is far from the SEP and may subside or emerge due to the onset of a homoclinic bifurcation [55]. Moreover, it has also been found that in a GFLC, a UPO encircling the SEP might emerge from a homoclinic bifurcation [90]. Notably, only one UPO and one homoclinic bifurcation have been reported in this prior work [90]. Up till now, the complex behaviors of GFLCs and GFMCs have not been thoroughly studied, including (1) the essential generation mechanism of aforementioned UPO and homoclinic bifurcation, (2) the possible existence of other periodic orbits and homoclinic bifurcations, (3) the fractal structure of basins of attraction, (4) the feasible method to identify the homoclinic bifurcation point, and (5) the similarity and differences between the homoclinic bifurcation behaviors observed in GFLCs and GFMCs.

To study the synchronization characteristics of converters with the grid, and to clarify the relationship between transient stability and basic nonlinear behavior [18, 41, 55, 90], both GFLC and GFMC are modeled as second-order systems in this chapter, where only the dynamics of essential synchronization controls are retained. Through these models, a general set of criteria is developed for studying the homoclinic bifurcation behavior of GFLCs and GFMCs, providing design guidelines to significantly enlarge the SEP's basin of attraction and enhance the transient stability of GFLCs and GFMCs.

4.1 Models of Grid-Following and Grid-Forming Converters

4.1.1 Model of Grid-Following Converters

As shown in Fig. 4.1a, the GFLC is connected to the power grid through the PCC. The power grid is represented by a stiff voltage v_s connected in series with a grid

J. Yang, C. K. Tse, *Complex Behavior of Grid-Connected Power Electronics Systems*, https://doi.org/10.1007/978-3-031-69340-3_4

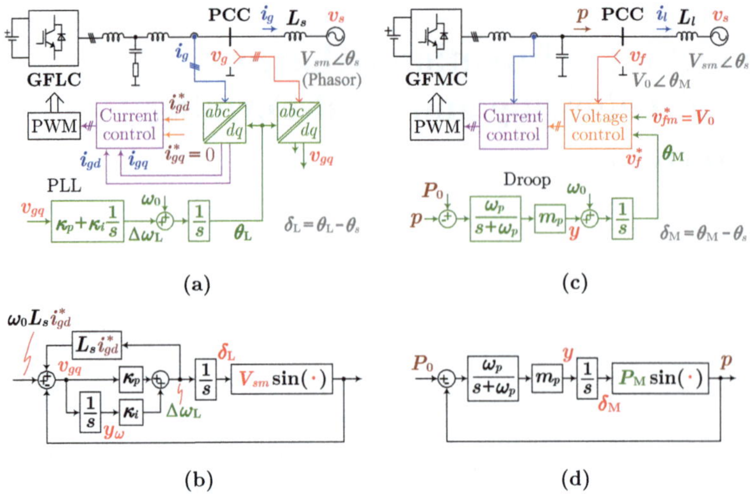

Fig. 4.1 GFLC: (**a**) circuit diagram and control strategy and (**b**) large-signal model. GFMC: (**c**) circuit diagram and control strategy and (**d**) large-signal model

inductance L_s. The magnitude, phase, and frequency of v_s are V_{sm}, θ_s, and ω_0, respectively. The PCC voltage is v_g, and the grid current is i_g. For the PLL, κ_p and κ_i are the proportional and integral parameters, respectively. Besides, $\Delta\omega_{\mathrm{L}}$ is the output variable of the PLL controller, and θ_{L} is the output phase angle of the PLL. We also define $\delta_{\mathrm{L}} = \theta_{\mathrm{L}} - \theta_s$. Therefore, in the dq-frame,

$$v_{gq} = v_{sq} + (\omega_0 + \Delta\omega_{\mathrm{L}})L_s i_{gd} , \tag{4.1}$$

where $v_{sd} + jv_{sq} = V_{sm}e^{j\theta_s}e^{-j\theta_{\mathrm{L}}} = V_{sm}e^{-j\delta_{\mathrm{L}}}$. For the current control, i_{gd}^* and i_{gq}^* are the references. In this chapter, $i_{gq}^* = 0$. The default parameters of the GFLC are given in Table 4.1. For the GFLC, the dynamics of the current control loop are much faster than the dynamics of the PLL. Therefore, it is reasonable to neglect the impact of current loops when studying the essential synchronization behavior, i.e., $i_{gd} \approx i_{gd}^*$, and $i_{gq} \approx i_{gq}^*$ [91, 92]. According to (4.1),

$$v_{gq} = -V_{sm}\sin\delta_{\mathrm{L}} + (\omega_0 + \Delta\omega_{\mathrm{L}})L_s i_{gd}^* . \tag{4.2}$$

Then, the large-signal model of the GFLC can be derived, as shown in Fig. 4.1b. It is noted that y_ω is the state variable of the PLL controller. Then, the GFLC can be modeled as

$$\begin{cases} \dot{\delta}_{\mathrm{L}} = \dfrac{-\kappa_p V_{sm}\sin\delta_{\mathrm{L}} + \kappa_i y_\omega + \omega_0}{1 - \kappa_p L_s i_{gd}^*} - \omega_0 \\ \dot{y}_\omega = -V_{sm}\sin\delta_{\mathrm{L}} + L_s i_{gd}^* \dfrac{-\kappa_p V_{sm}\sin\delta_{\mathrm{L}} + \kappa_i y_\omega + \omega_0}{1 - \kappa_p L_s i_{gd}^*} \end{cases} \tag{4.3}$$

Table 4.1 Default parameters of the GFLC

Parameter	Value	p.u.	Parameter	Value	p.u.
i_{gd}^*	30 A	1	ω_0	$100\,\pi$ rad/s	1
κ_p	0.413	1	κ_i	7.786	1

Note: As an example for initial analysis, the grid parameters are selected as $V_{sm} = 311$ V (i.e., 1 p.u.) and $L_s = 5$ mH (i.e., 0.214 p.u.). The impact of varying these parameters will be studied in Sect. 4.3.1

Table 4.2 Default parameters of the GFMC

Parameter	Value	p.u.	Parameter	Value	p.u.
V_0	311 V	1	P_0	25 kW	1
ω_p	$2\pi \times 0.1$	0.1	m_p	$0.04\,\omega_0/P_m$	0.04

Note: As an example for initial analysis, the grid parameters are selected as $V_{sm} = 311$ V (i.e., 1 p.u.) and $L_l = 15$ mH (i.e., 0.642 p.u.). The impact of varying these parameters will be studied in Sect. 4.3.1. Besides, $P_m = 50$ kW

It is worth noting that (4.3) is mathematically equivalent to a recently reported second-order model [62, 91, 92]. Here, the essential synchronization behavior (i.e., dynamics of δ_L and y_ω) is preserved.

4.1.2 *Model of Grid-Forming Converters*

As shown in Fig. 4.1c, the GFMC is connected to the power grid through the PCC, and the power grid is represented by v_s connected in series with a grid inductance L_l. The PCC voltage is v_f, and the instantaneous active power is p. The phase difference between v_f and v_s is defined as δ_M. For the droop control, ω_p is the cutoff frequency of the low-pass filter, m_p is the droop coefficient, P_0 is the nominal value of p, y is the output frequency difference, and θ_M is the output phase angle. The voltage magnitude reference is $v_{fm}^* = V_0$. Here, the voltage droop dynamics are neglected, because their influence on the synchronization behavior is insignificant [55]. Combining v_{fm}^* and θ_M, the voltage reference v_f^* is generated. The default parameters of the GFMC are given in Table 4.2.

For the GFMC, the dynamics of the voltage and current control loops are much faster than the dynamics of the power control loop. Therefore, we can assume that $v_f \approx v_f^*$ [18, 55]. Thus, $\delta_M = \theta_M - \theta_s$. From Fig. 4.1c, p is derived as

$$p = P_M \sin \delta_M = \frac{3}{2} \cdot \frac{V_{sm} V_0}{\omega_0 L_l} \sin \delta_M \,. \tag{4.4}$$

Then, the large-signal model of the GFMC can be derived, as shown in Fig. 4.1d. Thus, the GFMC can be modeled as

$$\begin{cases} \dot{\delta}_{\mathrm{M}} = y \\ \dot{y} \ = \omega_p \left(-y - m_p P_{\mathrm{M}} \sin \delta_{\mathrm{M}} + m_p P_0\right). \end{cases} \tag{4.5}$$

Here, the virtual inertia is considered (represented by ω_p [18]), and the essential synchronization behavior (i.e., dynamics of δ_{M} and y) is preserved.

4.2 Homoclinic Bifurcation Behaviors

4.2.1 Homoclinic Bifurcation of Grid-Following Converters

4.2.1.1 Equilibrium Points

Based on (4.3), the derivatives are set as zero, and the EQPs can be derived, namely,

$$\delta_{\mathrm{Lss1}} = \arcsin\left(\frac{\omega_0 L_s i_{gd}^*}{V_{sm}}\right), \quad y_{\omega \mathrm{ss1}} = 0 \tag{4.6}$$

for EQP1, and

$$\delta_{\mathrm{Lss2}} = \pi - \arcsin\left(\frac{\omega_0 L_s i_{gd}^*}{V_{sm}}\right), \quad y_{\omega \mathrm{ss2}} = 0 \tag{4.7}$$

for EQP2. Here, (4.3) is rewritten as

$$\begin{cases} \dot{\delta}_{\mathrm{L}} = f_1(\delta_{\mathrm{L}}, y_\omega) \\ \dot{y}_\omega = f_2(\delta_{\mathrm{L}}, y_\omega) \end{cases} \tag{4.8}$$

Then, its Jacobian can be derived as [5]

$$\mathbf{A}_{\mathrm{L}}(\delta_{\mathrm{L}}, y_\omega) = \begin{bmatrix} \frac{\partial f_1}{\partial \delta_{\mathrm{L}}} & \frac{\partial f_1}{\partial y_\omega} \\ \frac{\partial f_2}{\partial \delta_{\mathrm{L}}} & \frac{\partial f_2}{\partial y_\omega} \end{bmatrix}, \tag{4.9}$$

where

$$\frac{\partial f_1}{\partial \delta_{\mathrm{L}}} = \frac{-\kappa_p V_{sm} \cos \delta_{\mathrm{L}}}{1 - \kappa_p L_s i_{gd}^*},$$

$$\frac{\partial f_1}{\partial y_\omega} = \frac{\kappa_i}{1 - \kappa_p L_s i_{gd}^*},$$

$$\frac{\partial f_2}{\partial \delta_{\mathrm{L}}} = -V_{sm} \cos \delta_{\mathrm{L}} + L_s i_{gd}^* \frac{\partial f_1}{\partial \delta_{\mathrm{L}}},$$

$$\frac{\partial f_2}{\partial y_\omega} = L_s i_{gd}^* \frac{\partial f_1}{\partial y_\omega}$$

Therefore, the Jacobian at EQPℓ is $\mathbf{A}_{\mathbf{L}}(\delta_{\mathrm{Lss}\ell}, y_{\omega\mathrm{ss}\ell})$, where $\ell = 1, 2$. For each Jacobian, there are two eigenvalues, which are denoted as λ_1 and λ_2. Thus,

$$\lambda_1 \lambda_2 = \left(\frac{\partial f_1}{\partial \delta_{\mathrm{L}}} \frac{\partial f_2}{\partial y_\omega} - \frac{\partial f_1}{\partial y_\omega} \frac{\partial f_2}{\partial \delta_{\mathrm{L}}} \right) \bigg|_{\mathrm{EQP}\ell} = \frac{\kappa_i V_{sm} \cos \delta_{\mathrm{Lss}\ell}}{1 - \kappa_p L_s i_{gd}^*}, \tag{4.10}$$

$$\lambda_1 + \lambda_2 = \left(\frac{\partial f_1}{\partial \delta_{\mathrm{L}}} + \frac{\partial f_2}{\partial y_\omega} \right) \bigg|_{\mathrm{EQP}\ell} = \frac{\kappa_i L_s i_{gd}^* - \kappa_p V_{sm} \cos \delta_{\mathrm{Lss}\ell}}{1 - \kappa_p L_s i_{gd}^*} \tag{4.11}$$

It is noted that $1 - \kappa_p L_s i_{gd}^* > 0$ for the default parameter values shown in Table 4.1.

For EQP1, combining (4.6), (4.10), and (4.11),

$$\begin{cases} \lambda_1 \lambda_2 &= \dfrac{\kappa_i \sqrt{V_{sm}^2 - (\omega_0 L_s i_{gd}^*)^2}}{1 - \kappa_p L_s i_{gd}^*} \\ \lambda_1 + \lambda_2 &= \dfrac{\kappa_i L_s i_{gd}^* - \kappa_p \sqrt{V_{sm}^2 - (\omega_0 L_s i_{gd}^*)^2}}{1 - \kappa_p L_s i_{gd}^*} \end{cases} \tag{4.12}$$

Therefore, when $\kappa_i / \kappa_p < \sqrt{(V_{sm}/(L_s i_{gd}))^2 - \omega_0^2}$,

$$\lambda_1 \lambda_2 > 0 \quad \text{and} \quad \lambda_1 + \lambda_2 < 0,$$

implying that EQP1 is an SEP (i.e., $\mathrm{Re}(\lambda_1) < 0$, $\mathrm{Re}(\lambda_2) < 0$); when $\kappa_i / \kappa_p > \sqrt{(V_{sm}/(L_s i_{gd}))^2 - \omega_0^2}$,

$$\lambda_1 \lambda_2 > 0 \quad \text{and} \quad \lambda_1 + \lambda_2 > 0,$$

implying that EQP1 is a UEP (i.e., $\mathrm{Re}(\lambda_1) > 0$, $\mathrm{Re}(\lambda_2) > 0$). In this chapter, EQP1 is always an SEP unless otherwise specified.

For EQP2, combining (4.7), (4.10), and (4.11),

$$\begin{cases} \lambda_1 \lambda_2 &= -\dfrac{\kappa_i \sqrt{V_{sm}^2 - (\omega_0 L_s i_{gd}^*)^2}}{1 - \kappa_p L_s i_{gd}^*} \\ \lambda_1 + \lambda_2 &= \dfrac{\kappa_i L_s i_{gd}^* + \kappa_p \sqrt{V_{sm}^2 - (\omega_0 L_s i_{gd}^*)^2}}{1 - \kappa_p L_s i_{gd}^*} \end{cases} \tag{4.13}$$

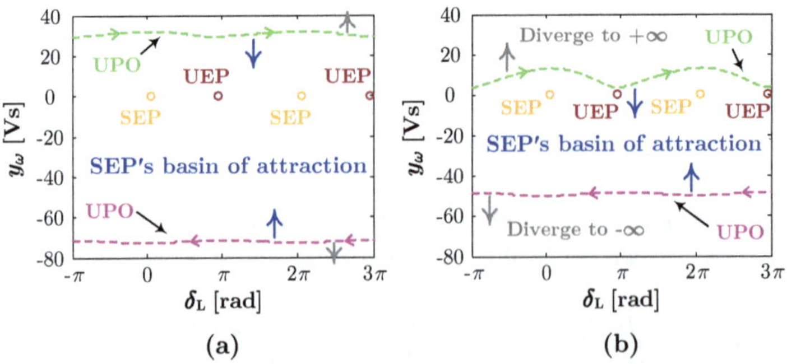

Fig. 4.2 UPOs of (4.3) with (**a**) $\kappa_p = 1$ p.u., and (**b**) $\kappa_p = 0.2$ p.u. The area between the top UPO and the bottom UPO is the SEP's basin of attraction. Other parameters follow the default set

Therefore, $\lambda_1\lambda_2 < 0$ and $\lambda_1 + \lambda_2 > 0$, implying that EQP2 is a UEP, i.e., a saddle (all two eigenvalues are real, and one is positive, while the other is negative).

4.2.1.2 Basins of Attraction

By using the stable manifolds of EQP2 (i.e., UEP), two UPOs can be found, as shown in Fig. 4.2. Here, the state variables δ_L and y_ω are plotted on the horizontal and vertical axes, respectively. In this plane, SEP/UEP is a point, while SPO/UPO is a curve with constant variation period. The derivation procedure of manifolds and UPOs will be explained later. From Fig. 4.2, none of these two UPOs surrounds the SEP, and they are quite different from the UPO reported previously, which surrounds the SEP [90]. For these new UPOs, δ_L has a period of 2π. As the UPOs repel the nearby trajectories, the area between these two UPOs forms the SEP's basin of attraction. From Fig. 4.2, as κ_p decreases, these two UPOs approach to the EQPs, especially for the top UPO (i.e., the green dashed curve), implying the shrinking of the SEP's basin of attraction.

For ease of understanding, the basins of attraction for different κ_p are derived, as shown in Fig. 4.3. There are three different basins of attraction. When the trajectory enters the blue area, it will converge to the SEP. When the trajectory enters the white area, it will diverge to $+\infty$. When the trajectory enters the gray area, it will diverge to $-\infty$. From Fig. 4.3a–c, the SEP's basin of attraction shrinks as κ_p decreases from 1 p.u. to 0.2 p.u., which is consistent with Fig. 4.2. On this basis, as κ_p keeps decreasing, the SEP's basin of attraction may be divided into infinite disconnected parts, as shown in Fig. 4.3d, e. The disconnected blue basins are separated by the white basins. Besides, the basin structure repeats itself infinitely near the gray area, resembling a fractal structure. When κ_p keeps decreasing, this fractal structure disappears, as shown in Fig. 4.3f. It is noted that if κ_p is further decreased, the SEP's basin of attraction will shrink, and eventually, the SEP will become unstable, as indicated by the eigenvalues shown in (4.12).

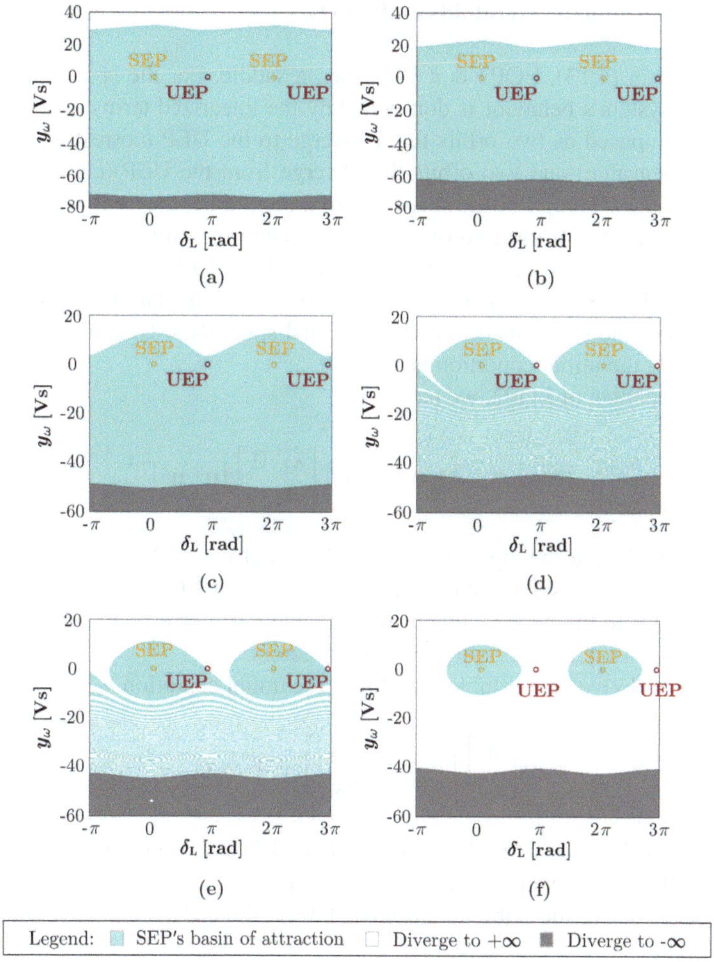

Fig. 4.3 Basins of attraction for (4.3), where **(a)** $\kappa_p = 1$ p.u., **(b)** $\kappa_p = 0.6$ p.u., **(c)** $\kappa_p = 0.2$ p.u., **(d)** $\kappa_p = 0.1$ p.u., **(e)** $\kappa_p = 0.06$ p.u., and **(f)** $\kappa_p = 0.015$ p.u. Other parameters follow the default set

From the perspective of the SEP's basin of attraction, there are two significant changes when κ_p decreases, as shown in Fig. 4.3c, d and in Fig. 4.3e, f. Thus, there are two corresponding bifurcations. At the onset of these two bifurcations, the stability status of EQPs remains unchanged, but the qualitative behavior significantly changes in the phase space, implying that these bifurcations are global bifurcations. For the global bifurcation, we may use the *stable and unstable manifolds* of EQP2 (UEP) to study the bifurcation behavior and to derive the bifurcation diagrams (i.e., the boundaries in the parameter space). The role of manifolds in a global bifurcation is similar to the role of eigenvalues in local bifurcations.

4.2.1.3 Derivation of Manifolds and UPOs

As indicated in (4.13), EQP2 is a UEP, i.e., a saddle. For the space close to the saddle, the system's behavior is dominated by the linearized terms of (4.3), which can be decomposed as two orbits that converge to the UEP (corresponding to the negative eigenvalue) and two orbits that diverge from the UEP (corresponding to the positive eigenvalue). For the space far from the saddle, the nonlinear terms play their roles, and the former two orbits are the stable manifolds of the UEP, while the latter two orbits are the unstable manifolds of the UEP, as shown in Fig. 4.4a. In other words, the manifold is a trajectory with a particular initial condition or final state. For the stable manifold W^s, the final state is the UEP. For the unstable manifold W^u, the initial condition is the UEP.

For the Jacobian of EQP2, we have

$$\mathbf{A}_{\mathrm{L}}(\delta_{\mathrm{Lss2}}, y_{\omega\mathrm{ss2}}) = \begin{bmatrix}\mathbf{u}^u & \mathbf{u}^s\end{bmatrix}\begin{bmatrix}\lambda_1 & 0\\ 0 & \lambda_2\end{bmatrix}\begin{bmatrix}\mathbf{u}^u & \mathbf{u}^s\end{bmatrix}^{-1}, \tag{4.14}$$

where λ_1 and λ_2 are the eigenvalues of EQP2, and $\lambda_1 > 0 > \lambda_2$ [54]. Besides, $\mathbf{u}^u$ and $\mathbf{u}^s$ are the unstable and stable eigenvectors of EQP2, respectively. Here, $\mathbf{u}^u$ corresponds to the positive eigenvalue λ_1, while $\mathbf{u}^s$ corresponds to the negative eigenvalue λ_2. They are all 2-dim real-value vectors.

In order to derive the unstable manifolds, the initial condition is selected as

$$\begin{bmatrix}\delta_{\mathrm{L}}\\ y_\omega\end{bmatrix}\Bigg|_{t=0} = \begin{bmatrix}\delta_{\mathrm{Lss2}}\\ y_{\omega\mathrm{ss2}}\end{bmatrix} + \mathbf{u}^u \times \varepsilon, \tag{4.15}$$

where $\varepsilon = 10^{-3}$ or $\varepsilon = -10^{-3}$. Then, based on (4.3), (4.7), and (4.15), the unstable manifolds W^u are derived, as shown in Fig. 4.4a.

In order to derive the stable manifolds, (4.3) is adjusted as

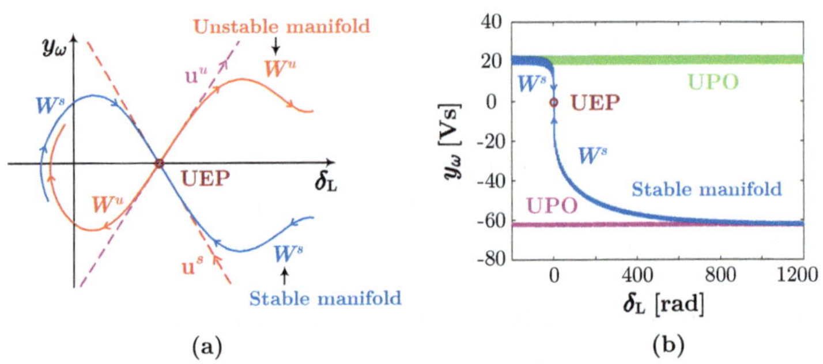

Fig. 4.4 (**a**) Stable manifolds W^s, unstable manifolds W^u, stable eigenvector $\mathbf{u}^s$, and unstable eigenvector $\mathbf{u}^u$ of the UEP. (**b**) Derivation of the UPOs by using the stable manifolds W^s, where $\kappa_p = 0.6$ p.u.

$$\begin{cases} \dot{\delta}_\mathrm{L} = -\dfrac{-\kappa_p V_{sm}\sin\delta_\mathrm{L} + \kappa_i y_\omega + \omega_0}{1-\kappa_p L_s i_{gd}^*} + \omega_0 \\ \dot{y}_\omega = V_{sm}\sin\delta_\mathrm{L} - L_s i_{gd}^* \dfrac{-\kappa_p V_{sm}\sin\delta_\mathrm{L} + \kappa_i y_\omega + \omega_0}{1-\kappa_p L_s i_{gd}^*} \end{cases}, \tag{4.16}$$

which is the time-reversal version of (4.3). Then, using the initial condition

$$\begin{bmatrix} \delta_\mathrm{L} \\ y_\omega \end{bmatrix}\Bigg|_{t=0} = \begin{bmatrix} \delta_\mathrm{Lss2} \\ y_{\omega\mathrm{ss2}} \end{bmatrix} + \mathbf{u}^s \times \varepsilon, \tag{4.17}$$

the stable manifolds W^s can be derived again by time reversal, as shown in Fig. 4.4a.

It is noted that the stable manifolds W^s may come from some repellers (e.g., UPOs) and converge to the UEP eventually. Therefore, we may use the stable manifolds (i.e., (4.16) and (4.17)) to find the UPOs, as shown in Fig. 4.4b. Thus, we can use the long-term static data of stable manifolds to plot the UPOs, as shown in Fig. 4.2.

4.2.1.4 Homoclinic Bifurcations

Here, the complex behavior of the aforementioned two bifurcations is explained through the stable and unstable manifolds. From (4.3), the nullclines and vector field can be derived, as shown in Fig. 4.5. Here, the δ-nullcline is defined as $y_\omega = f_\mathrm{L}(\delta_\mathrm{L})$ with $\dot{\delta}_\mathrm{L} = 0$, while the y_ω-nullcline is defined as $y_\omega = g_\mathrm{L}(\delta_\mathrm{L})$ with $\dot{y}_\omega = 0$.

In Fig. 4.6a, a UPO exists, and for $y_\omega > 0$, the stable manifold W^s is above the unstable manifold W^u. Besides, W^s is much closer to the UPO. As κ_p decreases, the distance between W^s and W^u becomes small, and finally, W^s may overlap with W^u to form a homoclinic orbit, as shown in Fig. 4.6b. In this case, a homoclinic bifurcation occurs. If we keep decreasing κ_p, W^s and W^u will separate, and W^s is underneath W^u, as shown in Fig. 4.6c. Here, the UPO does not exist anymore. The UPO, shown in Fig. 4.6a, is generated due to W^s and the upward-side vector field. According to the vector field shown in Fig. 4.5, any trajectory in the area where

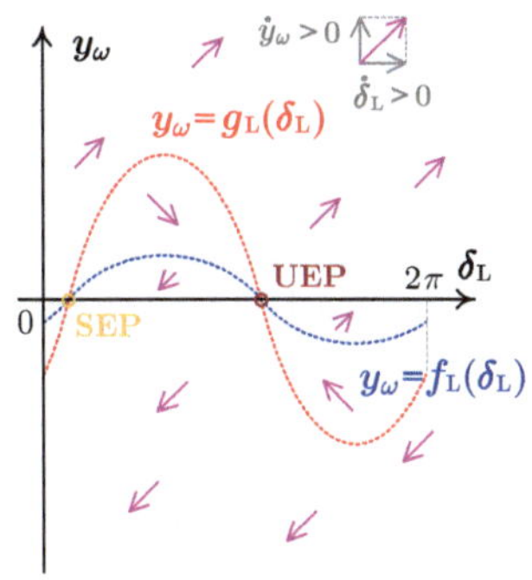

Fig. 4.5 Vector field of GFLC, which is drawn in purple

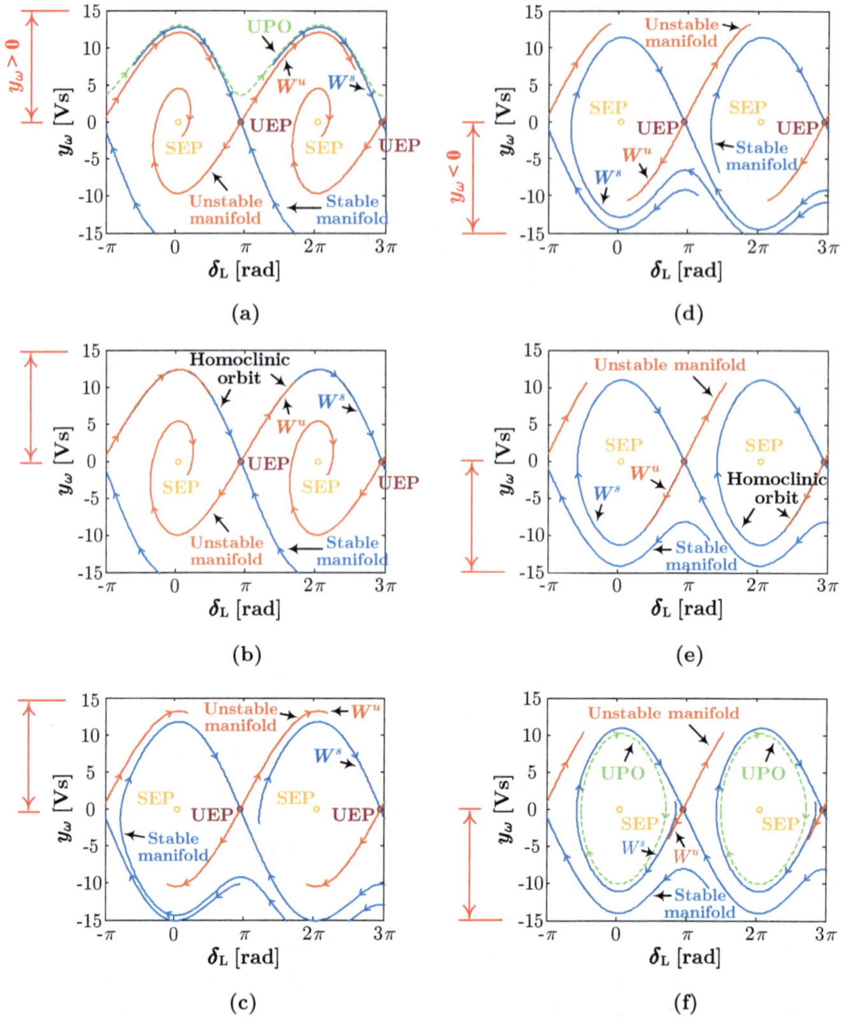

Fig. 4.6 For the GFLC, the stable manifolds and unstable manifolds of the UEP, where (**a**) $\kappa_p = 0.2$ p.u., (**b**) $\kappa_p = 0.166$ p.u., (**c**) $\kappa_p = 0.1$ p.u., (**d**) $\kappa_p = 0.06$ p.u., (**e**) $\kappa_p = 0.0188$ p.u., and (**f**) $\kappa_p = 0.015$ p.u. Other parameters follow the default set

$y_\omega > g_L(\delta_L)$ and $y_\omega > 0$ will diverge to $+\infty$. On the other hand, from Fig. 4.6a, for $y_\omega > 0$, the manifold W^s will converge to the UEP, but not diverge to $+\infty$. Therefore, a UPO must exist to divide the phase space $y_\omega > 0$ into two parts. However, in Fig. 4.6c, for $y_\omega > 0$, W^u becomes on the top of W^s. As this manifold W^u will diverge to $+\infty$, there is no UPO for $y_\omega > 0$. Moreover, in this case, the SEP is surrounded by W^s, which forms the SEP's basin of attraction, as shown in

Figs. 4.3d and 4.6c. It is noteworthy that the aforementioned homoclinic bifurcation is reported for the first time.

On this basis, if we keep decreasing κ_p, the stable and unstable manifolds with $y_\omega < 0$ may interact to generate another UPO. In Fig. 4.6d, there is no UPO, and for $y_\omega < 0$, the stable manifold W^s is below the unstable manifold W^u. As κ_p decreases, these two manifolds may overlap, resulting in another homoclinic orbit, as shown in Fig. 4.6e. When further decreasing κ_p, these two manifolds will separate, and W^s is above W^u, as shown in Fig. 4.6f. In this case, a new UPO exists, surrounding the SEP. Similarly, W^s is much closer to this UPO. This UPO is generated due to W^s and the vector field of the SEP: As shown in Fig. 4.6f, the trajectories near the SEP all converge to the SEP, while the manifold W^s and the trajectories outside W^s do not converge to the SEP. Therefore, a UPO must exist to separate these two different types of vector fields/trajectories. Besides, this UPO forms the SEP's basin of attraction, as shown in Figs. 4.3f and 4.6f. It is noteworthy that although this UPO and the second homoclinic bifurcation have been reported [90], their generation mechanism has not been seriously studied, especially from the perspective of the manifolds and vector field.

4.2.1.5 Example

As an example, the impact of a phase jump disturbance on the stability of GFLC is studied for different values of κ_p. In each case, the system operates at the SEP initially. For $\kappa_p = 0.2$ p.u., the area below the blue dashed curve (i.e., the UPO) is the SEP's basin of attraction, as shown in Fig. 4.7a. Therefore, after the phase jump disturbance, although the trajectory crosses the UEP, it is still within the SEP's basin of attraction. So, it converges to the SEP again after several cycles of oscillations, as shown in Fig. 4.7a, b. For $\kappa_p = 0.1$ p.u., the area inside the blue dashed curve (i.e., the stable manifolds of UEP) is the SEP's basin of attraction, as shown in Fig. 4.7c. In this case, after the disturbance, the trajectory passes the blue dashed curve and is out of the SEP's basin of attraction. As a result, it will not return to the SEP, but diverge to infinity, as shown in Fig. 4.7c, d.

4.2.2 *Homoclinic Bifurcation of Grid-Forming Converters*

4.2.2.1 Equilibrium Points

Similarly, based on (4.5), the values and eigenvalues of EQPs can be derived. For EQP1,

$$\delta_{\mathrm{Mss1}} = \arcsin(P_0/P_{\mathrm{M}}), \quad y_{\mathrm{ss1}} = 0 \tag{4.18}$$

and

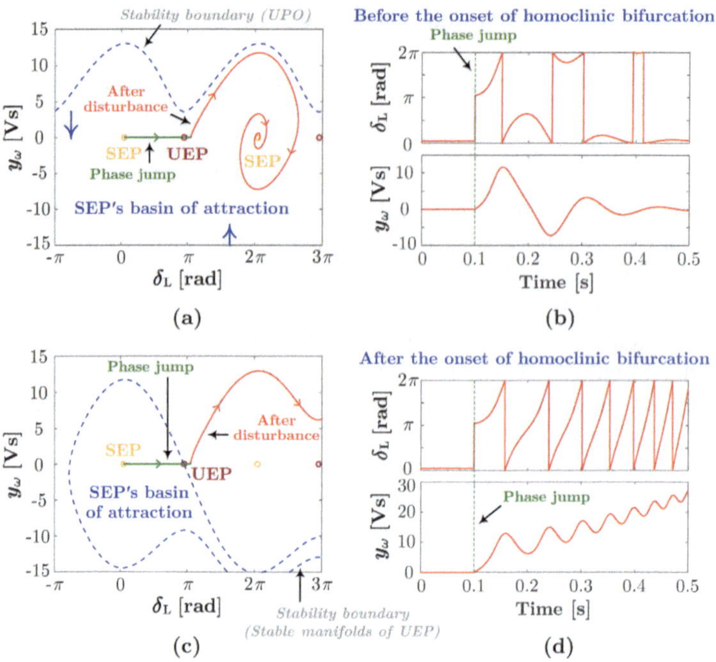

Fig. 4.7 For the GFLC, before and after a phase jump disturbance applied to v_s, the system behavior (**a**) in (δ_L, y_ω)-plane and (**b**) in the time domain for $\kappa_p = 0.2$ p.u. and the system behavior (**c**) in (δ_L, y_ω)-plane and (**d**) in the time domain for $\kappa_p = 0.1$ p.u. Other parameters follow the default set. For clarity, the variation range of δ_L is $0 \sim 2\pi$ in (**b**) and (**d**)

$$\begin{cases} \lambda_1\lambda_2 &= \omega_p m_p \sqrt{P_M^2 - P_0^2} \\ \lambda_1 + \lambda_2 &= -\omega_p \,. \end{cases} \tag{4.19}$$

Therefore, $\lambda_1\lambda_2 > 0$ and $\lambda_1 + \lambda_2 < 0$, implying that EQP1 is an SEP (i.e., $\mathrm{Re}(\lambda_1) < 0$, $\mathrm{Re}(\lambda_2) < 0$). For EQP2,

$$\delta_{Mss2} = \pi - \arcsin(P_0/P_M), \quad y_{ss2} = 0 \tag{4.20}$$

and

$$\begin{cases} \lambda_1\lambda_2 &= -\omega_p m_p \sqrt{P_M^2 - P_0^2} \\ \lambda_1 + \lambda_2 &= -\omega_p \end{cases} \tag{4.21}$$

Therefore, $\lambda_1\lambda_2 < 0$ and $\lambda_1 + \lambda_2 < 0$, implying that EQP2 is a UEP, i.e., a saddle.

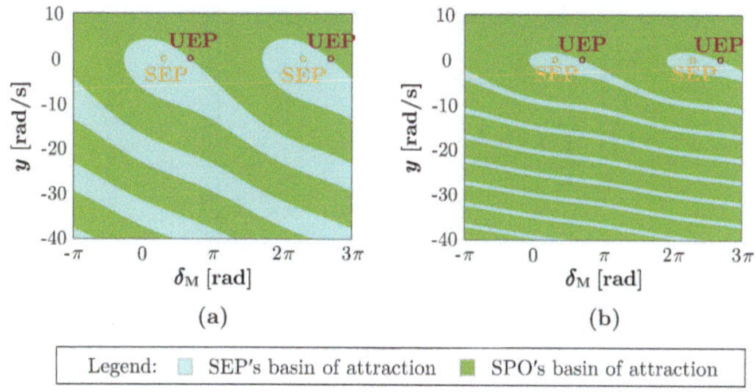

Fig. 4.8 Basins of attraction for (4.5) with **(a)** $\omega_p = 0.3$ p.u. and **(b)** $\omega_p = 0.1$ p.u. Other parameters follow the default set

Fig. 4.9 Vector field of the GFMC, which is drawn in purple

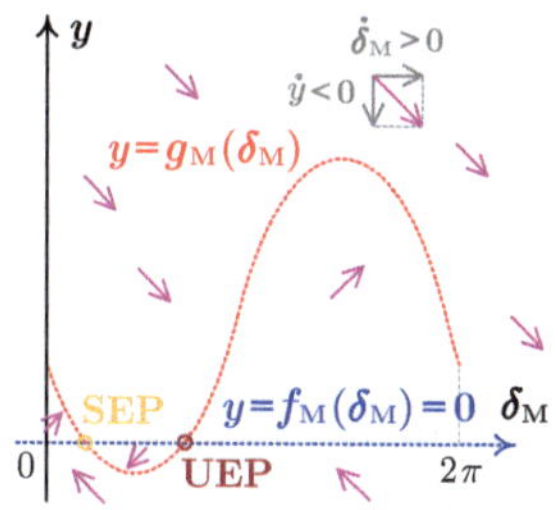

4.2.2.2 Basins of Attraction

For ease of understanding, the basins of attraction for small ω_p are derived first, as shown in Fig. 4.8. There are two different basins of attraction. When the trajectory enters the blue area, it will converge to the SEP. When the trajectory enters the green area, it will converge to the SPO. For $y \to -\infty$, the basin structure repeats itself infinitely, resembling a fractal structure. However, as there is no UPO, the basins of attraction are not squeezed into a small area. Besides, as ω_p decreases, the SEP's basin of attraction shrinks. But the SEP's basin of attraction does not become circular, however small ω_p is. Moreover, it is also found that when ω_p is significantly increased, the entire phase plane may become the SEP's basin of attraction. Therefore, there is a corresponding bifurcation that determines the change of the basins of attraction.

4.2.2.3 Homoclinic Bifurcation

Here, such bifurcation behavior is explained through the stable and unstable manifolds. From (4.5), the nullclines and vector field can be derived, as shown in Fig. 4.9. Here, the δ-nullcline and y-nullcline are defined as $y = f_M(\delta_M)$ and $y = g_M(\delta_M)$, respectively.

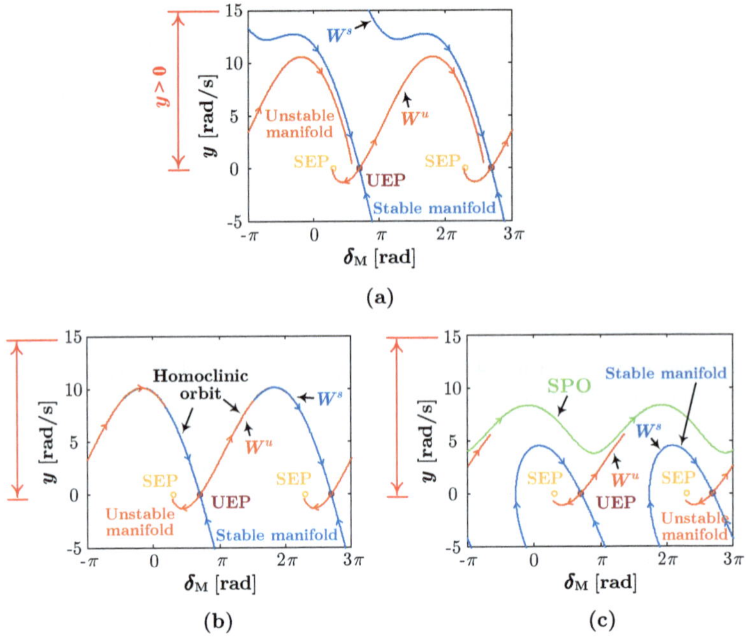

Fig. 4.10 For the GFMC, the stable manifolds and unstable manifolds of the UEP, where (**a**) $\omega_p = 0.8$ p.u., (**b**) $\omega_p = 0.6656$ p.u., and (**c**) $\omega_p = 0.3$ p.u. Other parameters follow the default set

In Fig. 4.10a, there is no SPO, and for $y > 0$, the stable manifold W^s is above the unstable manifold W^u. From Figs. 4.9 and 4.10a, the whole phase plane becomes the SEP's basin of attraction. As ω_p decreases, these two manifolds may overlap, resulting in a homoclinic orbit, as shown in Fig. 4.10b. In this case, a homoclinic bifurcation occurs. If we keep decreasing ω_p, these two manifolds will separate, and W^s is underneath W^u, as shown in Fig. 4.10c. Besides, W^u is much closer to the SPO, and the SEP is surrounded by W^s, forming the SEP's basin of attraction. It is noted that the SPO is generated due to W^u and the upward-side vector field, as shown in Figs. 4.9 and 4.10c.

Furthermore, as ω_p decreases, for $y < 0$, the unstable manifold W^u may approach to the stable manifold W^s. However, W^s and W^u do not overlap to trigger another homoclinic bifurcation, which will be explained later.

4.2.2.4 Example

The impact of a phase jump disturbance on the stability of GFMC is assessed for different values of ω_p. In each case, the system operates at the SEP initially. For $\omega_p = 0.8$ p.u., the whole phase plane is the SEP's basin of attraction, as shown in Fig. 4.11a. Therefore, the trajectory remains in the SEP's basin of attraction and

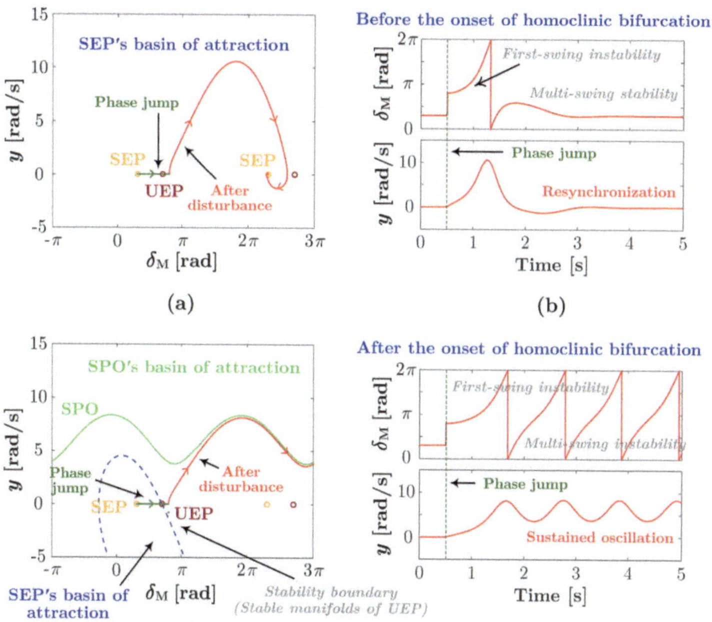

Fig. 4.11 For the GFMC, before and after a phase jump disturbance applied to v_s, the system behavior (**a**) in (δ_M, y)-plane and (**b**) in the time domain for $\omega_p = 0.8$ p.u.; the system behavior (**c**) in (δ_M, y)-plane and (**d**) in the time domain for $\omega_p = 0.3$ p.u. Other parameters follow the default set. For clarity, the variation range of δ_M is $0 \sim 2\pi$ in (**b**) and (**d**)

returns to the SEP eventually, as shown in Fig. 4.11a, b. For $\omega_p = 0.3$ p.u., the area inside the blue dashed curve (i.e., the stable manifolds of UEP) is the SEP's basin of attraction, as shown in Fig. 4.11c. In this case, after the disturbance, the trajectory passes the blue dashed curve and is out of the SEP's basin of attraction. As a result, it will not return to the SEP, but converge to the SPO, as shown in Fig. 4.11c, d.

4.2.3 General Criteria for Homoclinic Bifurcations

In this subsection, different homoclinic bifurcation behaviors of GFLC and GFMC are explained by using the Andronov-Leontovich theorem [54]. This theorem can be briefly indicated as follows. The sum of the two eigenvalues of the saddle (EQP2) is defined as the *saddle quantity*, i.e.,

$$\sigma = \lambda_1 + \lambda_2 \tag{4.22}$$

If $\sigma > 0$ on the homoclinic bifurcation point, a UPO will emerge before the onset of homoclinic bifurcation. If $\sigma < 0$ on the homoclinic bifurcation point, an SPO will emerge after the onset of homoclinic bifurcation.

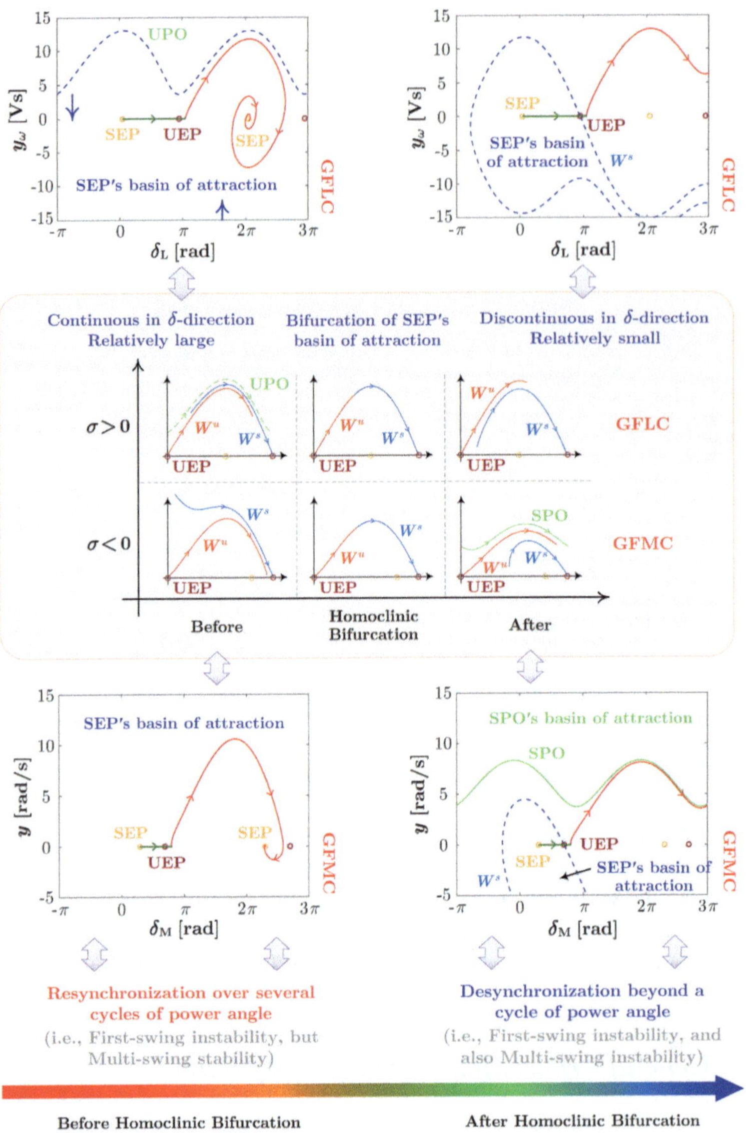

Fig. 4.12 General criteria for the first homoclinic bifurcation

Therefore, we can propose a general set of criteria based on analyzing the saddle quantity and manifolds to understand the behavior of GFLC and GFMC. As shown in Fig. 4.12, this general set of criteria is used to explain the first homoclinic bifurcation behavior. Here, the onset direction of the homoclinic bifurcation is defined as: W^s is on the top of W^u before the homoclinic bifurcation, while W^s is on the bottom of W^u after the homoclinic bifurcation. As analyzed above (i.e., (4.13) and (4.21)), $\sigma > 0$ for the GFLC, and $\sigma < 0$ for the GFMC, all the time. Hence, Fig. 4.12 indicates the duality property between the GFLC and GFMC. Specifically, for the GFLC, a UPO emerges before the onset of homoclinic bifurcation, and no UPO exists after the onset of homoclinic bifurcation. Moreover, for the GFMC, no SPO exists before the onset of homoclinic bifurcation, and an SPO emerges after the onset of homoclinic bifurcation.

From Fig. 4.12, before the onset of the first homoclinic bifurcation, the SEP's basin of attraction is continuous in the power-angle direction. Thus, GFLCs and GFMCs are able to resynchronize with the grid within several cycles of power angle. In this case, GFLCs and GFMCs seem to lose the synchronization in the first swing, but eventually regain the synchronization in the multiple swings, as shown in Figs. 4.7b, 4.11b, and 4.12. On the other hand, after the onset of the first homoclinic bifurcation, the SEP's basin of attraction is discontinuous in the power-angle direction. Thus, GFLCs and GFMCs may desynchronize with the grid beyond one cycle of power angle. In this case, GFLCs and GFMCs lose synchronization in the first swing and also the multiple swings, as shown in Figs. 4.7d, 4.11d, and 4.12.

Furthermore, these general criteria can be used to explore the onset possibility and/or the behavior of the second homoclinic bifurcation. For the GFLC, as $\sigma > 0$ all the time, another UPO may emerge through the second homoclinic bifurcation. However, the second homoclinic bifurcation cannot occur in the GFMC, which is proved by contradiction as follows. There are two different onset possibilities:

1. *Possible Existence of Another SPO:* As $\sigma < 0$ for all time, an SPO surrounding EQP1 may exist. However, EQP1 must become unstable in this case (i.e., EQP1 becomes a UEP and an SPO), which is contradicted by the fact that EQP1 is always stable, as reflected from (4.19).
2. *Possible Existence of a New UPO:* It is assumed that a new UPO surrounding the SEP (i.e., EQP1) exists. Then, according to the general criteria, the saddle quantity σ must become positive during the onset of the second homoclinic bifurcation, which, however, is contradicted by the fact that $\sigma < 0$ for the GFMC all the time, as reflected in (4.21).

4.3 Transient Stability Enhancement

In this chapter, we propose a new method to enhance the transient synchronization stability by avoiding the onset of the first homoclinic bifurcation.

4.3.1 Design-Oriented Analysis

In order to derive the stability boundaries in the parameter space (i.e., bifurcation diagrams), the onset of the homoclinic bifurcation should be identified numerically.

This chapter proposes a new method based on the unstable manifold to identify the homoclinic bifurcation point. Here, we focus on the unstable manifold with $\dot{\delta}_{\text{L}}|_{t=0} > 0$ and $\dot{y}_{\omega}|_{t=0} > 0$, or $\dot{\delta}_{\text{M}}|_{t=0} > 0$ and $\dot{y}|_{t=0} > 0$. For the GFLC, this unstable manifold will converge to the SEP before the onset of the first homoclinic bifurcation, as shown in Fig. 4.13a, while it will diverge to infinity after the onset of the first homoclinic bifurcation, as shown in Fig. 4.13b. For the GFMC, the unstable manifold will converge to the SEP before the onset of the homoclinic bifurcation, as shown in Fig. 4.13c, while it will converge to the SPO after the onset of the homoclinic bifurcation, as shown in Fig. 4.13d.

On this basis, the related bifurcation diagrams can be derived, as shown in Fig. 4.14a for the GFLC and Fig. 4.14b for the GFMC. As usual, the bifurcation diagrams are used to show the influence of specific parameters on the bifurcation behavior in question [5, 54]. In Fig. 4.14a, for each solid curve, the homoclinic bifurcation occurs on the solid curve, the top UPO (i.e., the green dashed curve in Fig. 4.2) exists above the solid curve, and the top UPO subsides below the solid curve. In order to enhance the transient stability, it is better to put the operating point in the area where the top UPO exists. Hence, from Fig. 4.14a, we can increase κ_p or V_{sm} to enhance the transient stability. Besides, we can enlarge the top UPO regime by decreasing L_s. In Fig. 4.14b, for each solid curve, the homoclinic bifurcation

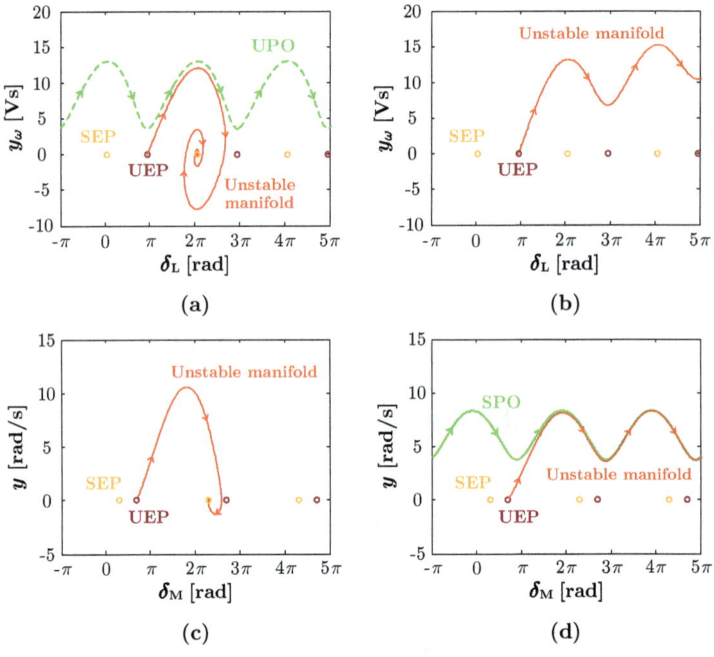

Fig. 4.13 Unstable manifold of the UEP (**a**) before and (**b**) after the onset of the first homoclinic bifurcation for the GFLC, (**c**) before and (**d**) after the onset of the homoclinic bifurcation for the GFMC

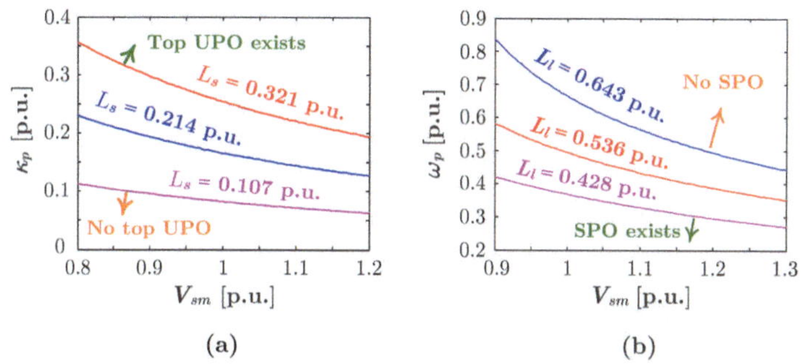

Fig. 4.14 (**a**) For the GFLC, the codimension-2 homoclinic bifurcation diagrams on V_{sm}-κ_p plane with varying L_s. (**b**) For the GFMC, the codimension-2 homoclinic bifurcation diagrams on V_{sm}-ω_p plane with varying L_l. Other parameters follow the default set

Table 4.3 Qualitative design guideline

GFLC	κ_p	κ_i	i_{gd}^*	
	↑	↓	↓	
GFMC	ω_p	m_p	P_0	V_0
	↑	↓	↓	↑
Grid	V_{sm}	L_s	L_l	SCR
	↑	↓	↓	↑

Note: "↑" ("↓") denotes that the transient stability can be enhanced by increasing (decreasing) the corresponding parameter to avoid the first homoclinic bifurcation and significantly enlarge the SEP's basin of attraction. The grid strength is reflected by the short-circuit ratio (SCR), which is defined as $V_{sm}/(\omega_0 L_s I_n)$ or $V_{sm}/(\omega_0 L_l I_n)$. Here, $I_n = 30$ A. Besides, it is noteworthy that the usual EQP is stable all the time when adjusting these parameters

occurs on the solid curve, no SPO exists above the solid curve, and the SPO exists below the solid curve. In order to enhance the transient stability, we can select suitable parameters to let the operating point in the area where no SPO exists. Hence, from Fig. 4.14b, we can increase ω_p or V_{sm} to enhance the transient stability. Besides, we can shrink the SPO regime to improve the transient stability by decreasing L_l. Similarly, we can derive other bifurcation diagrams, and collect all these analytical results, as shown in Table 4.3.

4.3.2 Practical Values and Advantages

In much of the prior work, the stability boundaries of GFLC and GFMC in the phase plane are derived through an energy-based approach (e.g., equivalent potential energy and Lyapunov function). The derived SEP's basins of attraction (which are surrounded by the aforementioned stability boundaries) are circular and disconnected, which do not contain the UEP (i.e., EQP2), as shown in Fig. 4.15a. In

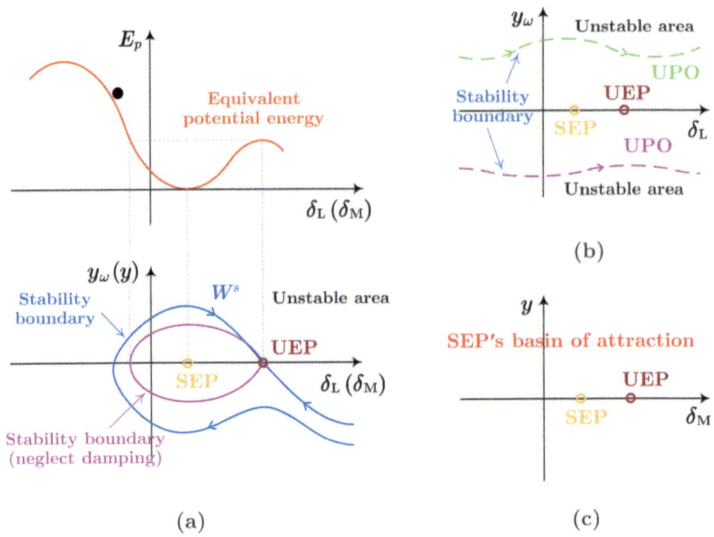

Fig. 4.15 (**a**) Based on the equivalent potential energy, the stability boundaries of GFLC and GFMC. Here, the area inside the stability boundaries is the SEP's basin of attraction. Based on our proposed method, the SEP's basin of attraction is significantly enlarged, for (**b**) the GFLC and (**c**) the GFMC

this case, sufficient damping is needed to keep the trajectories far from the UEP so that the transient stability can be maintained [18, 62, 90–94].

In this chapter, the transient stability is enhanced from a new perspective, where the SEP's basins of attraction are significantly enlarged to contain the UEP, as shown in Fig. 4.15b, c. In this case, the trajectory will finally return to the SEP when it crosses over the UEP, even for the system with poor damping. In practice, the GFLCs and GFMCs can resynchronize with the grid after a few cycles of oscillations. Such a resynchronization mechanism is effective for various types of transient disturbances. In this chapter, we also find that the aforementioned SEP's basins of attraction can be achieved by avoiding the onset of the first homoclinic bifurcation and be identified with the help of unstable manifolds, as shown in Fig. 4.13. On this basis, the influence of adjusting control and circuit parameters on the first homoclinic bifurcation is studied, and a qualitative design guideline (i.e., Table 4.3) is proposed to enhance the transient stability, where the parameters are selected or designed in advance.

4.4 Experimental Verification

A series of *hardware-in-loop* experiments have been carried out to verify the theoretical findings. The power circuit of the grid and the grid-connected converters was

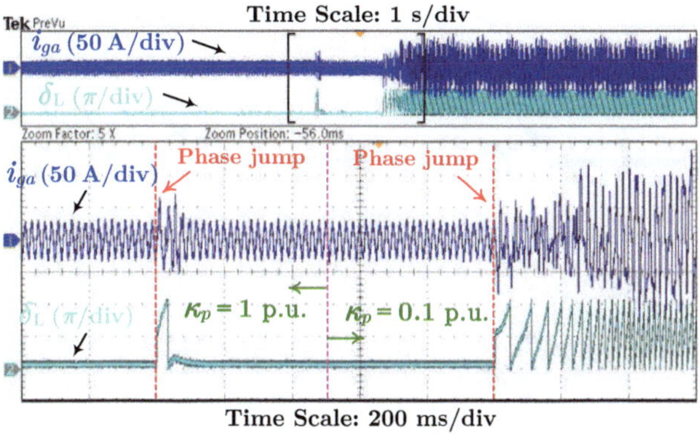

Fig. 4.16 Experimental waveforms of δ_L and i_{ga} before and after the phase jump disturbances and the adjustment of κ_p. Here, i_{ga} is the phase-A component of i_g. Other parameters follow the default set

emulated by the OPAL-RT equipment OP5600, while the PWM signal generation and the control algorithms were implemented using DSP TMS320F28335. The default parameters are given in Tables 4.1 and 4.2, unless otherwise specified.

The experimental result of GFLC is shown in Fig. 4.16. Initially, $\kappa_p = 1$ p.u., and the system operates normally. Then, a sudden phase jump is applied to v_s. After this disturbance, δ_L and i_{ga} return to the normal state within a cycle of oscillation, implying that the trajectory passes the UEP but remains in the SEP's basin of attraction. In other words, the UEP is contained in the SEP's basin of attraction. Then, κ_p is set to 0.1 p.u., and the same phase disturbance is applied. After the disturbance, i_{ga} is found to be highly distorted, and δ_L oscillates fiercely, implying that the trajectory passes the UEP as well, but the SEP's basin of attraction does not contain the UEP this time. This experimental result is consistent with Fig. 4.7 and confirms that the transient stability of the GFLC can be significantly improved when the UEP is contained in the SEP's basin of attraction, which can be achieved by adjusting κ_p to avoid the onset of the first homoclinic bifurcation.

The experimental result of the GFMC is shown in Fig. 4.17. In the beginning, $\omega_p = 0.8$ p.u., and the system operates normally. Then, a sudden phase jump is applied to v_s. After the phase jump, δ_M, p and i_{la} experience a cycle of oscillation and finally return to their normal operating state. In particular, δ_M increases and crosses 2π, implying that the trajectory passes the UEP. Besides, as there is no sustained oscillation, the trajectory remains in the SEP's basin of attraction. Then, ω_p is adjusted to 0.3 p.u., and the same phase jump disturbance is applied to v_s. In this case, sustained sub-synchronous oscillations emerge in δ_M, p, and i_{la}, implying that the trajectory passes the UEP as well, but the SEP's basin of attraction does not contain the UEP this time. This experimental result is consistent with Fig. 4.11 and confirms that the transient stability of the GFMC can be significantly improved when the UEP is contained in the SEP's basin of attraction, which can be achieved by adjusting ω_p to avoid the onset of the first homoclinic bifurcation.

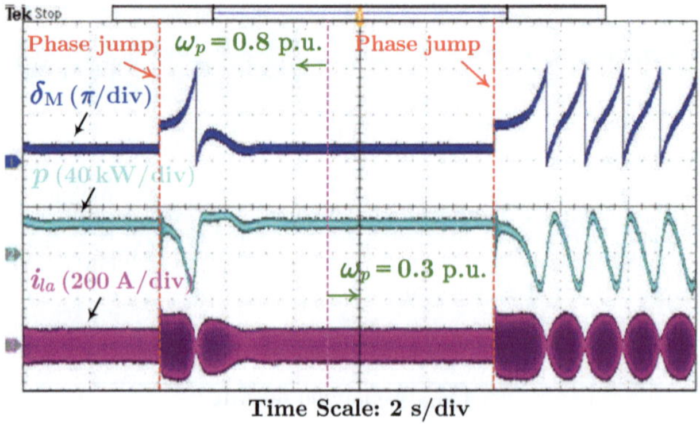

Fig. 4.17 Experimental waveforms of δ_M, p, and i_{la} before and after the phase jump disturbances and the adjustment of ω_p. Here, i_{la} is the phase-A component of i_l. Other parameters follow the default set

4.5 Recapitulation

In this chapter, a general set of criteria based on the manifolds and saddle quantity is established for studying the homoclinic bifurcation behavior of GFLCs and GFMCs. It is shown that there are three UPOs and two homoclinic bifurcations in the GFLC, while there are only one SPO and one homoclinic bifurcation in the GFMC. According to Fig. 4.12, the relation between the homoclinic bifurcation behavior and the transient synchronization stability can be summarized as follows:

1. Before the onset of the first homoclinic bifurcation, the SEP's basin of attraction contains the UEP and is continuous in the power-angle direction. In this case, the trajectory is able to return to the SEP even when it crosses over the UEP. In practice, the GFLCs and GFMCs can resynchronize with the grid after a few cycles of oscillations, where the power angle increases at least 2π. Therefore, the first-swing instability here does not imply multi-swing instability. The converter-based system shows higher robustness against the transient disturbance.
2. After the onset of the first homoclinic bifurcation, the SEP's basin of attraction does not contain the UEP and is discontinuous in the power-angle direction. In this case, the trajectory cannot return to the SEP once it crosses over the UEP. In practice, the GFLCs and GFMCs will desynchronize with the grid, and sustained oscillation may emerge. Therefore, the first-swing instability here is exactly the multi-swing instability. The converter-based system shows lower robustness against the transient disturbance.

Chapter 5
Impact of Outer Loops of Grid-Following Converters on Transient Stability

In Chap. 4, an ideal current loop has been assumed, i.e., the AC-side current perfectly follows the control reference. Hence, the slow-scale dynamics of the PLL can be considered dominant, and the PLL-based second-order model has been adopted for the transient synchronization stability analysis. While this approach can approximately explain the behavior of GLFCs, neglecting the current transients inevitably leads to inaccurate transient stability assessment. Moreover, according to Shil'nikov theorem, the essential bifurcation of GFLCs (i.e., homoclinic bifurcation) can be considerably complex if the system's dimension exceeds two [54]. Hence, the assumption of an ideal current loop could bury the essential qualitative changes. An example is shown in Fig. 5.1. The blue area is the SEP's basin of attraction predicted by using the second-order model. From Fig. 5.1, the trajectory (i.e., red curve) once the disturbance is removed is still in the blue area, implying that the trajectory will eventually return to SEP according to the second-order model. However, based on the full-circuit cycle-by-cyle simulation, the trajectory diverges to infinity. This discrepancy indicates that neglecting the dynamics of the nonideal current loop leads to over-optimistic prediction of SEP's basin of attraction.

On the other hand, the renewable resources may use other DC/DC or AC/DC converters as intermediaries to connect the GLFCs, and constant DC power may be injected into the DC-link by other converters, such as under maximum power tracking control. In this case, constant power, but not constant current, may be injected by GLFCs to the power grid [95, 96]. However, this factor is not considered in the aforementioned second-order model. Besides, although the DC-link voltage control (DVC) that embeds a current limit is commonly used in GLFCs and may have a significant impact on the transient synchronization stability, the DC-link dynamics have been neglected in the aforementioned second-order model.

Based on the results presented in Chap. 4, this chapter attempts to investigate the influence of the nonideal current loops and the DVC with current limit on the transient synchronization stability.

J. Yang, C. K. Tse, *Complex Behavior of Grid-Connected Power Electronics Systems*, https://doi.org/10.1007/978-3-031-69340-3_5

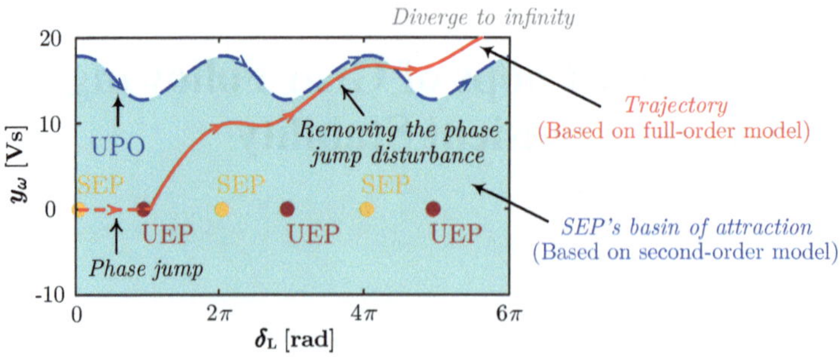

Fig. 5.1 SEP's basin of attraction (i.e., the blue area) predicted by the second-order model and the trajectory (i.e., the red curve) based on the full-circuit cycle-by-cycle simulation with $\kappa_p = 0.4$ p.u.

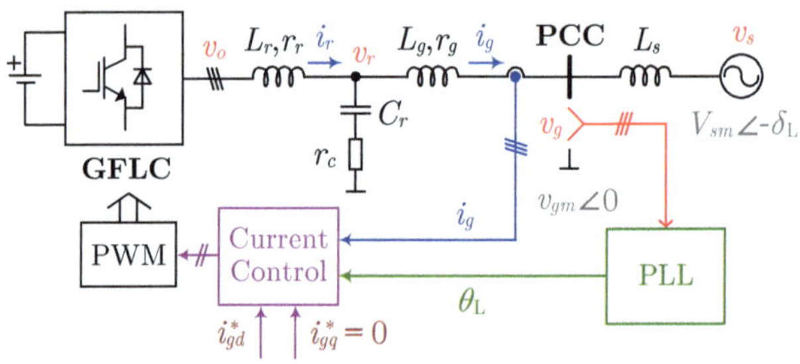

Fig. 5.2 Circuit diagram and control of the GFLC

5.1 Impact of Current Control Loop on Transient Stability

In this section, nonideal current loops are included in the established model, and the impact of these nonideal current loops on the transient synchronization stability is studied.

5.1.1 Full-Order Model Considering Current-Loop Dynamics

As shown in Fig. 5.2, the GFLC connects the power grid via an LCL filter. The power grid is modeled as a voltage source v_s and a grid impedance, where the amplitude and frequency of v_s are V_{sm} and ω_0, respectively, while the grid impedance includes an inductance L_s and a resistance r_s. For the GFLC, the AC-side voltage is v_g, and the AC-side current is i_g. The default parameters in Table 5.1 are designed using Bode plots to ensure small-signal stability, where the bandwidth of the PLL and the bandwidth of the current loop are $f_c^{\text{pll}} = 23.36$ Hz and $f_c^{\text{cl}} = 126.29$ Hz, respectively.

Table 5.1 Default parameters of GFLC

Parameter	Value	p.u.	Parameter	Value	p.u.
V_{sm}	311 V	1	L_r	2 mH	0.061
L_s	5 mH	0.152	r_r	0.1 Ω	0.010
r_s	0.1 Ω	0.010	C_r	60 μF	0.195
i_{gd}^*	30 A	1	r_c	1 Ω	0.096
i_{gq}^*	0	0	L_g	0.6 mH	0.018
κ_p	0.413	1	r_g	0.05 Ω	0.005
κ_i	7.786	1	ω_0	100π	1
β_p	1.648	1	T_{ctr}	0.25 ms	0.125
β_i	51.779	1	f_{sw}	2 kHz	40

According to Chap. 4, the dynamics of the PLL can be described by

$$\frac{\mathrm{d}y_\omega}{\mathrm{d}t} = v_{gq} \tag{5.1}$$

$$\frac{\mathrm{d}\delta_\mathrm{L}}{\mathrm{d}t} = \omega_\mathrm{L} - \omega_0 \tag{5.2}$$

$$\omega_\mathrm{L} = \kappa_p v_{gq} + \kappa_i y_\omega + \omega_0 \tag{5.3}$$

where y_ω is the state variable, δ_L is the phase difference between v_g and v_s, ω_L is the frequency of PLL and κ_p and κ_i are the proportional and integral parameters, respectively.

The block diagram of the nonideal current loop is shown in Fig. 5.3. The current loop generates the PWM reference of the converter terminal voltage v_o^*, i.e.,

$$\frac{\mathrm{d}}{\mathrm{d}t}\begin{bmatrix} y_{id} \\ y_{iq} \end{bmatrix} = \begin{bmatrix} i_{gd}^* \\ i_{gq}^* \end{bmatrix} - \begin{bmatrix} i_{gd} \\ i_{gq} \end{bmatrix} \tag{5.4}$$

$$\begin{bmatrix} v_{od}^* \\ v_{oq}^* \end{bmatrix} = \beta_p \left(\begin{bmatrix} i_{gd}^* \\ i_{gq}^* \end{bmatrix} - \begin{bmatrix} i_{gd} \\ i_{gq} \end{bmatrix} \right) + \beta_i \begin{bmatrix} y_{id} \\ y_{iq} \end{bmatrix} \tag{5.5}$$

where y_{id} and y_{iq} are the state variables of the current loop, and β_p and β_i are the proportional and integral parameters, respectively.

The converter terminal voltage v_o is affected by the time delay, which can be modeled as

$$\frac{\mathrm{d}}{\mathrm{d}t}\begin{bmatrix} v_{od} \\ v_{oq} \end{bmatrix} = \frac{1}{T_d} \left(\begin{bmatrix} v_{od}^* \\ v_{oq}^* \end{bmatrix} - \begin{bmatrix} v_{od} \\ v_{oq} \end{bmatrix} \right) \tag{5.6}$$

where $T_d = 1.5T_{\text{ctr}}$ is the equivalent time constant and T_{ctr} is the control period.

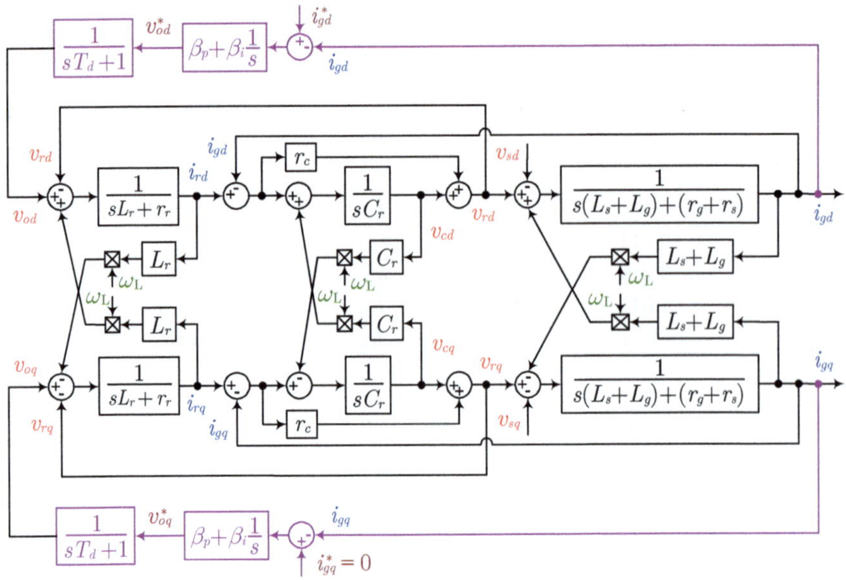

Fig. 5.3 Block diagram of the nonideal current loop of GFLC

Finally, the grid-side current i_g is determined by the dynamics of the LCL filter and grid impedance. The LCL filter includes a converter-side inductance L_r, a capacitance C_r, a damping resistance r_c, and a grid-side inductance L_g, where the stray resistances of L_r and L_g are denoted by r_r and r_g, respectively. Thus, the dynamics of the filter and grid impedance are described by

$$\frac{\mathrm{d}}{\mathrm{d}t}\begin{bmatrix} i_{rd} \\ i_{rq} \end{bmatrix} = \frac{1}{L_r}\left(\begin{bmatrix} v_{od} \\ v_{oq} \end{bmatrix} - \begin{bmatrix} v_{rd} \\ v_{rq} \end{bmatrix} - \begin{bmatrix} r_r & -\omega_{\mathrm{L}} L_r \\ \omega_{\mathrm{L}} L_r & r_r \end{bmatrix}\begin{bmatrix} i_{rd} \\ i_{rq} \end{bmatrix}\right) \tag{5.7}$$

$$\frac{\mathrm{d}}{\mathrm{d}t}\begin{bmatrix} v_{cd} \\ v_{cq} \end{bmatrix} = \frac{1}{C_r}\left(\begin{bmatrix} i_{rd} \\ i_{rq} \end{bmatrix} - \begin{bmatrix} i_{gd} \\ i_{gq} \end{bmatrix} - \begin{bmatrix} 0 & -\omega_{\mathrm{L}} C_r \\ \omega_{\mathrm{L}} C_r & 0 \end{bmatrix}\begin{bmatrix} v_{cd} \\ v_{cq} \end{bmatrix}\right) \tag{5.8}$$

$$\begin{aligned}\frac{\mathrm{d}}{\mathrm{d}t}\begin{bmatrix} i_{gd} \\ i_{gq} \end{bmatrix} = \frac{1}{L_s + L_g}\Bigg(&\begin{bmatrix} v_{rd} \\ v_{rq} \end{bmatrix} - \begin{bmatrix} v_{sd} \\ v_{sq} \end{bmatrix} \\ &- \begin{bmatrix} r_s + r_g & -\omega_{\mathrm{L}}(L_s + L_g) \\ \omega_{\mathrm{L}}(L_s + L_g) & r_s + r_g \end{bmatrix}\begin{bmatrix} i_{gd} \\ i_{gq} \end{bmatrix}\Bigg)\end{aligned} \tag{5.9}$$

and

$$\begin{bmatrix} v_{rd} \\ v_{rq} \end{bmatrix} = \begin{bmatrix} v_{cd} \\ v_{cq} \end{bmatrix} + r_c\left(\begin{bmatrix} i_{rd} \\ i_{rq} \end{bmatrix} - \begin{bmatrix} i_{gd} \\ i_{gq} \end{bmatrix}\right) \tag{5.10}$$

where $v_{sd} = V_{sm} \cos \delta_{\mathrm{L}}$, $v_{sq} = -V_{sm} \sin \delta_{\mathrm{L}}$, i_{rd} and i_{rq} are the converter-side currents, v_{cd} and v_{cq} are the capacitance voltages, and v_{rd} and v_{rq} are the voltages across C_r and r_c.

Also, a more numerically stable representation of v_{gq} can be found as follows:

$$v_{gq} = \frac{L_s L_g}{L_s + L_g} \left(\frac{1}{L_s} v_{sq} + \frac{1}{L_g} v_{rq} + \left(\frac{r_s}{L_s} - \frac{r_g}{L_g} \right) i_{gq} \right) \tag{5.11}$$

Combining (5.1)–(5.11) yields a 12th-order nonlinear system, which can be represented as

$$\frac{\mathrm{d}\boldsymbol{x}}{\mathrm{d}t} = f(\boldsymbol{x}) \tag{5.12}$$

where

$$\boldsymbol{x} = \left[y_\omega \; \delta_{\mathrm{L}} \; i_{rd} \; i_{rq} \; v_{cd} \; v_{cq} \; i_{gd} \; i_{gq} \; y_{id} \; y_{iq} \; v_{od} \; v_{oq} \right]^T \tag{5.13}$$

is the state vector.

5.1.2 *Numerical Continuation for Bifurcation Analysis*

5.1.2.1 Homoclinic Bifurcation

The homoclinic bifurcation occurs when a homoclinic orbit connects a saddle [54]. The saddle is a UEP with both positive and negative eigenvalues. As a parameter changes, the split of the homoclinic orbit can generate an SPO or UPO, depending on the polarity of the saddle quantity. For a second-order system, the saddle quantity refers to the algebraic sum of the saddle's two eigenvalues. This feature of the homoclinic bifurcation results in a dual characteristic between the GFMC and the GFLC [55, 90, 97], as presented in Chap. 4. It is reasonable to expect this bifurcation to remain in the full-order model since the slow dynamics of the PLL are dominating for the GFLC.

However, the homoclinic bifurcation can have distinctive features in high-order systems. From the homoclinic central manifold theorem [54], the leading eigenvalues, instead of all eigenvalues, determine the property of the homoclinic bifurcation. The leading eigenvalues are defined as the eigenvalues that are closest to the imaginary axis. For a saddle with eigenvalues satisfying

$$\mathrm{Re}(\lambda_a) < \cdots < \mathrm{Re}(\lambda_1) < 0 < \mathrm{Re}(\lambda_2) < \cdots < \mathrm{Re}(\lambda_b), \tag{5.14}$$

the leading stable and unstable eigenvalues are λ_1 and λ_2, respectively. Then, the saddle quantity is given as

$$\sigma = \mathrm{Re}(\lambda_1) + \mathrm{Re}(\lambda_2). \tag{5.15}$$

As in the case of the second-order system, $\sigma < 0$ in high-order systems means an SPO emerging from bifurcation. In contrast, Shil'nikov theorem [54] predicts that $\sigma > 0$ in high-order systems can have two different results:

1. *One Simple UPO:* This case occurs when the leading eigenvalues are both real values, i.e., $\mathrm{Im}(\lambda_1) = \mathrm{Im}(\lambda_2) = 0$. This saddle is referred to as the Type-1 saddle in the following.
2. *Infinite UPOs and One Chaotic Attractor:* This case occurs when at least a pair of leading eigenvalues are complex, i.e., $\mathrm{Im}(\lambda_1) \neq 0$ or $\mathrm{Im}(\lambda_2) \neq 0$. This saddle is referred to as the Type-2 saddle in the sequel.

5.1.2.2 Numerical Continuation

The adoption of the full-order model presents a crucial challenge. The analysis method for the second-order model is no longer applicable. For the second-order model, the stable and unstable manifolds of the saddle can be calculated by integrating the model along the saddle's eigenvectors. A homoclinic orbit is found when the two manifolds overlap. However, the full-order model as a high-order system can be much stiffer in the inverse direction of time. Therefore, the calculation of the stable manifold easily fails as the integration errors blow up instantaneously [54].

In this chapter, we try to find the UPO emerging from the homoclinic bifurcation. To this end, the numerical continuation of UPOs is applied in a periodic boundary-value problem. Let the parametric representation of (5.12) be $f(\boldsymbol{x}, \alpha)$, where α is the selected continuation parameter. The time $t \in [0, T]$ can be normalized to $\tau \in [0, 1]$, where T is the period of the UPO $\boldsymbol{x}_t$. The normalized system can be represented as

$$\frac{\mathrm{d}\boldsymbol{\mu}}{\mathrm{d}\tau} = T f(\boldsymbol{\mu}, \alpha) \tag{5.16}$$

where $\boldsymbol{\mu}_\tau$ is equivalent to $\boldsymbol{x}_t$. Hence, $\boldsymbol{\mu}_\tau$ is a UPO with period 1, and the periodic boundary condition is given as

$$\boldsymbol{\mu}_0 = \boldsymbol{\mu}_1. \tag{5.17}$$

The discretization of (5.16) and (5.17) defines a curve consisting of UPOs, i.e.,

$$\mathcal{F}(\boldsymbol{\xi}, \alpha) = 0,\ \mathcal{F} : \mathbb{R}^{N+1} \to \mathbb{R}^N \tag{5.18}$$

With an additional phase condition, the periodic boundary-value problem defined by (5.18) can be solved via Newton-like iterations. Some well-studied test functions

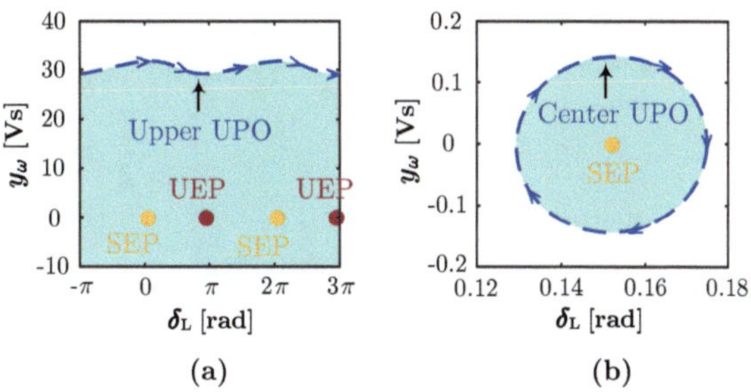

Fig. 5.4 UPOs emerging from the homoclinic bifurcations in the second-order model, where (**a**) $\kappa_p = 1$ p.u., and (**b**) $\kappa_p = 0.0092$ p.u.

can be used to locate the bifurcation points along this curve. For implementation details, refer to Kuznetsov [54].

As the numerical continuation relies on iterations, the initial point to start the continuation is of great significance. By studying the second-order model, the rough features of the UPOs generated from the homoclinic bifurcations are already known. For the upper homoclinic bifurcation, the associated UPO behaves close to a straight line at relatively large κ_p, as shown in Fig. 5.4a. For the center homoclinic bifurcation, the associated UPO converges to the SEP at relatively small κ_p, eventually leading to a subcritical Hopf bifurcation, as shown in Fig. 5.4b. These features of the UPOs in the second-order model provide an effective estimation for the initial points. Therefore, for the full-order model, the continuation of the upper and center UPOs here starts from an approximate straight line and the subcritical Hopf bifurcation, respectively. The corresponding homoclinic orbits can be found when the periods of these UPOs approach infinity.

5.1.3 Complex Behavior

5.1.3.1 Basin of Attraction

The full-order model is applied to derive the SEP's basin of attraction numerically. Specifically, the trajectory starting from each initial state in the phase plane is calculated by the Runge-Kutta method. Convergence or divergence of the numerical process can be used to identify the basin of attraction. As the system normally operates at the SEP, the cross-section containing the SEP is analyzed. Figure 5.5 shows that the SEP's basin of attraction is connected at $\kappa_p = 1$ p.u., implying two potential UPOs in the upper and bottom sides. As κ_p decreases to 0.26 p.u., the SEP's basin of attraction splits into a fractal structure. Furthermore, the SEP's basin of attraction shrinks to an ellipse at $\kappa_p = 0.22$ p.u., suggesting a potential UPO near the SEP. If κ_p decreases further, the SEP eventually becomes a UEP via the subcritical Hopf bifurcation. These rough features are consistent with that of the second-order model, indicating the primary role of the PLL in the dynamics of the GFLC.

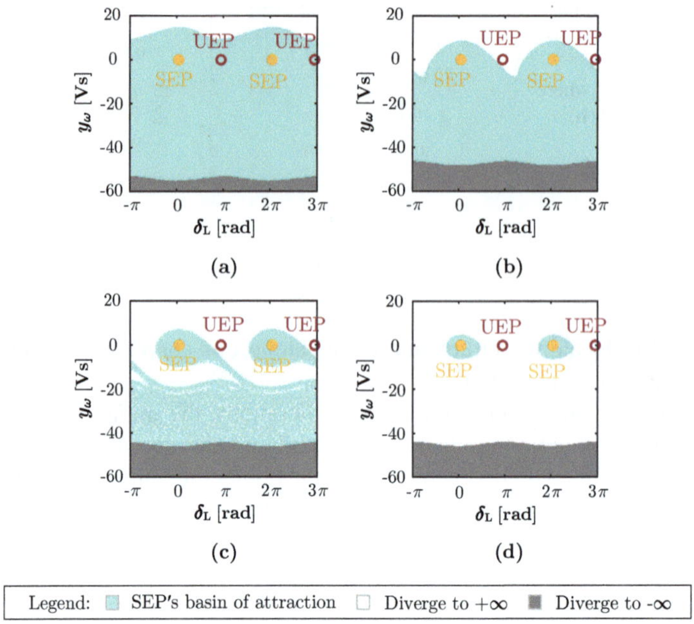

Fig. 5.5 SEP's basin of attraction of the full-order model where (**a**) $\kappa_p = 1$ p.u., (**b**) $\kappa_p = 0.4$ p.u., (**c**) $\kappa_p = 0.26$ p.u., and (**d**) $\kappa_p = 0.22$ p.u. Other parameters adopt the default values

However, the comparison between Figs. 5.1 and 5.5b indicates that the SEP's basin of attraction shrinks much faster in the full-order model. Consequently, the projection of the UEP in Fig. 5.5b becomes out of the SEP's basin of attraction at relatively large κ_p. This significant difference can be attributed to the adverse effects of current transients. The result demonstrates that the second-order model overestimates the GLFC's robustness to phase jump disturbances, highlighting the importance of the nonideal current loop in the transient synchronization stability.

5.1.3.2 Bifurcation Behavior

The numerical continuation method is utilized to identify the bifurcation behavior behind the change of the SEP's basin of attraction. The continuation results of the upper and center UPOs are shown in Figs. 5.6 and 5.7, respectively. The manifolds in Figs. 5.6a and 5.7a show the trajectory variations of the corresponding UPOs with different κ_p. The periods of the UPOs in Figs. 5.6b and 5.7b approach infinity near $\kappa_p = 0$, implying homoclinic bifurcation. The saddle quantities at the upper and center homoclinic bifurcations are 24.52 and 27.44, respectively, with both the saddles having complex leading eigenvalues. Hence, as Shil'nikov theorem predicts, infinite UPOs will emerge near the homoclinic bifurcations. This phenomenon is verified by the repetitive small "wiggles" near the homoclinic bifurcations illustrated in Figs. 5.6b and 5.7b.

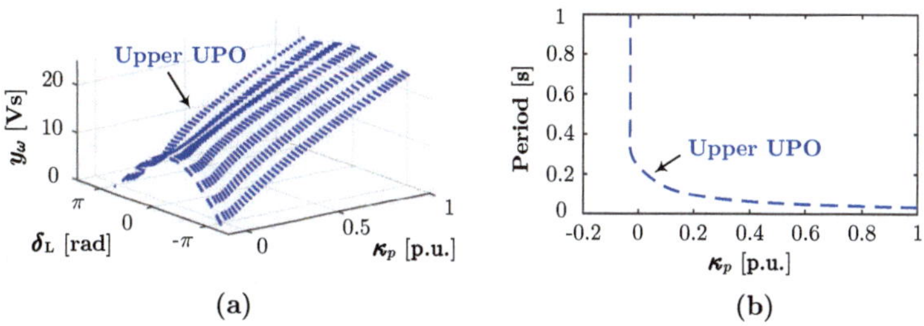

Fig. 5.6 (**a**) Projection of the upper UPO manifold and (**b**) periods of the upper UPOs

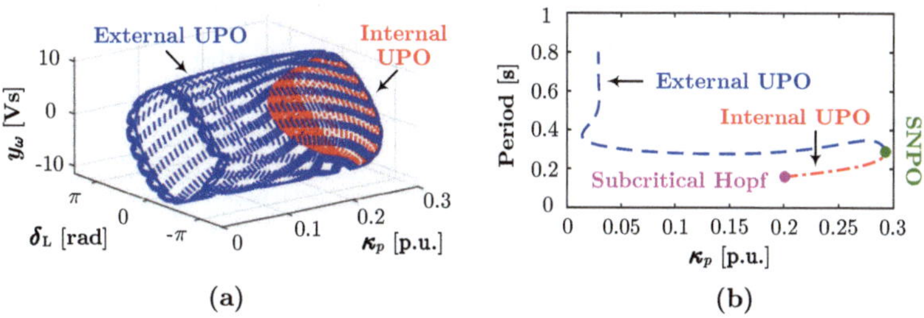

Fig. 5.7 (**a**) Projection of the center UPO manifold and (**b**) periods of the center UPOs. Dashed and dash-dotted lines represent external and internal branches, respectively

Figure 5.6 shows that, within the normal parameter range (i.e., $\kappa_p > 0$), there exists at least an upper UPO whose stable manifold can affect the SEP's basin of attraction. From (5.12), for any values of κ_p, the SEP lies in the same position as shown in Fig. 5.5. Thus, Fig. 5.6 indicates that as κ_p decreases from 1 p.u., the upper UPO gets closer to the UEP and, simultaneously, the SEP, resulting in the contraction of the SEP's basin of attraction.

Moreover, Fig. 5.7 shows two main branches of UPOs in the center UPO manifold. For clarity, a codim-1 bifurcation diagram illustrating the center UPOs is presented in Fig. 5.8. The bifurcation diagram shows that the external and internal branches of the center UPOs emerge from the homoclinic and Hopf bifurcations, respectively. The internal branch makes the SEP's basin of attraction significantly sensitive to parameter choices. As the internal branch approaches the SEP, the SEP's basin of attraction shrinks rapidly from a connected area ($\kappa_p = 0.292$ p.u.) to zero ($\kappa_p = 0.201$ p.u.) until a subcritical Hopf bifurcation sets in. To enhance the transient synchronization stability of the GLFC, the internal branch's existence should be avoided. This objective can be achieved by triggering the SNPO at $\kappa_p = 0.292$ p.u., where the two branches collide and annihilate. Notably, two UPOs merge here, which is different from the SNPO shown in Chap. 2.

The significance of the SNPO is further illustrated in Fig. 5.9a, where the magnitudes of the critical phase jumps $\Delta\theta_s$ leading to instability are presented. As κ_p decreases from the SNPO, the critical phase jumps drop to zero in a narrow parameter range. Therefore, the full-order model suggests triggering the SNPO to

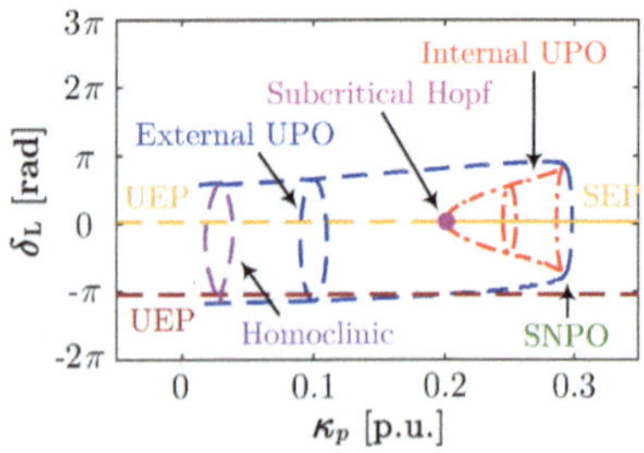

Fig. 5.8 Codim-1 bifurcation diagram regarding the center UPOs, where solid line represents the SEP, and dashed and dash-dotted lines represent the UEPs and UPOs. All parameters except κ_p adopt the default values

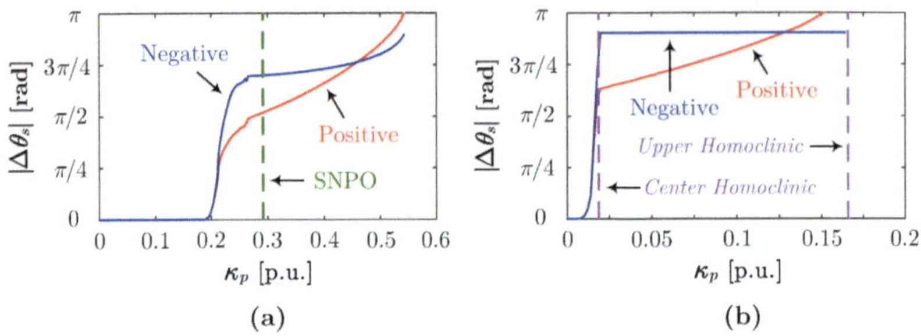

Fig. 5.9 Magnitudes of the critical phase jumps of (**a**) full-order model and (**b**) second-order model

enhance the synchronization stability. In contrast, the critical phase jumps of the second-order model in Fig. 5.9b suggest triggering the upper homoclinic bifurcation at $\kappa_p = 0.166$ p.u. With this design principle derived from the second-order model, the real critical phase jumps have decreased to zero, indicating no SEP's basin of attraction, as shown in Fig. 5.9a. Therefore, comparing Fig. 5.9a and b validates the full-order model and the proposed design principle.

5.1.4 *Effects of Current Loop Bandwidth*

Studies have shown that a faster current loop can reduce the current transient and mitigate its adverse impact on the transient synchronization stability [98, 99]. In this subsection, not only the qualitative effects of the current loop bandwidth f_c^{cl} but also its quantitative effects are studied in terms of the SEP's basin of attraction and bifurcation behavior. With $f_c^{\text{cl}} \leq 0.2 f_{\text{sw}}$ [100], where $f_{\text{sw}} = 2$ kHz is the switching frequency, four cases with $f_c^{\text{cl}} = 100$, 200, 300, and 400 Hz are presented in Fig. 5.10. The current loop bandwidth is tuned while the ratio β_i/β_p remains unchanged. Figure 5.10a shows that increasing f_c^{cl} enlarges the SEP's basin of attraction considerably. However, even with a very high bandwidth of 400 Hz, the discrepancy between the second-order model and full-order model is still significant, as shown in Figs. 5.9b and 5.10a. Therefore, while the second-

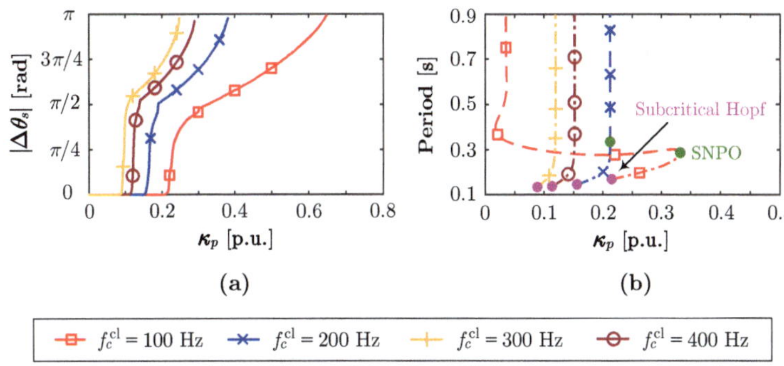

Fig. 5.10 (**a**) Positive critical phase jumps with different f_c^{cl} and (**b**) periods of center UPOs with different f_c^{cl}, where dashed and dash-dotted lines represent external and internal branches, respectively

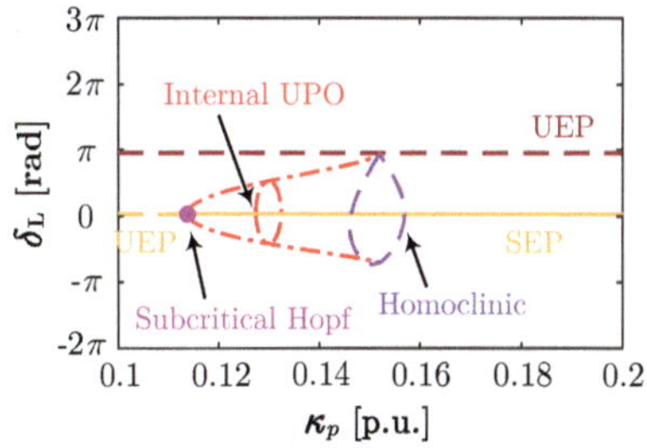

Fig. 5.11 Codim-1 bifurcation diagram illustrating the center UPO with $f_c^{\mathrm{cl}} = 300$ Hz, where solid line represents the SEP and dashed and dash-dotted lines represent the UEPs and UPO. Other parameters adopt the default values

order model provides an approximate description of the dynamical behavior, the current transient is important for accurate estimation of the SEP's basin of attraction. The changes in bifurcation behavior regarding the center UPOs are illustrated in Fig. 5.10b. With $f_c^{\mathrm{cl}} = 100$ Hz, the external branch of the center UPOs emerges from a homoclinic bifurcation at small κ_p (nearly zero). As f_c^{cl} increases to 200 Hz, another homoclinic bifurcation occurs at relatively large κ_p (over 0.2 p.u.), of which the generated UPO forms the external branch. If f_c^{cl} increases further, the SNPO and the external branch disappear. This phenomenon can be explained by applying Shil'nikov theorem. When $f_c^{\mathrm{cl}} = 200$ Hz, the saddle's leading eigenvalues are $-21.52 \pm j16.78$ and 57.07, indicating a Type-2 saddle. When $f_c^{\mathrm{cl}} = 300$ Hz, the leading eigenvalues change to -26.02 and 55.63, indicating a Type-1 saddle. Hence, the transformation of the saddle means that the homoclinic bifurcation generates only a simple UPO for very high f_c^{cl}. A codim-1 bifurcation diagram with $f_c^{\mathrm{cl}} = 300$ Hz is presented in Fig. 5.11, where the only center UPO is directly connected by the subcritical Hopf bifurcation and the homoclinic bifurcation. Indeed, such bifurcation behavior of the full-order model at very high f_c^{cl} is similar to that of the second-order model, which can be regarded as a full-order model with infinite current loop bandwidth.

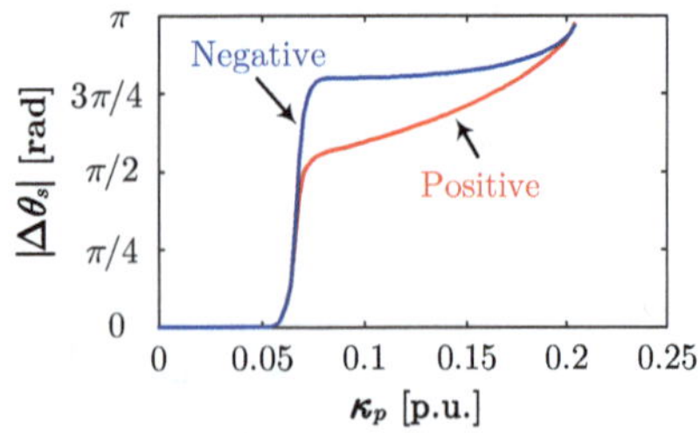

Fig. 5.12 Magnitudes of the critical phase jumps with $f_c^{\mathrm{cl}} = 900\,\mathrm{Hz}$

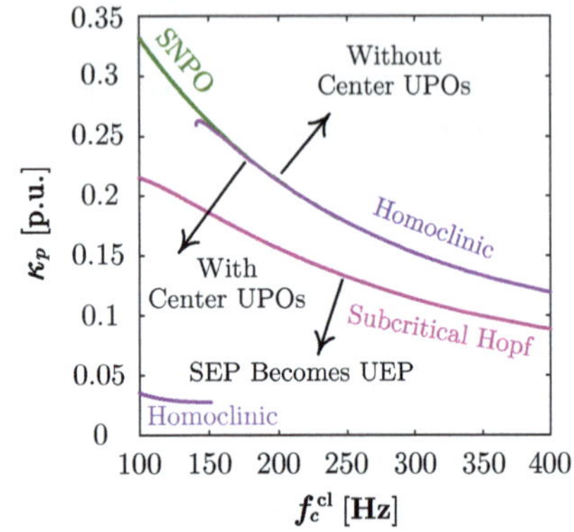

Fig. 5.13 Codim-2 bifurcation diagram in the (f_c^{cl},κ_p)-plane

Notably, the available bandwidth of the current loop is limited by small-signal stability as well as the switching frequency. The extreme bandwidth of the current loop here is nearly 900 Hz in terms of small-signal stability. With such an extreme setting, the critical phase jumps of the full-order model are presented in Fig. 5.12, which is close to, though not exactly same as, Fig. 5.9b. Therefore, the second-order model cannot fully substitute the full-order model with only the current loop bandwidth being tuned. Besides, tuning the current loop bandwidth to adapt to the second-order model may cause other problems, e.g., reducing the small-signal stability margin. Thus, the appropriate model should be chosen according to the specific setting.

To clarify the bifurcation behavior with varying current loop bandwidth, a codim-2 bifurcation diagram is presented in Fig. 5.13. It can be observed that, at relatively low f_c^{cl}, the SNPO serves as the boundary between the existence and disappearance of the center UPOs. With f_c^{cl} exceeding 140.85 Hz, a center homoclinic bifurcation occurs near the SNPO and eventually overlaps with the SNPO at $f_c^{\mathrm{cl}} = 244.52\,\mathrm{Hz}$. Simultaneously, the saddle connected by the homoclinic orbit transforms from a Type-2 saddle to a Type-1 saddle. To enhance the transient synchronization stability, the existence of the center UPOs should be avoided. This objective can be achieved by triggering the SNPO for relatively low current loop bandwidth or avoiding the center homoclinic bifurcation for very high current loop bandwidth.

5.1.5 Effects of Voltage Feedforward Scheme

Although the PCC voltage is neglected in the derivation of the current loop bandwidth, it can significantly impact current regulation [101]. To address the current harmonics that result from voltage distortion, grid voltage can be feedforwarded into the current loop. This scheme is also crucial for transient stability, as it enhances the speed of the current loop. Thus, the output of the current controller in (5.5) can change to

$$\begin{bmatrix} v_{od}^* \\ v_{oq}^* \end{bmatrix} = \beta_p \left(\begin{bmatrix} i_{gd}^* \\ i_{gq}^* \end{bmatrix} - \begin{bmatrix} i_{gd} \\ i_{gq} \end{bmatrix} \right) + \beta_i \begin{bmatrix} y_{id} \\ y_{iq} \end{bmatrix} + k \begin{bmatrix} v_{gd} \\ v_{gq} \end{bmatrix} \tag{5.19}$$

where

$$v_{gd} = \frac{L_s L_g}{L_s + L_g} \left(\frac{1}{L_s} v_{sd} + \frac{1}{L_g} v_{rd} + \left(\frac{r_s}{L_s} - \frac{r_g}{L_g} \right) i_{gd} \right) \tag{5.20}$$

Here, a feedforward weight $k \in [0, 1]$ is adopted because the unweighted voltage feedforward scheme will increase the risk of small-signal instability of converters with LCL filters, especially under weak grids [102].

Figure 5.14 shows the positive critical phase jumps of the full-order model adopting the voltage feedforward scheme. As k varies from 0 to 1, the SEP's basin of attraction of the full-order model gradually becomes close to that of the second-order model. Furthermore, the bifurcation behavior with varying k is illustrated by the codim-2 bifurcation diagram in Fig. 5.15. Similar to increasing f_c^{cl}, increasing k leads to the change of the bifurcation behavior from Figs. 5.8, 5.9, 5.10 and 5.11. The reason is that both increasing f_c^{cl} and k can accelerate current regulation.

Furthermore, the full-order model gives the same SEP's basin of attraction and bifurcation behavior as the second-order model for an extreme k (i.e., 1), but not for an extreme f_c^{cl} (i.e., 800 Hz). Thus, the simplification of the second-order model is valid for the GFLCs using the unweighted voltage forward scheme, though this scheme can lead to oscillations under a high grid impedance [102]. Besides,

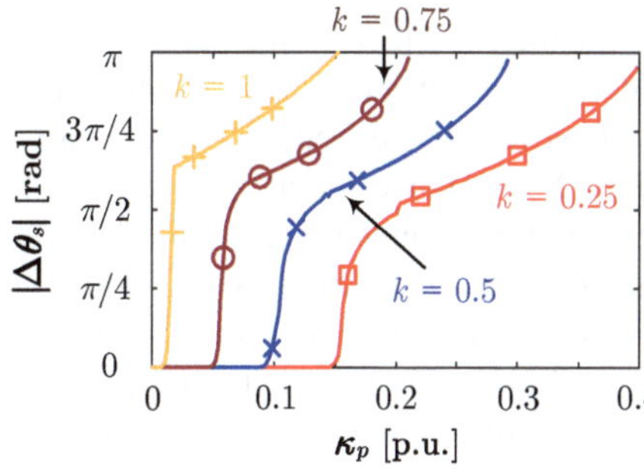

Fig. 5.14 Positive critical phase jumps with the voltage feedforward scheme

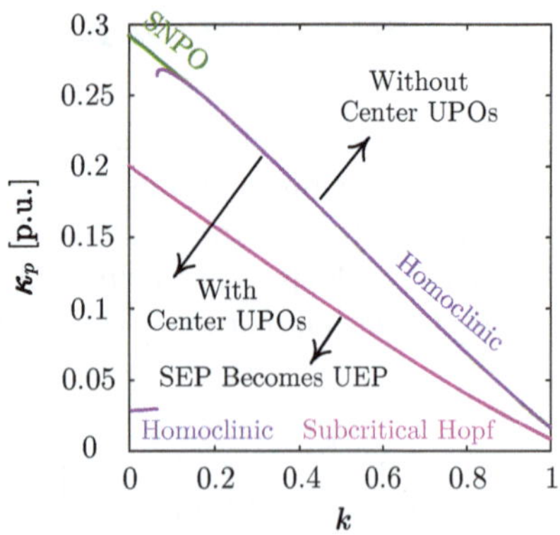

Fig. 5.15 Codim-2 bifurcation diagram in the (k,κ_p)-plane

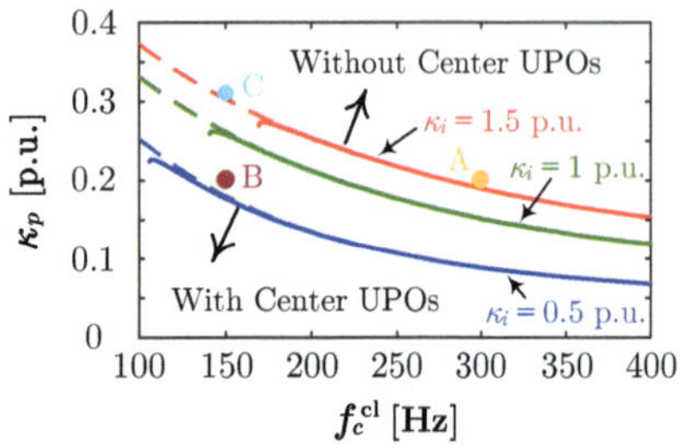

Fig. 5.16 Bifurcation diagrams in the ($f_c^{\rm cl}$,κ_p)-plane with varying κ_i. Dashed and solid lines represent the SNPO and homoclinic bifurcations, respectively

regardless of the choice of $f_c^{\rm cl}$ and k, the full-order model and the numerical continuation are always valid.

5.1.6 *Design-Oriented Analysis*

As a classical application of bifurcation diagrams, design-oriented analysis can be conveniently performed with the full-order model. The bifurcation diagrams corresponding to the SNPO and the center homoclinic bifurcation with different control and grid parameters are provided in Fig. 5.16. As summarized in Table 5.2, the qualitative effects of the parameters on the SEP's basin of attraction, except $f_c^{\rm cl}$ and k, in the full-order model are the same as in the second-order model [56].

However, the desired parameters vary significantly with the current loop bandwidth. For instance, when $f_c^{\rm cl}$ decreases from 300 to 150 Hz with other parameters remaining unchanged, the minimum required κ_p gets nearly doubled (from 0.152 p.u. to 0.262 p.u.). Suppose κ_p above the bifurcation boundary by 0.05 p.u. is adopted. Then, system A ($\kappa_p = 0.202$ p.u.) in Fig. 5.16 should be chosen for $f_c^{\rm cl} = 300$ Hz. On the other hand, the conventional methods neglecting the current transients recommend the same κ_p for different current loop bandwidth, which may

Table 5.2 Qualitative effects of parameters

	κ_p	κ_i	
PLL	↑	↓	
Current loop	i_{gd}^*	f_c^{cl}	k
	↓	↑	↑
Grid	V_{sm}	L_s, r_s	
	↑	↓	

Note: "↑" ("↓") denotes that the transient synchronization stability can be enhanced by increasing (decreasing) the corresponding parameter

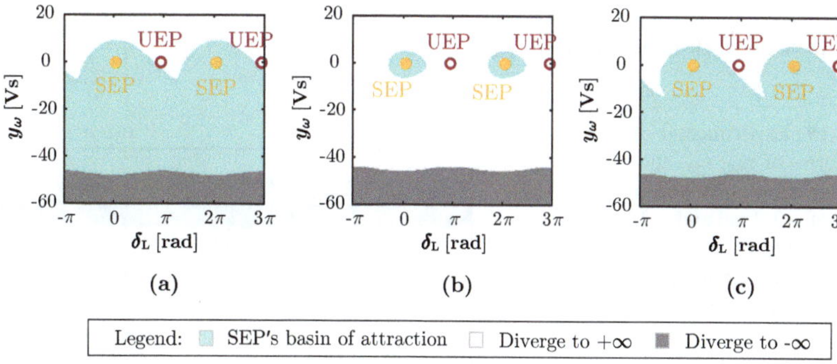

Fig. 5.17 SEP's basin of attraction where (**a**) $\kappa_p = 0.202$ p.u., $f_c^{\mathrm{cl}} = 300$ Hz for system A, (**b**) $\kappa_p = 0.202$ p.u., $f_c^{\mathrm{cl}} = 150$ Hz for system B, and (**c**) $\kappa_p = 0.312$ p.u., $f_c^{\mathrm{cl}} = 150$ Hz for system C. Other parameters adopt the default values

lead to system B ($\kappa_p = 0.202$ p.u., $f_c^{\mathrm{cl}} = 150$ Hz) in Fig. 5.16. For such lower f_c^{cl}, the proposed method suggests choosing system C ($\kappa_p = 0.312$ p.u.) in Fig. 5.16 to ensure transient synchronization stability.

The SEP's basin of attraction of systems A, B, and C are provided in Fig. 5.17. Clearly, the SEP's basin of attraction of systems A and C is significantly larger than that of system B. The robust performance of systems A and C can be further reflected by the critical clearing time t_c against voltage dips and SCFs. The voltage dip is modeled as the temporary drop of the amplitude of the grid voltage to V_{sm}^f. Likewise, the SCF is modeled as the temporary increase of the inductance and resistance of the grid impedance to L_s^f and r_s^f, respectively. After the fault is cleared, the grid voltage and impedance are assumed to recover to the default values. The critical clearing time to remain stable is presented in Fig. 5.18. For a voltage dip with $V_{sm}^f = 0.35$ p.u., systems A, B, and C must clear the fault within 0.30 s, 0.05 s, and 0.09 s, respectively. Such difference is also significant for an SCF with $L_s^f = 1.8$ p.u. and $r_s^f = 1.8$ p.u., where t_c of systems A, B, and C are 1.01 s, 0.47 s, and 0.88 s, respectively. The results show that systems A and C have more time to clear the relatively severe faults. In contrast, system B, designed without considering the

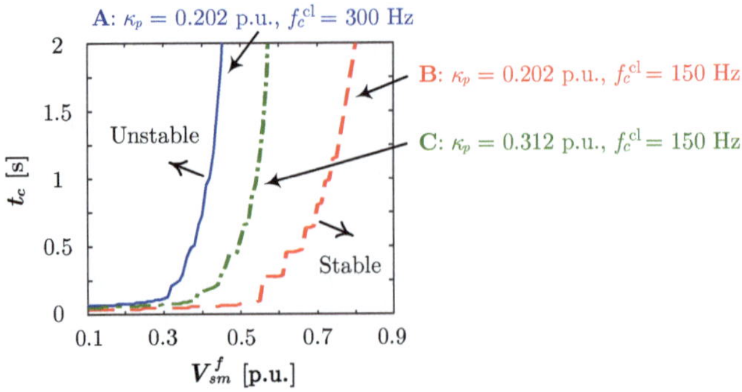

Fig. 5.18 Critical clearing time for voltage dips

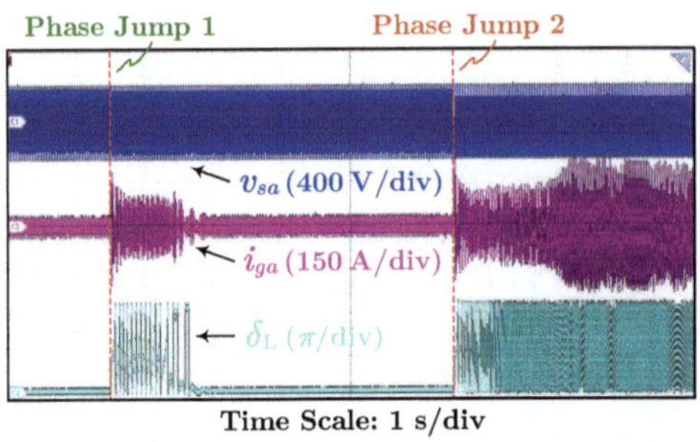

Fig. 5.19 Experimental waveforms of v_{sa}, i_{ga}, and δ_L after two subsequent phase jumps of −2.2 rad and −2.4 rad applied to v_s, with κ_p = 0.4 p.u. Other parameters adopt the default values

current transients, is more likely to become unstable with a short duration of grid faults. Hence, the bifurcation analysis with the full-order model enables a more accurate design reflecting the adverse effects of the nonideal current loop and is applicable to different current loop parameters.

5.1.7 Experimental Verification

The *control hardware-in-loop* experiments have been performed to verify the analytical findings. The grid and the GFLC are emulated by the OPAL-RT equipment OP4512, while the control signals are generated using DSP TMS320F28377D. The default parameters in Table 5.1 are adopted unless otherwise stated, where the short-circuit ratio of the grid is 6.6.

First, the phase jump disturbances are applied to v_s to validate the accuracy of the full-order model. As shown in Fig. 5.19, a phase jump of −2.2 rad cannot destabilize a GFLC with κ_p = 0.4 p.u. and other parameters adopting the default values. When the phase jump increases to −2.4 rad, the GFLC loses the synchronization, with δ_L diverging and i_{ga} soaring to a very high value. Despite the different stability results,

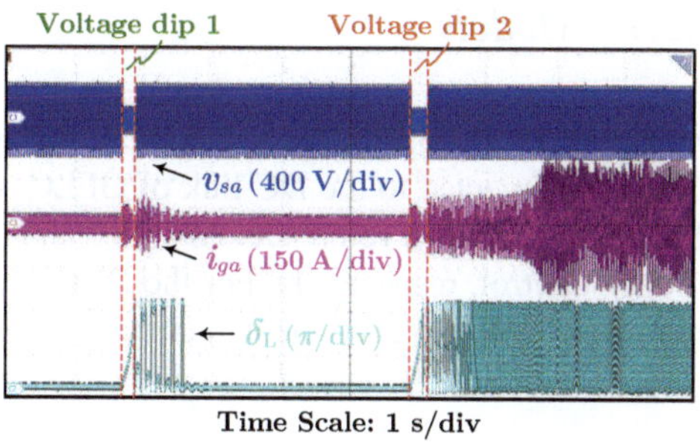

Fig. 5.20 Experimental waveforms of v_{sa}, i_{ga}, and δ_L after two subsequent voltage dips, with $V_{sm}^f = 0.35$ p.u., $\kappa_p = 0.202$ p.u., and $f_c^{cl} = 300$ Hz. Other parameters adopt the default values. The first and second voltage dips last for 0.2 s and 0.25 s, respectively

the amplitude of i_{ga} is far from the reference value (30 A) during the transient process, which justifies the consideration of the nonideal current loop. These experimental results are consistent with the numerical result given in Fig. 5.9a, where the full-order model predicts that the GFLC with $\kappa_p = 0.4$ p.u. has a critical phase jump of −2.3 rad.

Figure 5.20 shows the response of system A under two subsequent voltage dips with $V_{sm}^f = 0.35$ p.u. While the fault depth is considerable, system A recovers the normal current and phase angle with the first voltage dip being cleared 0.2 s after its occurrence. Then, after the second voltage dip, which lasts for 0.25s, system A falls into continuous oscillation. These experimental results are close to the numerical critical clearing time (0.30 s) given in Fig. 5.18.

The stability conclusions of the experiments are consistent with those of the numerical analysis. Hence, the experiments on systems A and C demonstrate the robust performance of the proposed design principle. Since the two systems have distinctive current loops, the experiments also corroborate the broad applicability of the proposed design principle to GFLCs with different current loop bandwidth.

5.2 Impact of DC-Link Voltage Control Loop on Transient Stability

In this section, the impact of the DC-link voltage control loop on the transient synchronization stability is studied. For the sake of analysis, it is assumed that the current inner loops are faster than the PLL and the DC-link voltage control loop; therefore, the dynamics of current loops are neglected here, i.e.,

$$i_{gd} \approx i_{gd}^*, \quad i_{gq} \approx i_{gq}^*, \tag{5.21}$$

which is ensured by using high-bandwidth current loops and/or weighted voltage feedforward scheme [103], as indicated in the last section.

5.2.1 Models

As shown in Fig. 5.21a, C_d is the DC-link capacitor of GFLC, $P_{\rm dc}$ is the constant DC power injected to the DC-link of GFLC, $v_{\rm dc}$ is the DC-link voltage, and $v_{\rm dc}^*$ is the DC-link voltage reference. Also, y_ω and y_v are the state variables of PLL and voltage control, respectively. For the DVC, α_p, α_i, and $I_{gd,{\rm lim}}$ are the proportional coefficient, integral coefficient, and current limit, respectively.

For the DVC,

$$\frac{{\rm d}v_{\rm dc}^2}{{\rm d}t} = \frac{2}{C_d}\left(P_{\rm dc} - \frac{3}{2}(v_{gd}i_{gd} + v_{gq}i_{gq})\right), \tag{5.22}$$

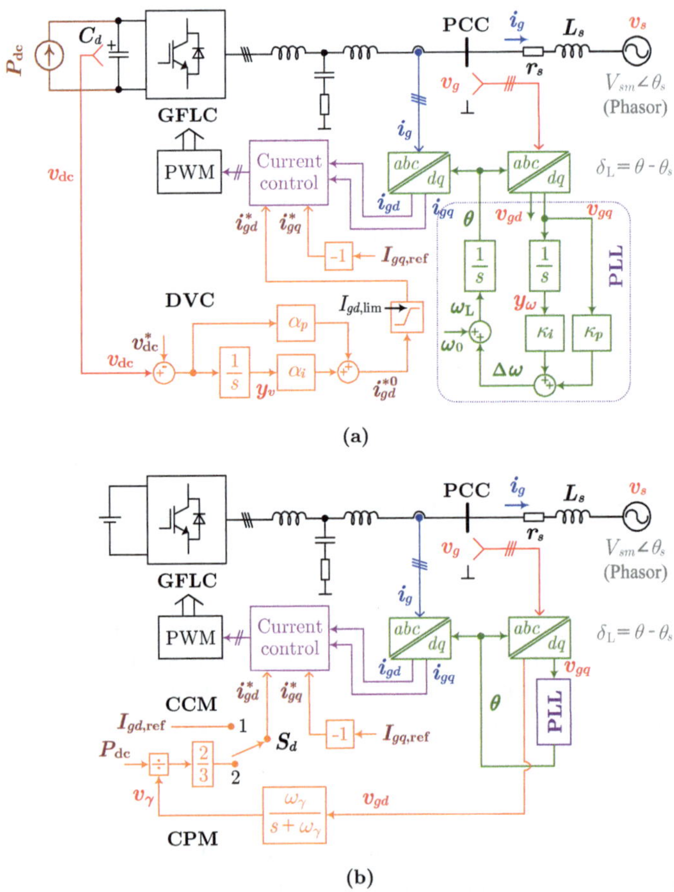

Fig. 5.21 (**a**) GFLC with DC-link voltage control loop. (**b**) DC-link dynamics omitted to simplify the models for analyzing transient synchronization stability

$$\frac{\mathrm{d}y_v}{\mathrm{d}t} = v_{\mathrm{dc}} - v_{\mathrm{dc}}^*, \tag{5.23}$$

and

$$i_{gd}^{*0} = \alpha_p(v_{\mathrm{dc}} - v_{\mathrm{dc}}^*) + \alpha_i y_v. \tag{5.24}$$

$$i_{gd}^* = \begin{cases} I_{gd,\mathrm{lim}}, & i_{gd}^{*0} \geqslant I_{gd,\mathrm{lim}} \\ i_{gd}^{*0}, & |i_{gd}^{*0}| < I_{gd,\mathrm{lim}} \\ -I_{gd,\mathrm{lim}}, & i_{gd}^{*0} \leqslant -I_{gd,\mathrm{lim}} \end{cases} \tag{5.25}$$

$$i_{gq}^* = -I_{gq,\mathrm{ref}} \tag{5.26}$$

For the network and grid,

$$\begin{bmatrix} v_{gd} \\ v_{gq} \end{bmatrix} = \begin{bmatrix} v_{sd} \\ v_{sq} \end{bmatrix} + \begin{bmatrix} r_s & -\omega_{\mathrm{L}} L_s \\ \omega_{\mathrm{L}} L_s & r_s \end{bmatrix} \begin{bmatrix} i_{gd} \\ i_{gq} \end{bmatrix} \tag{5.27}$$

where $v_{sd} = V_{sm} \cos \delta_{\mathrm{L}}$, and $v_{sq} = -V_{sm} \sin \delta_{\mathrm{L}}$. Combining (5.1)–(5.3) and (5.21)–(5.27), we get the model of the GFLC adopting DVC. The state variables are y_ω, δ_{L}, v_{dc}, and y_v.

For Fig. 5.21b, the DC-link dynamics are neglected, and i_{gd}^* is determined by switch S_d. Specifically, $S_d = 1$ and $S_d = 2$ represent the constant current mode (CCM) and the constant power mode (CPM), respectively.

For the CCM (i.e., $S_d = 1$),

$$i_{gd}^* = I_{gd,\mathrm{ref}} \tag{5.28}$$

Combining (5.1)–(5.3), (5.21), and (5.26)–(5.28), we get the model for CCM. The state variables are y_ω and δ_{L}.

For the CPM (i.e., $S_d = 2$),

$$\frac{\mathrm{d}v_\gamma}{\mathrm{d}t} = \omega_\gamma (v_{gd} - v_\gamma), \tag{5.29}$$

$$i_{gd}^* = \frac{2}{3} \frac{P_{\mathrm{dc}}}{v_\gamma} \tag{5.30}$$

where ω_γ is the cutoff frequency of the low-pass filter. Combining (5.1)–(5.3), (5.21), (5.26)–(5.27), (5.29), and (5.30), we get the model for CPM. The state variables are y_ω, δ_{L}, and v_γ.

In the following, we will study the impact of the DVC (i.e., varying α_p, α_i, and $I_{gd,\mathrm{lim}}$) on the transient synchronization stability. On this basis, we will explore

Table 5.3 Circuit and control parameters of a GFLC

Parameter	Value	p.u.	Parameter	Value	p.u.
C_d	3 mF	/	r_s	0.1 Ω	/
ω_0	100π	/	v_{dc}^*	1 kV	/
κ_p	0.413	1	κ_i	7.786	1
α_p	1.491	1	α_i	28.106	1

Note: As an example for initial analysis, the grid parameters are selected as $V_{sm} = 311$ V (i.e., 1 p.u.) and $L_s = 5$ mH (i.e., 0.214 p.u.)

Table 5.4 Other parameters for DVC, CCM, and CPM

	Parameter	Value	p.u.
DVC	P_{dc}	16.1 kW	1
	$I_{gd,\text{lim}}$	50 A	1.67
	$I_{gq,\text{ref}}$	30 A	1
CCM	$I_{gd,\text{ref}}$	30 A	1
	$I_{gq,\text{ref}}$	30 A	1
CPM	P_{dc}	16.1 kW	1
	ω_y	8π	4
	$I_{gq,\text{ref}}$	30 A	1

under what conditions the DVC model can be simplified as the CCM model or the CPM model when studying the transient synchronization stability.

The default circuit and control parameters are given in Tables 5.3 and 5.4. Using the parameters shown in Table 5.4, the output active/reactive currents are identical in DVC, CCM, and CPM.

5.2.2 *Stability Boundary and Equivalent Distance*

Here, two new tools are proposed to assess the transient synchronization stability, including the upper-half-plane stability boundary and the equivalent distance. On this basis, we will study the impact of varying α_p and α_i on the transient stability. For the sake of analysis, it is assumed that $I_{gd,\text{lim}} \to +\infty$ in this subsection. The impact of varying $I_{gd,\text{lim}}$ on the transient stability will be investigated in the next subsection.

5.2.2.1 Role of Upper-Half-Plane Stability Boundary

Based on the derived model for DVC, we get some typical trajectories by using the Runge-Kutta numerical method, as shown in Fig. 5.22a. Then, using these trajectories, we can identify the upper-half-plane stability boundary in (δ, y_ω)-plane. The detailed derivation procedure will be presented in the next subsection.

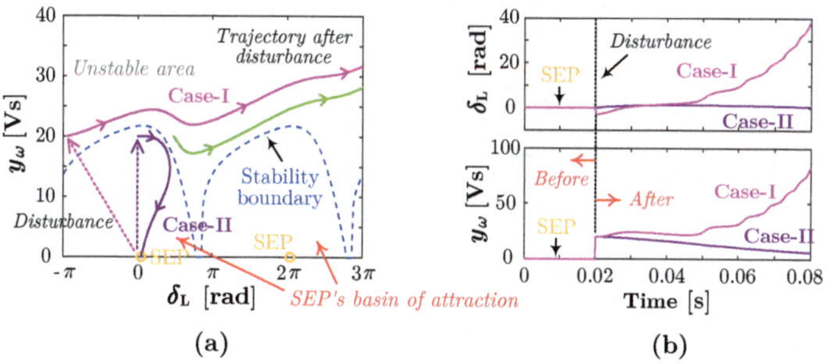

Fig. 5.22 For DVC with $I_{gd,\lim} \to +\infty$, (**a**) phase portraits in (δ_L, y_ω)-plane, showing the upper-half-plane stability boundary and some typical trajectories, (**b**) time-domain waveforms of Case-I and Case-II trajectories

In Fig. 5.22a, the dashed blue curve is the upper-half-plane stability boundary. The area below the stability boundary is the SEP's basin of attraction, while the area above the stability boundary is the unstable area. Case-I and Case-II trajectories are also presented in Fig. 5.22 for further illustrations. Usually, the system stays at the SEP. However, the trajectory may move away from the SEP under a transient disturbance. If the trajectory crosses the stability boundary to escape the SEP's basin of attraction when the disturbance is removed, it will diverge to infinity, as indicated by Case-I trajectory. Otherwise, it will converge to the SEP, as indicated by Case-II trajectory. So, if the stability boundary is far from the SEP, only a highly intense disturbance can drive the trajectory to escape the SEP's basin of attraction, implying that the system robustness against the transient disturbance is strong.

In contrast to existing research, we only focus on the upper-half-plane part of the stability boundary, i.e., $y_\omega \geqslant 0$. This upper-half-plane stability boundary is important because the trajectory usually crosses over such a boundary under a transient disturbance to destabilize the system. For example, during the phase jump disturbance, the phase of v_s changes suddenly, leading to a phase jump in δ_L and zero variation in y_ω. Moreover, during the voltage dip disturbance, V_{sm} drops, and v_{gq} increases according to (5.27); hence, from (5.1), y_ω becomes large. For the disconnection of a transmission line, L_s may be enlarged [96], increasing v_{gq} and y_ω.

5.2.2.2 Derivation Procedure

The upper-half-plane stability boundary can be determined point by point using the flow chart shown in Fig. 5.23. To derive each boundary point, multiple trajectories with different initial conditions are generated. Here, only $y_\omega(0)$ varies between $c_{\min}$ and $c_{\max}$, and the initial values of other state variables remain unchanged. (It is also noted that $\delta_L(0)$ is different for different boundary points.) Using the flow chart, two trajectories with similar $y_\omega(0)$ values are identified, one diverging and the other

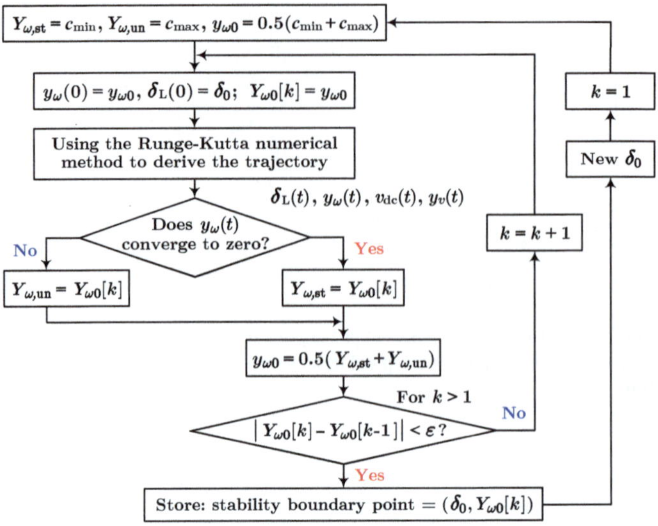

Fig. 5.23 Main flow chart to generate upper-half-plane stability boundary in $(\delta_\mathrm{L}, y_\omega)$-plane. Initial conditions of $v_\mathrm{dc}(t)$ and $y_v(t)$ follow the SEP value

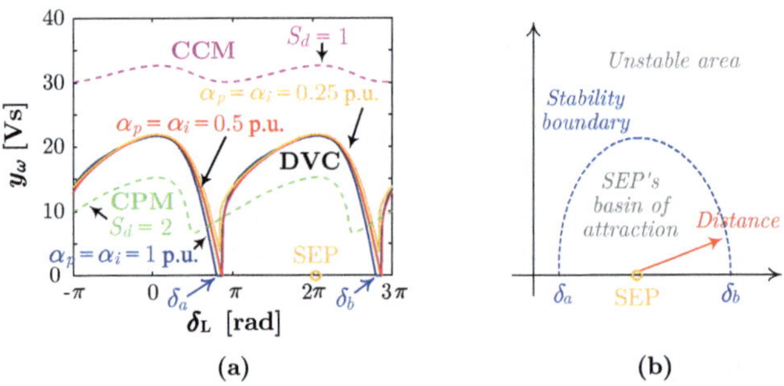

Fig. 5.24 (**a**) Upper-half-plane stability boundaries of DVC, CCM, and CPM. Here, $I_{gd,\lim} \to +\infty$, and $\omega_\gamma \to +\infty$. (**b**) Distance between the SEP and the upper-half-plane stability boundary

converging, with an error less than ε. Besides, the bisection method is adopted to accelerate the derivation of critical y_ω value. For each iteration, the variable $Y_{\omega,\mathrm{st}}$ (or $Y_{\omega,\mathrm{un}}$) is updated with the current $y_\omega(0)$ value if the trajectory converges to (or diverges from) the SEP. Then, the initial value of y_ω for next iteration is set as $0.5 \times (Y_{\omega,\mathrm{st}} + Y_{\omega,\mathrm{un}})$. In Fig. 5.23, the variable $y_{\omega 0}$ and the array $Y_{\omega 0}[k]$ are used to record $y_\omega(0)$ values. Here, we choose $\varepsilon = 0.01$, $c_{\min} = 0$, and $c_{\max} = 50$.

5.2.2.3 Equivalent Distance

The upper-half-plane stability boundaries for different modes are plotted in Fig. 5.24a. For DVC, the solid curves are the stability boundaries. Each solid curve corresponds to specific values of α_p and α_i. From Fig. 5.24a, the influence of α_p and α_i is relatively small, although small α_p and α_i (i.e., slow response) may increase the SEP's basin of attraction slightly. For CCM, the dashed purple curve is the stability boundary. For CPM, the dashed green curve is the stability boundary. From Fig. 5.24a, the dashed purple curve is on the top of and far from the other curves, implying that the constant power injection and the extra DVC may significantly shrink the SEP's basin of attraction and weaken the transient stability.

Furthermore, from Figs. 5.22 and 5.24b, the distance between the stability boundary and the SEP represents the probability of instability under a transient disturbance. For the sake of quantitative analysis, we use the power-norm to provide an overall measure of the equivalent distance, i.e.,

$$d_e = \sqrt{\frac{1}{2\pi}\int_{\delta_a}^{\delta_b}\left(\left|\frac{\delta_{\mathrm{L}} - \delta_{\mathrm{L,SEP}}}{\delta_{\mathrm{L,base}}}\right|^2 + \left|\frac{y_\omega - y_{\omega,\mathrm{SEP}}}{y_{\omega,\mathrm{base}}}\right|^2\right)\mathrm{d}\delta}, \tag{5.31}$$

where the point $(\delta_{\mathrm{L}}, y_\omega)$ is in the stability boundary, $\delta_{\mathrm{L,SEP}}$ and $y_{\omega,\mathrm{SEP}}$ are the values of SEP, and $\delta_b - \delta_a = 2\pi$, as shown in Fig. 5.24. Besides, it is selected that $\delta_{\mathrm{L,base}} = 2\pi$, and $y_{\omega,\mathrm{base}} = 40$. Here, the equivalent distance d_e indicates the average probability of losing stability for various transient disturbances. With increasing values of d_e, the average probability of losing stability decreases, indicating the strong robustness against transient disturbances.

From (5.31), $d_e = 0.5406$ for the solid blue curve (i.e., stability boundary of DVC), $d_e = 0.8506$ for the dashed purple curve (i.e., stability boundary of CCM), and $d_e = 0.4411$ for the dashed green curve (i.e., stability boundary of CPM). Hence, in terms of d_e, the influence of DVC and CPM on the transient stability is similar, but is distinct from the influence of CCM on the transient stability.

5.2.3 *Effect of the Current Limit*

5.2.3.1 Influential Mechanism

In order to clarify the impact of varying the current limit of DVC (i.e., varying $I_{gd,\mathrm{lim}}$), three typical trajectories are given:

- Case 1: $I_{gd,\mathrm{lim}} \to +\infty$, and the trajectory starts from point A.
- Case 2: $I_{gd,\mathrm{lim}} = 1.33$ p.u., and the trajectory starts from point B.
- Case 3: $I_{gd,\mathrm{lim}} = 1.33$ p.u., and the trajectory starts from point C.

In Fig. 5.25a, the blue and red dashed curves represent the upper-half-plane stability boundaries with $I_{gd,\mathrm{lim}} \to +\infty$ and $I_{gd,\mathrm{lim}} = 1.33$ p.u., respectively.

For Case 1, point A is above the blue dashed curve, and the trajectory diverges to infinity eventually, as shown in Fig. 5.25. In Case 2, point B is near point A, and the trajectory is initially similar to that in Case 1. Once i_{gd}^* reaches its limit ($I_{gd,\mathrm{lim}}$), the

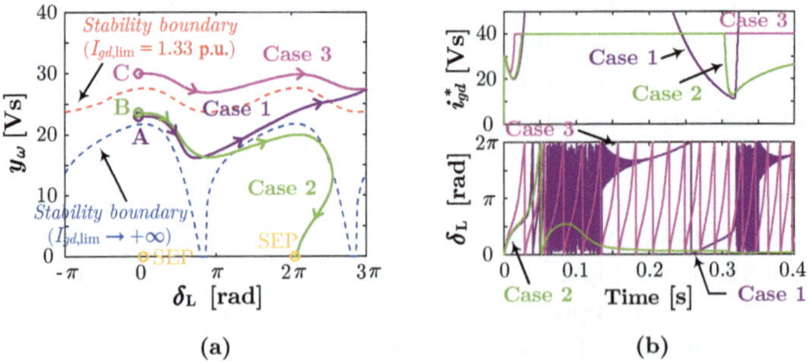

Fig. 5.25 For DVC, (**a**) three typical trajectories and two upper-half-plane stability boundaries with $I_{gd,\lim} \to +\infty$ and $I_{gd,\lim} = 1.33$ p.u., respectively. (**b**) time-domain waveforms for the aforementioned three trajectories. For clarity, δ_L varies from 0 to 2π in (**b**)

upper-half-plane stability boundary becomes the red dashed curve. As the trajectory remains below such a boundary, it will finally converge to the SEP. In Case 3, point C is above both dashed curves, causing all state variables to diverge as i^*_{gd} quickly reaches its limit. Combining (5.21), (5.22), and (5.27),

$$v^2_{dc}(t_2) - v^2_{dc}(t_1) = \int_{t_1}^{t_2} \frac{2}{C_d}\left(P_{dc} - \frac{3}{2}(V_{sm} i^*_{gm} \cos(\delta_L + \varphi) + r_s i^{*2}_{gm})\right) dt \qquad (5.32)$$

where $(t_2 - t_1)$ represents a period of δ_L, i.e., δ_L increases from 0 to 2π. Besides, $\cos(\varphi) = i^*_{gd}/i^*_{gm}$, and $i^*_{gm} = \sqrt{i^{*2}_{gd} + i^{*2}_{gq}}$. As $P_{dc} \gg 1.5 r_s i^{*2}_{gm}$, $v^2_{dc}(t_2) > v^2_{dc}(t_1)$, implying that v_{dc} keeps increasing. From (5.23) to (5.25), i^{*0}_{gd} keeps increasing, and $i^*_{gd} = I_{gd,\lim}$ in the following time, implying that the upper-half-plane stability boundary remains unchanged (i.e., the red dashed curve). Therefore, the trajectory in Case 3 will diverge to infinity, as it is on the top of the aforementioned boundary all the time.

Hence, the current limit may reshape the upper-half-plane stability boundary significantly by bypassing the DC-link voltage control loop and restricting the variation of i^*_{gd}. In other words, the current limit has a significant impact on the transient synchronization stability.

5.2.3.2 Quantitative Analysis

Here, the impact of varying $I_{gd,\lim}$ on the upper-half-plane stability boundary is investigated, as shown in Fig. 5.26a. Here, $I_{gd,\lim} \to +\infty$ for the dashed blue curve, while $I_{gd,\lim}$ = 1.33 p.u./1.67 p.u./2 p.u./3.33 p.u. for the solid purple/green/red/brown curves. From Fig. 5.26a, the solid brown curve is close to the dashed blue curve, while the other solid curves are largely distinct from the

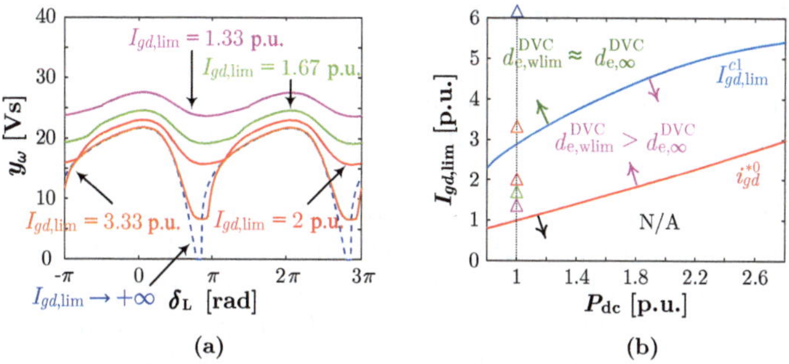

Fig. 5.26 For DVC, (**a**) upper-half-plane stability boundaries for varying $I_{gd,\lim}$ (**b**) the critical current limit $I_{gd,\lim}^{cl}$ for varying P_{dc}

dashed blue curve. This finding implies that the reshaping effect of the current limit on stability boundaries is weak for large $I_{gd,\lim}$, but strong for small $I_{gd,\lim}$.

Furthermore, we use d_e to quantitatively show the impact of the current limit. The value of d_e for $I_{gd,\lim} \to +\infty$ is denoted as $d_{e,\infty}^{\mathrm{DVC}}$, while the value of d_e for specific $I_{gd,\lim}$ is denoted as $d_{e,\mathrm{wlim}}^{\mathrm{DVC}}$. Based on $d_{e,\infty}^{\mathrm{DVC}}$ and $d_{e,\mathrm{wlim}}^{\mathrm{DVC}}$, the critical current limit $I_{gd,\lim}^{cl}$ can be found, as shown in Fig. 5.26b. The solid red curve represents the injection current i_{gd}^{*0} (see Fig. 5.21a) for varying P_{dc}, while the solid blue curve represents the critical current limit $I_{gd,\lim}^{cl}$ for varying P_{dc}. Between these two curves, $d_{e,\mathrm{wlim}}^{\mathrm{DVC}} > d_{e,\infty}^{\mathrm{DVC}}$, implying that the stability boundary for given $I_{gd,\lim}$ is distinct from the one for $I_{gd,\lim} \to +\infty$. Above the solid blue curve, $d_{e,\mathrm{wlim}}^{\mathrm{DVC}} \approx d_{e,\infty}^{\mathrm{DVC}}$, implying that the stability boundary for given $I_{gd,\lim}$ is close to the one for $I_{gd,\lim} \to +\infty$. Here, $d_{e,\mathrm{wlim}}^{\mathrm{DVC}} \approx d_{e,\infty}^{\mathrm{DVC}}$ is guaranteed by the condition

$$\frac{|d_{e,\mathrm{wlim}}^{\mathrm{DVC}} - d_{e,\infty}^{\mathrm{DVC}}|}{d_{e,\mathrm{wlim}}^{\mathrm{DVC}}} < 0.01 \tag{5.33}$$

Usually, the current limit is slightly larger than the rated current. However, in practice, the actual DC-link power P_{dc} may vary from zero to the rated value (or even a little larger than the rated value). Therefore, the actual current i_{gd}^{*0} before a transient disturbance (i.e., pre-fault current) may be far from or close to the limit. As shown in Fig. 5.26), if the pre-fault current is much smaller than $I_{gd,\lim}$, the system behaves as if there is no current limit. In this case, we can preserve the DC-link dynamics and use the DVC-based model for analysis; or neglect the DC-link dynamics, adjust the value of ω_γ, and use the CPM-based model as the alternative. On the other hand, if the pre-fault current is close to $I_{gd,\lim}$, the current limit, but not the DC-link voltage controller, has a significant impact on the transient stability. In this case, the DC-link dynamics can be neglected, and the model based on the CCM can be adopted (i.e., only the PLL dynamics are considered).

5.2.4 Experimental Verification

We adopt hardware-in-the-loop experiments for verification. The default parameters given in Tables 5.3 and 5.4 are adopted. Two sets of experimental results are presented as follows.

In the first set, $I_{gd,\lim} = 1.67$ p.u., and $P_{dc} = 1$ p.u. for DVC, implying that the pre-fault current (1 p.u.) is close to the current limit. The experimental results for DVC are shown in Fig. 5.27a. The grid voltage v_s exhibits two voltage dip disturbances with different durations Δt_{dip} and constant magnitude (i.e., $V_{sm} = 10$ V). After removing the first disturbance $\Delta t_{dip} = 0.14$ s, we see that i_g, δ_L, and v_{dc} eventually recover to their normal values. However, after removing the second disturbance $\Delta t_{dip} = 0.2$ s, i_g is highly distorted, δ_L oscillates between $-\pi$ and π, and v_{dc} deviates from its normal value. Then, we use the circuit shown in Fig. 5.21b and set $S_d = 1$ (i.e., the CCM is adopted) and $I_{gd,ref} = 1.67$ p.u. The experimental results for CCM are shown in Fig. 5.27b. Under the same voltage dip disturbances, the waveforms of i_{ga} and δ_L in Fig. 5.27b are similar to those in Fig. 5.27a, confirming the effectiveness of using the CCM-based model as the equivalent model when the pre-fault current is close to the current limit. Moreover, we set $S_d = 2$, i.e., the CPM is adopted. The experimental results for CPM are shown in Fig. 5.27c. Under the voltage dip disturbance $\Delta t_{dip} = 0.14$ s, we see that

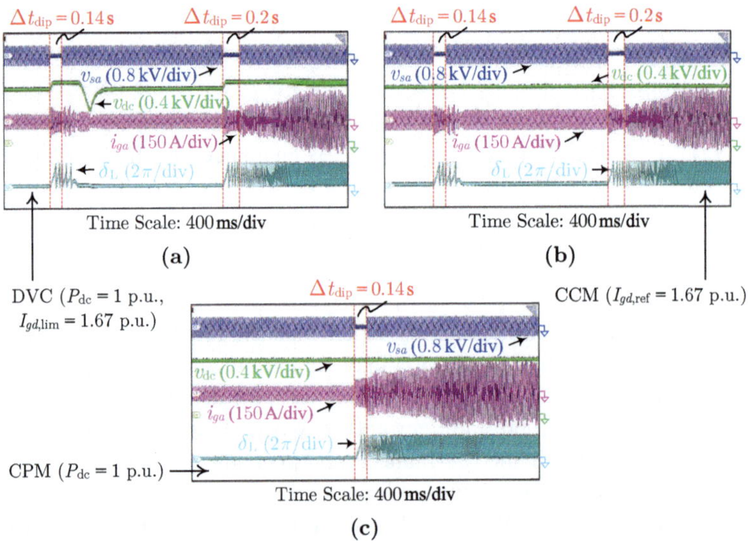

Fig. 5.27 Experimental waveforms of v_{sa}, δ_L, i_{ga}, and v_{dc} before and after the voltage dip disturbances, where $V_{sm} = 10$ V during the disturbances, and $V_{sm} = 311$ V otherwise. (**a**) The DVC is adopted with $I_{gd,\lim} = 1.67$ p.u. and $P_{dc} = 1$ p.u.; (**b**) the CCM is adopted with $I_{gd,ref} = 1.67$ p.u.; and (**c**) the CPM is adopted with $P_{dc} = 1$ p.u. Besides, v_{sa} and i_{ga} are the phase A variables of v_s and i_g, respectively

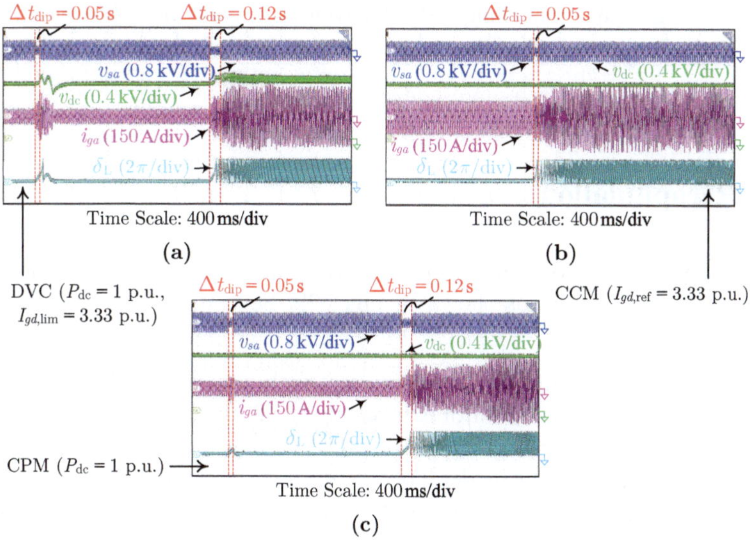

Fig. 5.28 Experimental waveforms of v_{sa}, δ_L, i_{ga}, and v_{dc} before and after the voltage dip disturbances, where $V_{sm} = 100$ V during the disturbances, and $V_{sm} = 311$ V otherwise. (**a**) The DVC is adopted with $I_{gd,\text{lim}} = 3.33$ p.u. and $P_{dc} = 1$ p.u.; (**b**) the CCM is adopted with $I_{gd,\text{ref}} = 3.33$ p.u.; and (**c**) the CPM is adopted with $P_{dc} = 1$ p.u. Besides, v_{sa} and i_{ga} are the phase A variables of v_s and i_g, respectively

i_g is distorted, and δ_L oscillates between $-\pi$ and π, implying that the robustness against transient disturbances for CPM is weaker than the one for DVC or CCM. In other words, it may be inaccurate to use the CPM-based model to approximate the DVC system if the pre-fault current is close to the current limit.

In the second set, $I_{gd,\text{lim}} = 3.33$ p.u., and $P_{dc} = 1$ p.u. for DVC, implying that the pre-fault current (1 p.u.) is much smaller than the current limit. The experimental results for DVC are shown in Fig. 5.28a. The grid voltage v_s exhibits two voltage dip disturbances with different durations Δt_{dip} and constant magnitude (i.e., $V_{sm} = 100$ V). After removing the first disturbance $\Delta t_{\text{dip}} = 0.05$ s, we see that i_g, δ_L, and v_{dc} recover to their normal values at last. But for the second disturbance $\Delta t_{\text{dip}} = 0.12$ s, i_g is highly distorted, δ_L oscillates between $-\pi$ and π, and v_{dc} deviates from its normal value. Then, we use the circuit shown in Fig. 5.21b and set $S_d = 1$ (i.e., the CCM is adopted) and $I_{gd,\text{ref}} = 3.33$ p.u. The experimental results for CCM are shown in Fig. 5.28b. Under the voltage dip disturbance $\Delta t_{\text{dip}} = 0.05$ s, we see that i_g is distorted, and δ_L oscillates between $-\pi$ and π, implying that the robustness against transient disturbances for CCM is weaker than the one for DVC or CPM. In other words, it may be inaccurate to use the CCM-based model to approximate the DVC system if the pre-fault current is much smaller than the current limit. Furthermore, we set $S_d = 2$ (i.e., the CPM is adopted). The experimental results for CPM are shown in Fig. 5.28c. Under the same voltage dip disturbances, the waveforms of i_{ga} and δ_L in Fig. 5.28c are similar to those in Fig. 5.28a, confirming the effectiveness of using the CPM-based model as the equivalent model when the pre-fault current is much smaller than the current limit.

5.3 Recapitulation

The nonideal current loop in the GFLC plays a decisive role in determining the dynamics of the system, as shown in Fig. 5.29. Several observations are worth noting:

1. If the bandwidth of the current loop is relatively low, and/or the coefficient of the voltage feedforward control is relatively small, the complex behavior of the full-order system (considering the nonideal current loop) is distinct from the PLL-based second-order model. For the full-order system, there are two UPOs which surround the SEP. The blue UPO generates from the onset of a homoclinic bifurcation, while the red UPO generates from the onset of a subcritical Hopf bifurcation. These two UPOs merge and disappear through the onset of an SNPO, leading to a significant enlargement of the SEP's basin of attraction. Therefore, in this scenario, the dynamics of current loops should be retained to guarantee the accuracy of transient stability prediction. Notably, as two UPOs merge here, the SNPO is different from that in Chap. 2.
2. If the bandwidth of the current loop is relatively wide, and/or the coefficient of the voltage feedforward control is relatively large, the complex behavior of the full-order system (considering the nonideal current loop) is similar to the PLL-based second-order model. In this case, there is only one UPO surrounding the SEP. This UPO generates from the onset of a subcritical Hopf bifurcation and subsides via the onset of a homoclinic bifurcation. After the onset of the homoclinic

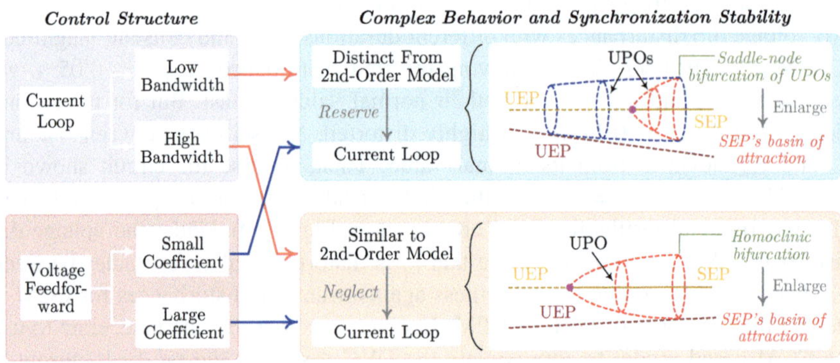

Fig. 5.29 Impact of the nonideal current loops on the complex behavior and transient synchronization stability

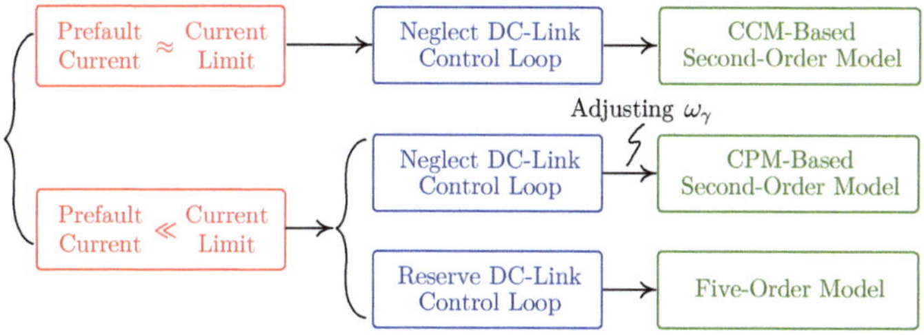

Fig. 5.30 Impact of the DVC and current limit on the transient synchronization stability

bifurcation, the SEP's basin of attraction enlarges significantly. Therefore, in this scenario, the dynamics of current loops can be neglected.

On the other hand, it is shown that the transient synchronization stability is less affected by the controller parameters of DVC, but profoundly influenced by the output limit of DVC (i.e., current limit). Therefore, we can develop some equivalent models by omitting the DC-link dynamics and focusing solely on the AC-side behavior. As shown in Fig. 5.30, we can adopt the CCM-based second-order model (i.e., the PLL-based second-order model, where the current limit serves as the injection current) if the pre-fault current is close to the current limit, while we can use the CPM-based second-order model if the pre-fault current is much smaller than the current limit. These equivalent models provide a quick route to study the transient synchronization stability of GFLCs, as only the AC-side dynamics are considered.

Part II
Stability of Islanded Multi-Converter Systems

Chapter 6
Reduced-Order Models of Islanded Microgrid with Multiple Grid-Forming Converters

For a power-electronic-dominated power system (e.g., the converter-based microgrid [3, 37]), a reduced-order model would present an advantage of providing the necessary dynamic information at a reasonable computational cost. This will benefit future research on the robustness assessment and cascading failure mechanism of power systems [1, 104, 105]. Besides, the study of order reduction of islanded microgrid models may help find the dominant structure that determines the nonlinear dynamics and develop an optimal mitigation method to enhance the grid's robustness.

For an islanded microgrid comprising GFMCs, a full-order state-space model has been proposed previously [106–108]. It is worth noting that a virtual resistor is adopted to help model the power network with a complex structure, which may, however, lead to inaccurate stability assessment [107, 109]. Hence, a first-order Taylor expansion was applied to simplify the power network [110], which will be compared with other network modeling methods in this chapter. Focusing on the slow-scale dynamics, some reduced-order models of GFMCs have been derived using the singular perturbation method [109, 111, 112]. However, analysis based on such models naturally omits fast-scale states, resulting in unrealistic analysis or inconsistent observations since fast-scale states do interact with slow-scale states [113]. To address this issue, a reduced-order model considering the integral coefficients of voltage loops has been proposed [113]. Furthermore, considering the slow-scale Hopf and saddle-node bifurcations, some reduced-order models of the islanded microgrid were derived using the bifurcation theory [114]. However, very little work has been devoted to the development of reduced-order models of the complete islanded system of converters including the essential power network. In addition, the effects of inner loops (especially the voltage loop) and virtual inductors have seldom been explored.

In this chapter, we focus on the islanded microgrid comprising GFMCs, simple/complex power networks, and constant impedance loads. Based on this microgrid, we consider three types of slow-scale smooth bifurcations (i.e., Hopf, saddle-

J. Yang, C. K. Tse, *Complex Behavior of Grid-Connected Power Electronics Systems*, https://doi.org/10.1007/978-3-031-69340-3_6

node, and homoclinic bifurcations) and attempt to find adequate reduced-order models to describe the essential nonlinear behaviors. It is noteworthy that the slow-scale Hopf bifurcation being studied in this work is associated with a sub-synchronous oscillation.

6.1 Islanded Microgrid

An islanded microgrid with a complex power network and multiple local buses is shown in Fig. 6.1a. Each local bus is connected to a GFMC and a constant impedance load. Here, $L_{f\ell}$ and $C_{f\ell}$ are the inductance and capacitance of the ℓth GFMC's output filter, respectively, where $\ell = 1, 2, \cdots, N$. The ℓth load consists of resistance $R_{b\ell}$ and inductance $L_{b\ell}$. The local buses are connected to each other through the connecting lines. The ρth connecting line contains inductance $L_{c\rho}$ and resistance $r_{c\rho}$, where $\rho = 1, 2, \cdots, (N-1)$. The inductance of the connection line between the ℓth local bus and the ℓth GFMC is denoted by $L_{l\ell}$, which has an equivalent series resistance (ESR) of $r_{l\ell}$. Likewise, $r_{f\ell}$ is the ESR of $L_{f\ell}$. Furthermore, let $i_{f\ell}$ denote the current of $L_{f\ell}$, $i_{l\ell}$ be the current of $L_{l\ell}$, $v_{f\ell}$ be the voltage of $C_{f\ell}$, and $v_{b\ell}$ be the voltage of the ℓth load. These variables are expressed in the *abc*-frame, as shown in Fig. 6.1a. For example, $i_{f\ell}$ denotes $i_{fa\ell}$, $i_{fb\ell}$, and $i_{fc\ell}$. Likewise, we represent the corresponding variables in the *dq*-frame as $i_{fd\ell}$ and $i_{fq\ell}$.

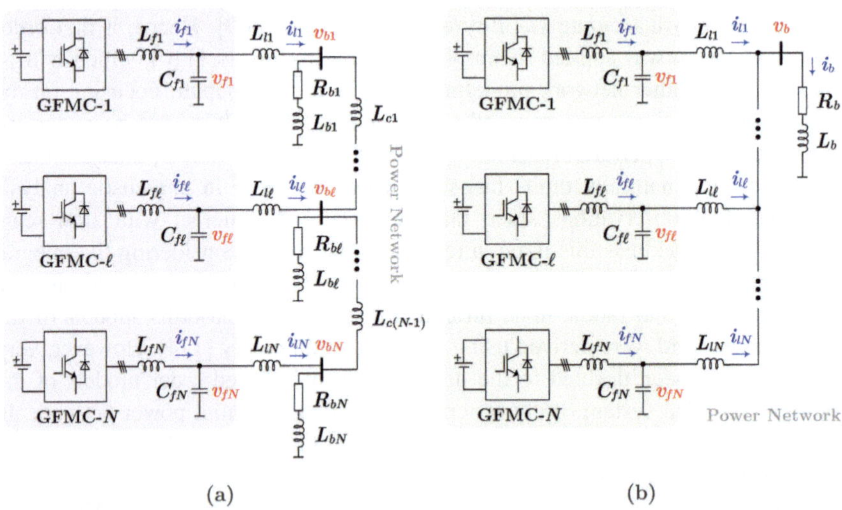

Fig. 6.1 Islanded microgrid (**a**) with a complex power network and multiple local buses and (**b**) with a simple power network. The ESR of $L_{f\ell}$, $L_{l\ell}$, and $L_{c\rho}$ are $r_{f\ell}$, $r_{l\ell}$, and $r_{c\rho}$, respectively, where $\ell = 1, 2, \cdots, N$, and $\rho = 1, 2, \cdots, (N-1)$

For ease of analysis, another islanded microgrid with a simple power network is also provided here, as shown in Fig. 6.1b. Specifically, N GFMCs are connected in parallel to a common bus. Also, R_b and L_b are the resistance and inductance of the load, respectively. Similarly, let i_b denote the load current and v_b be the load voltage. Comparing these two topologies shown in Fig. 6.1, it is straightforward to derive the differential equations of the power network and load for Fig. 6.1b, while it may be hard to derive them for Fig. 6.1a. Hence, for clarity and analytical convenience, the reduced-order models for the microgrid shown in Fig. 6.1b will be investigated first in the following. Subsequently, the analysis of reduced-order models will be extended to the microgrid with a complex power network, as shown in Fig. 6.1a.

The control structure in the dq-frame for each GFMC is shown in Fig. 6.2. There are three control loops, namely, the power loop, the voltage loop, and the current loop. For the ℓth GFMC, the instantaneous active power p_ℓ and reactive power q_ℓ are

$$p_\ell = \frac{3}{2}(v_{fd\ell}i_{ld\ell} + v_{fq\ell}i_{lq\ell}), \quad q_\ell = \frac{3}{2}(-v_{fd\ell}i_{lq\ell} + v_{fq\ell}i_{ld\ell}). \tag{6.1}$$

The average active power P_ℓ and reactive power Q_ℓ can be found, respectively, from p_ℓ and q_ℓ after applying low-pass filtering to $(P_{0\ell} - p_\ell)$ and $(Q_{0\ell} - q_\ell)$, where $P_{0\ell}$ and $Q_{0\ell}$ are the nominal values of active and reactive power, respectively. The

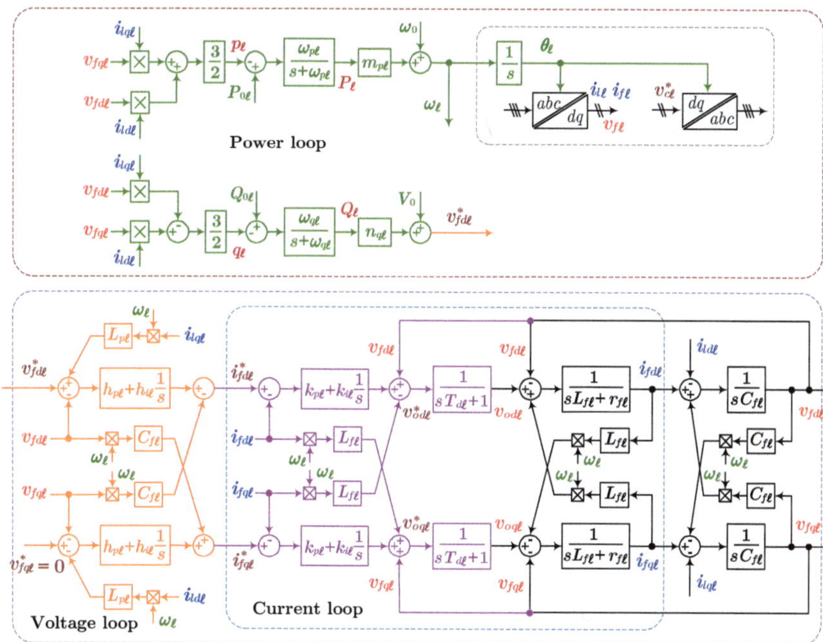

Fig. 6.2 Control block diagram in dq-frame of ℓth GFMC

cutoff frequencies of the low-pass filters are $\omega_{p\ell}$ and $\omega_{q\ell}$, respectively. These low-pass filters provide the virtual inertia as well. According to Fig. 6.2, the angular frequency (ω_ℓ) and the voltage references ($v^*_{fd\ell}$ and $v^*_{fq\ell}$) are generated as

$$\omega_\ell = \omega_0 + m_{p\ell} P_\ell, \quad v^*_{fd\ell} = V_0 + n_{q\ell} Q_\ell, \quad v^*_{fq\ell} = 0, \tag{6.2}$$

where $m_{p\ell}$ and $n_{q\ell}$ are the droop coefficients, ω_0 is the frequency of $v_{f\ell}$ when $p_\ell = P_{0\ell}$, and V_0 is the magnitude of $v_{f\ell}$ when $q_\ell = Q_{0\ell}$.

The system regulates $v_{fd\ell}$ and $v_{fq\ell}$ according to $v^*_{fd\ell}$ and $v^*_{fq\ell}$, respectively. For the voltage controller, $h_{p\ell}$ and $h_{i\ell}$ are the proportional coefficient and integral coefficient, respectively. Also, $L_{p\ell}$ is the virtual inductance, which is chosen to satisfy $\omega_\ell(L_{p\ell} + L_{l\ell}) \gg r_{l\ell}$, i.e., active power control loop and the reactive power control loop are decoupled [106, 115]. For the current controller, $i^*_{fd\ell}$ and $i^*_{fq\ell}$ are the references, and $k_{p\ell}$ and $k_{i\ell}$ are the proportional and integral coefficients, respectively. As far as slow-scale dynamics are concerned, the time delay can be simplified by a first-order approximation, i.e., $\exp(-sT_{d\ell}) \approx 1/(sT_{d\ell} + 1)$, where $T_{d\ell} = 1.5T_{ctrl\ell}$, and $T_{ctrl\ell}$ is the control period of the ℓth GFMC [107]. Finally, $v^*_{cd\ell}$ and $v^*_{cq\ell}$ are the inputs of the time delay, and $v_{cd\ell}$ and $v_{cq\ell}$ are the outputs of the time delay.

For the GFMCs, the circuit parameters are given in Table 6.1, and the typical control parameters are given in Tables 6.2 and 6.3. As an example, three converters' parameters are provided here. These converters share the same set of voltage and current control parameters. For comparison, two cases with different voltage and current control parameters are provided in Table 6.3. The dynamical response of the voltage and current loops in Case 1 is slower than that in Case 2.

Table 6.1 Circuit parameters of GFMCs

Parameter	Value	Parameter	Value	Parameter	Value
$L_{f\ell}$	1.8 mH	$r_{f\ell}$	0.1 Ω	$C_{f\ell}$	300 μF

Table 6.2 Typical parameters of power control loops of GFMCs

Parameter	Value	p.u.	Parameter	Value	p.u.
P_m	50 kW	2	Q_m	15 kVar	0.6
ω_0	100π	1	V_0	311 V	1
$\omega_{p\ell}$	$2\pi \times 0.1$	0.1	$\omega_{q\ell}$	$2\pi \times 0.1$	0.1
L_{p1}	3 mH	/	L_{p2}	2 mH	/
L_{p3}	2 mH	/			
P_{01}	50 kW	2	P_{02}	25 kW	1
P_{03}	25 kW	1	$Q_{0\ell}$	0	0
m_{p1}	$0.02\omega_0/P_m$	0.02	m_{p2}	$0.04\omega_0/P_m$	0.04
m_{p3}	$0.04\omega_0/P_m$	0.04	n_{q1}	$0.025V_0/Q_m$	0.025
n_{q2}	$0.05V_0/Q_m$	0.05	n_{q3}	$0.05V_0/Q_m$	0.05

Table 6.3 Typical parameters of voltage and current control loops of GFMCs

Case	$h_{p\ell}$	$h_{i\ell}$	$k_{p\ell}$	$k_{i\ell}$	$f_{cv\ell}$ [Hz]	$\mathrm{PM}_{v\ell}$ [°]	$f_{ci\ell}$ [Hz]	$\mathrm{PM}_{i\ell}$ [°]
1	0.11	6.72	1.31	123	59.3	51.3	117	69.1
2	0.21	13.43	2.62	247	120	54.7	232	56.1

Note: The time delay model $\exp(-sT_{d\ell})$ is adopted here to derive $f_{cv\ell}$, $\mathrm{PM}_{v\ell}$, $f_{ci\ell}$, and $\mathrm{PM}_{i\ell}$. Here, $f_{cv\ell}$ and $\mathrm{PM}_{v\ell}$ represent the gain crossover frequency and phase margin of the voltage-loop loop gain, respectively, and $f_{ci\ell}$ and $\mathrm{PM}_{i\ell}$ represent the gain crossover frequency and phase margin of the current-loop loop gain, respectively. Besides, the switching frequency is $f_{sw\ell} = 2$ kHz, and the control frequency is $f_{ctrl} = 4$ kHz, where $f_{ctr\ell} = 1/T_{ctr\ell}$

6.2 Reduced-Order Models

In this section, several reduced-order models of the islanded microgrid will be presented, and their effectiveness in representing the system's slow-scale dynamics will be compared.

6.2.1 *Full-Order Model for Grid-Forming Converters*

Referring to Fig. 6.2, for the ℓth GFMC, the power loop can be modeled as

$$\frac{\mathrm{d}}{\mathrm{d}t}\begin{bmatrix} P_\ell \\ Q_\ell \end{bmatrix} = \begin{bmatrix} \omega_{p\ell} & 0 \\ 0 & \omega_{q\ell} \end{bmatrix}\left(-\begin{bmatrix} P_\ell \\ Q_\ell \end{bmatrix} + \begin{bmatrix} P_{0\ell} \\ Q_{0\ell} \end{bmatrix} - \begin{bmatrix} p_\ell \\ q_\ell \end{bmatrix}\right) \quad . \tag{6.3}$$

The dynamics of the voltage control are represented by

$$\frac{\mathrm{d}}{\mathrm{d}t}\begin{bmatrix} x_{vd\ell} \\ x_{vq\ell} \end{bmatrix} = \begin{bmatrix} v^*_{fd\ell} \\ v^*_{fq\ell} \end{bmatrix} - \begin{bmatrix} v_{fd\ell} \\ v_{fq\ell} \end{bmatrix} - \omega_\ell L_{p\ell}\begin{bmatrix} 0 & -1 \\ 1 & 0 \end{bmatrix}\begin{bmatrix} i_{ld\ell} \\ i_{lq\ell} \end{bmatrix} \tag{6.4}$$

$$\begin{aligned} \begin{bmatrix} i^*_{fd\ell} \\ i^*_{fq\ell} \end{bmatrix} &= h_{i\ell}\begin{bmatrix} x_{vd\ell} \\ x_{vq\ell} \end{bmatrix} + \omega_\ell C_{f\ell}\begin{bmatrix} 0 & -1 \\ 1 & 0 \end{bmatrix}\begin{bmatrix} v_{fd\ell} \\ v_{fq\ell} \end{bmatrix} \\ &\quad + h_{p\ell}\left(\begin{bmatrix} v^*_{fd\ell} \\ v^*_{fq\ell} \end{bmatrix} - \begin{bmatrix} v_{fd\ell} \\ v_{fq\ell} \end{bmatrix} - \omega_\ell L_{p\ell}\begin{bmatrix} 0 & -1 \\ 1 & 0 \end{bmatrix}\begin{bmatrix} i_{ld\ell} \\ i_{lq\ell} \end{bmatrix}\right) \end{aligned} \tag{6.5}$$

where $x_{vd\ell}$ and $x_{vq\ell}$ are the state variables of the voltage controller. Moreover, the dynamics of the current control are described by

$$\frac{\mathrm{d}}{\mathrm{d}t}\begin{bmatrix} x_{id\ell} \\ x_{iq\ell} \end{bmatrix} = \begin{bmatrix} i^*_{fd\ell} \\ i^*_{fq\ell} \end{bmatrix} - \begin{bmatrix} i_{fd\ell} \\ i_{fq\ell} \end{bmatrix} \tag{6.6}$$

$$\begin{bmatrix} v_{cd\ell}^{*} \\ v_{cq\ell}^{*} \end{bmatrix} = k_{p\ell}\left(\begin{bmatrix} i_{fd\ell}^{*} \\ i_{fq\ell}^{*} \end{bmatrix} - \begin{bmatrix} i_{fd\ell} \\ i_{fq\ell} \end{bmatrix}\right) + k_{i\ell}\begin{bmatrix} x_{id\ell} \\ x_{iq\ell} \end{bmatrix} + \omega_\ell L_{f\ell}\begin{bmatrix} 0 & -1 \\ 1 & 0 \end{bmatrix}\begin{bmatrix} i_{fd\ell} \\ i_{fq\ell} \end{bmatrix} + \begin{bmatrix} v_{fd\ell} \\ v_{fq\ell} \end{bmatrix} \tag{6.7}$$

where $x_{id\ell}$ and $x_{iq\ell}$ are the state variables of the current controller. The dynamics of the time delay and the output filter are described by

$$\frac{\mathrm{d}}{\mathrm{d}t}\begin{bmatrix} v_{cd\ell} \\ v_{cq\ell} \end{bmatrix} = \frac{1}{T_{d\ell}}\left(\begin{bmatrix} v_{cd\ell}^{*} \\ v_{cq\ell}^{*} \end{bmatrix} - \begin{bmatrix} v_{cd\ell} \\ v_{cq\ell} \end{bmatrix}\right) \tag{6.8}$$

$$\frac{\mathrm{d}}{\mathrm{d}t}\begin{bmatrix} i_{fd\ell} \\ i_{fq\ell} \end{bmatrix} = \frac{1}{L_{f\ell}}\left(\begin{bmatrix} v_{cd\ell} \\ v_{cq\ell} \end{bmatrix} - \begin{bmatrix} v_{fd\ell} \\ v_{fq\ell} \end{bmatrix} - \begin{bmatrix} r_{f\ell} & -\omega_\ell L_{f\ell} \\ \omega_\ell L_{f\ell} & r_{f\ell} \end{bmatrix}\begin{bmatrix} i_{fd\ell} \\ i_{fq\ell} \end{bmatrix}\right) \tag{6.9}$$

$$\frac{\mathrm{d}}{\mathrm{d}t}\begin{bmatrix} v_{fd\ell} \\ v_{fq\ell} \end{bmatrix} = \frac{1}{C_{f\ell}}\left(\begin{bmatrix} i_{fd\ell} \\ i_{fq\ell} \end{bmatrix} - \begin{bmatrix} i_{ld\ell} \\ i_{lq\ell} \end{bmatrix} - \omega_\ell C_{f\ell}\begin{bmatrix} 0 & -1 \\ 1 & 0 \end{bmatrix}\begin{bmatrix} v_{fd\ell} \\ v_{fq\ell} \end{bmatrix}\right) \tag{6.10}$$

For the ℓth GFMC, each set of equations shown above is represented in its dq-frame with angular frequency ω_ℓ. One particular dq-frame is chosen as the common coordinate frame, i.e., the first GFMC's coordinate frame. For convenience, we denote this common frame as the DQ-frame with angular frequency ω_1. The angle between the $d_\ell q_\ell$-frame and the DQ-frame is δ_ℓ, as shown in Fig. 6.3 [106]. Therefore,

$$\frac{\mathrm{d}\delta_\ell}{\mathrm{d}t} = \omega_\ell - \omega_1, \quad \delta_1 = 0. \tag{6.11}$$

It should be noted that $\delta_\ell = \theta_\ell - \theta_1$, where θ_ℓ is the phase angle for the Park transformation, as shown in Fig. 6.2. As shown in Fig. 6.2, for each GFMC, $i_{ld\ell}$ and $i_{lq\ell}$ are the only input variables, which are determined by the power network and the load. As the state equations of the power network and the load are represented in the DQ-frame, for GFMCs, a transformation from $[i_{lD\ell}\ i_{lQ\ell}]^T$ to $[i_{ld\ell}\ i_{lq\ell}]^T$ is needed, i.e.,

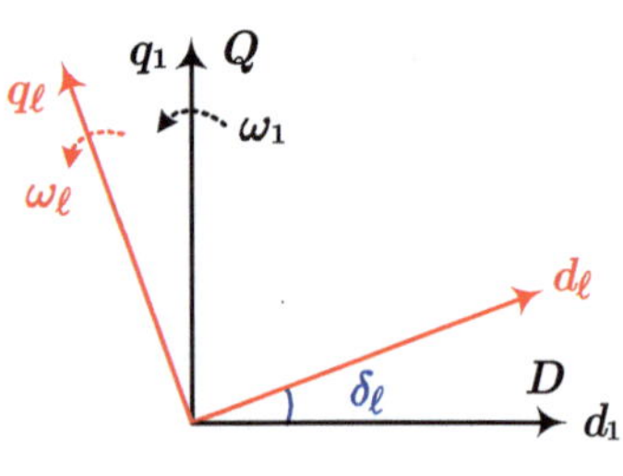

Fig. 6.3 Relation between the $d_\ell q_\ell$-frame and the common DQ-frame

$$\begin{bmatrix} i_{ld\ell} \\ i_{lq\ell} \end{bmatrix} = \begin{bmatrix} \cos\delta_\ell & \sin\delta_\ell \\ -\sin\delta_\ell & \cos\delta_\ell \end{bmatrix} \begin{bmatrix} i_{lD\ell} \\ i_{lQ\ell} \end{bmatrix} \triangleq \mathbf{T}^{-1}(\delta_\ell) \begin{bmatrix} i_{lD\ell} \\ i_{lQ\ell} \end{bmatrix}. \tag{6.12}$$

For the power network and the load, $v_{fd\ell}$ and $v_{fq\ell}$ are the input variables, as shown in Fig. 6.1, which have to be transformed via

$$\begin{bmatrix} v_{fD\ell} \\ v_{fQ\ell} \end{bmatrix} = \begin{bmatrix} \cos\delta_\ell & -\sin\delta_\ell \\ \sin\delta_\ell & \cos\delta_\ell \end{bmatrix} \begin{bmatrix} v_{fd\ell} \\ v_{fq\ell} \end{bmatrix} \triangleq \mathbf{T}(\delta_\ell) \begin{bmatrix} v_{fd\ell} \\ v_{fq\ell} \end{bmatrix}. \tag{6.13}$$

6.2.2 *Reduced-Order Models for Power Network and Loads*

Mathematical models for describing the dynamics of an islanded microgrid are discussed in the following subsections. For ease of referral, we adopt the following convention:

- $\mathrm{L_V}$ and $\mathrm{L_C}$ denote the voltage loop and current loop, respectively.
- NL denotes the power network and the load.
- Superscript p denotes partial inclusion in the model and "overline" denotes omission in the model.

Specifically, we consider full inclusion, partial inclusion, and omission of the dynamics of the power network and load, giving the $\mathrm{L_VL_C}$–NL, $\mathrm{L_VL_C}$–NL^p, and $\mathrm{L_VL_C}$–$\overline{\mathrm{NL}}$ models, respectively. Here, $\mathrm{L_VL_C}$ denotes the full inclusion of the voltage and current loops, while NL, NL^p, and $\overline{\mathrm{NL}}$ denote full inclusion, partial inclusion, and omission, respectively, of the dynamics of the connecting line and load.

In this subsection, the islanded microgrid shown in Fig. 6.1b will be investigated first. In this microgrid, there are two GFMCs, i.e., $\ell = 1, 2$ and $N = 2$. The parameters of these converters are given in Tables 6.1 to 6.3, and the parameters of the power network and the load are provided in Table 6.4.

6.2.2.1 $\mathrm{L_VL_C}$–NL Model

According to Fig. 6.1b,

$$\frac{\mathrm{d}\mathbf{i}_{\mathrm{lDQ}}}{\mathrm{d}t} = \mathbf{A}_{\mathrm{N}}^{-1}\left(-\mathbf{B}_{\mathrm{N}}\mathbf{i}_{\mathrm{lDQ}} + \mathbf{v}_{\mathrm{fDQ}}\right) \tag{6.14}$$

Table 6.4 Circuit parameters of the simple power network

Parameter	Value	Parameter	Value
L_{l1}	3 mH	r_{l1}	0.8 Ω
L_{l2}	2 mH	r_{l2}	0.5 Ω
R_b	2.1 Ω	L_b	1 mH

where

$$\mathbf{i}_{\mathbf{lDQ}} = \left[\mathbf{i}_{\mathbf{lDQ1}}\ \mathbf{i}_{\mathbf{lDQ2}} \cdots \mathbf{i}_{\mathbf{lDQ}N}\right]^T,$$

$$\mathbf{v}_{\mathbf{fDQ}} = \left[\mathbf{v}_{\mathbf{fDQ1}}\ \mathbf{v}_{\mathbf{fDQ2}} \cdots \mathbf{v}_{\mathbf{fDQ}N}\right]^T,$$

$$\mathbf{i}_{\mathbf{lDQ}\ell} \triangleq \left[i_{lD\ell}\ i_{lQ\ell}\right]^T, \quad \mathbf{v}_{\mathbf{fDQ}\ell} \triangleq \left[v_{fD\ell}\ v_{fQ\ell}\right]^T,$$

$$\mathbf{A_N} = \begin{bmatrix} \mathbf{A_{l1}} + \mathbf{A_b} & \mathbf{A_b} & \cdots & \mathbf{A_b} \\ \mathbf{A_b} & \mathbf{A_{l2}} + \mathbf{A_b} & \cdots & \mathbf{A_b} \\ \vdots & \vdots & \ddots & \vdots \\ \mathbf{A_b} & \mathbf{A_b} & \cdots & \mathbf{A}_{\mathbf{l}N} + \mathbf{A_b} \end{bmatrix},$$

$$\mathbf{B_N} = \begin{bmatrix} \mathbf{B_{l1}} + \mathbf{B_b} & \mathbf{B_b} & \cdots & \mathbf{B_b} \\ \mathbf{B_b} & \mathbf{B_{l2}} + \mathbf{B_b} & \cdots & \mathbf{B_b} \\ \vdots & \vdots & \ddots & \vdots \\ \mathbf{B_b} & \mathbf{B_b} & \cdots & \mathbf{B}_{\mathbf{l}N} + \mathbf{B_b} \end{bmatrix},$$

$$\mathbf{A}_{\mathbf{l}\ell} = \begin{bmatrix} L_{l\ell} & 0 \\ 0 & L_{l\ell} \end{bmatrix}, \quad \mathbf{B}_{\mathbf{l}\ell} = \begin{bmatrix} r_{l\ell} & -\omega_1 L_{l\ell} \\ \omega_1 L_{l\ell} & r_{l\ell} \end{bmatrix},$$

$$\mathbf{A_b} = \begin{bmatrix} L_b & 0 \\ 0 & L_b \end{bmatrix}, \quad \mathbf{B_b} = \begin{bmatrix} R_b & -\omega_1 L_b \\ \omega_1 L_b & R_b \end{bmatrix}.$$

Equations (6.1) to (6.14) represent the $L_V L_C$–NL model. The number of state equations is $N_1 = 15N - 1$, where $\dot{\delta}_1 = 0$ is neglected. The state vector is

$$\mathbf{x} = \left[\mathbf{x}_{\mathbf{GFMC1}} \cdots \mathbf{x}_{\mathbf{GFMC}N}\ \boldsymbol{\delta}_{\mathbf{GFMC}}\ \mathbf{i}_{\mathbf{lDQ1}} \cdots \mathbf{i}_{\mathbf{lDQ}N}\right]^T \triangleq \left[x_1\ x_2 \cdots x_{N_k}\right]^T,$$

where

$$\boldsymbol{\delta}_{\mathbf{GFMC}} = \left[\delta_2 \cdots \delta_N\right]^T,$$

$$\mathbf{x}_{\mathbf{GFMC}\ell} = \left[\mathbf{S}_\ell\ \mathbf{x}_{\mathbf{vdq}\ell}\ \mathbf{x}_{\mathbf{idq}\ell}\ \mathbf{v}_{\mathbf{cdq}\ell}\ \mathbf{i}_{\mathbf{fdq}\ell}\ \mathbf{v}_{\mathbf{fdq}\ell}\right]^T,$$

$$\mathbf{S}_\ell = \begin{bmatrix} P_\ell \\ Q_\ell \end{bmatrix}, \quad \mathbf{x}_{\mathbf{vdq}\ell} = \begin{bmatrix} x_{vd\ell} \\ x_{vq\ell} \end{bmatrix}, \quad \mathbf{x}_{\mathbf{idq}\ell} = \begin{bmatrix} x_{id\ell} \\ x_{iq\ell} \end{bmatrix},$$

$$\mathbf{v}_{\mathbf{cdq}\ell} = \begin{bmatrix} v_{cd\ell} \\ v_{cq\ell} \end{bmatrix}, \quad \mathbf{i}_{\mathbf{fdq}\ell} = \begin{bmatrix} i_{fd\ell} \\ i_{fq\ell} \end{bmatrix}, \quad \mathbf{v}_{\mathbf{fdq}\ell} = \begin{bmatrix} v_{fd\ell} \\ v_{fq\ell} \end{bmatrix}.$$

Then, the L_VL_C–NL model can be represented as

$$\frac{d\mathbf{x}}{dt} = \mathbf{f}(\mathbf{x}) = \left[f_1(x_1, \cdots, x_{N_k}) \ \cdots \ f_{N_k}(x_1, \cdots, x_{N_k})\right]^T \tag{6.15}$$

where $k = 1$. The system stability of this L_VL_C–NL model can be determined by the eigenvalues λ_σ of the Jacobian of (6.15), where $\sigma = 1, 2, \cdots, N_k$. If all λ_σ satisfy $\mathrm{Re}(\lambda_\sigma) < 0$, the system is stable. Otherwise, the system is unstable [5, 89]. It is noted that the eigenvalues imply the stability status of specific EQPs; thereby, they can be used to predict the Hopf and saddle-node bifurcations. However, for the homoclinic bifurcation, the stability of EQPs remains unvaried, while the qualitative behavior significantly changes in the entire phase space instead. Hence, simply using the eigenvalues may not be able to predict the homoclinic bifurcation.

6.2.2.2 L_VL_C–NL^p Model

We apply a system reduction procedure (introduced in [110]) to (6.14) to get

$$\frac{d\mathbf{v_{fDQ}}}{dt} = \mathbf{Y_{N1}^{-1}}\left(-\mathbf{Y_{N0}}\mathbf{v_{fDQ}} + \mathbf{i_{lDQ}}\right) \tag{6.16}$$

where

$$\mathbf{Y_{N0}} = \mathbf{B_N^{-1}}, \quad \mathbf{Y_{N1}} = \frac{\partial}{\partial s}\left((s\mathbf{A_N} + \mathbf{B_N})^{-1}\right)\bigg|_{s=0}.$$

Combining (6.1) to (6.13) and (6.16), we get the L_VL_C–NL^p model. The number of state equations is $N_2 = 13N - 1$, and the state vector becomes $\mathbf{x} = [\mathbf{x_{GFMC1}} \cdots \mathbf{x_{GFMC\mathit{N}}} \ \boldsymbol{\delta}_{\mathbf{GFMC}}]^T$. The reduced-order system is represented as (6.15) with $k = 2$.

6.2.2.3 L_VL_C–$\overline{\text{NL}}$ Model

Omitting the dynamics of the power network and impedance load, (6.14) becomes

$$\mathbf{i_{lDQ}} = \mathbf{B_N^{-1}}\mathbf{v_{fDQ}}. \tag{6.17}$$

Combining (6.1) to (6.13) and (6.17), we get the L_VL_C–$\overline{\text{NL}}$ model. The number of state equations is $N_3 = 13N - 1$, and the state vector becomes $\mathbf{x} = [\mathbf{x_{GFMC1}} \cdots \mathbf{x_{GFMC\mathit{N}}} \ \boldsymbol{\delta}_{\mathbf{GFMC}}]^T$. The reduced-order system is represented by (6.15) with $k = 3$.

6.2.2.4 Comparison of the Models in Detecting Hopf Bifurcation

In order to compare $\text{L}_\text{V}\text{L}_\text{C}$–NL, $\text{L}_\text{V}\text{L}_\text{C}$–$\text{NL}^p$, and $\text{L}_\text{V}\text{L}_\text{C}$–$\overline{\text{NL}}$ models, the root loci of λ_σ are derived through adjusting $m_{p\ell}$, as shown in Fig. 6.4, where Fig. 6.4a corresponds to Case 2 of Table 6.3, and Fig. 6.4b corresponds to Case 1 of Table 6.3. Specifically, when a pair of complex eigenvalues crosses the imaginary axis as $m_{p\ell}$ increases, Hopf bifurcation occurs. (It can be numerically verified that the Hopf bifurcation here occurs in supercritical mode, implying that there is only one basin of attraction before and after the Hopf bifurcation [5].) The imaginary part of the eigenvalues reveals the oscillation frequency, whereas the real part reveals the convergence rate of the solution trajectories. Therefore, when we derive the reduced-order model, accurate slow-scale eigenvalues are necessary. From Fig. 6.4, the eigenvalues of the $\text{L}_\text{V}\text{L}_\text{C}$–$\text{NL}^p$ model almost coincide with those of the $\text{L}_\text{V}\text{L}_\text{C}$–NL model. Moreover, the eigenvalues of the $\text{L}_\text{V}\text{L}_\text{C}$–$\overline{\text{NL}}$ model are close to those of the $\text{L}_\text{V}\text{L}_\text{C}$–NL model, especially in the vicinity of the imaginary axis. Furthermore, slow dynamic response of the voltage loop (i.e., small f_{cvj}) helps reduce the difference between the $\text{L}_\text{V}\text{L}_\text{C}$–$\overline{\text{NL}}$ and $\text{L}_\text{V}\text{L}_\text{C}$–NL models. The above conclusion is still valid for other values of ω_{pj}, ω_{qj}, and $r_{l\ell}/L_{l\ell}$, where the root loci of λ_σ are similar to Fig. 6.4.

Although the $\text{L}_\text{V}\text{L}_\text{C}$–$\text{NL}^p$ model has higher fidelity than the $\text{L}_\text{V}\text{L}_\text{C}$–$\overline{\text{NL}}$ model, the $\text{L}_\text{V}\text{L}_\text{C}$–$\text{NL}^p$ model has two significant drawbacks. First, false right-half-plane (RHP) eigenvalues may exist in the $\text{L}_\text{V}\text{L}_\text{C}$–$\text{NL}^p$ model, as shown in Fig. 6.5. It is also found that the RHP eigenvalues can be eliminated only when $C_{f\ell}$ is very large, which is not economical or practical for grid-connected converter implementation. Second, the GFMCs and the power network strongly couple due to (6.10) and (6.16), which increases the derivation complexity and the computation cost. Therefore, in practice, the $\text{L}_\text{V}\text{L}_\text{C}$–$\overline{\text{NL}}$ model can be used for predicting the Hopf bifurcation

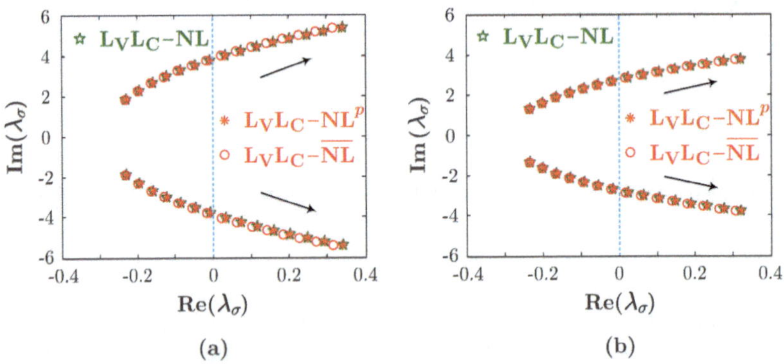

Fig. 6.4 Root loci of eigenvalues λ_σ of $\text{L}_\text{V}\text{L}_\text{C}$–NL, $\text{L}_\text{V}\text{L}_\text{C}$–$\text{NL}^p$, and $\text{L}_\text{V}\text{L}_\text{C}$–$\overline{\text{NL}}$ models. Parameter values are given in Tables 6.1, 6.2, and 6.4. The black arrow indicates increasing $m_{p\ell}$ direction, and $m_{p1} = 0.5m_{p2}$. (**a**) Case 2 of Table 6.3 is adopted, and m_{p1} varies from 0.01 p.u. to 0.08 p.u.; (**b**) Case 1 of Table 6.3 is adopted, and m_{p1} varies from 0.005 p.u. to 0.04 p.u.

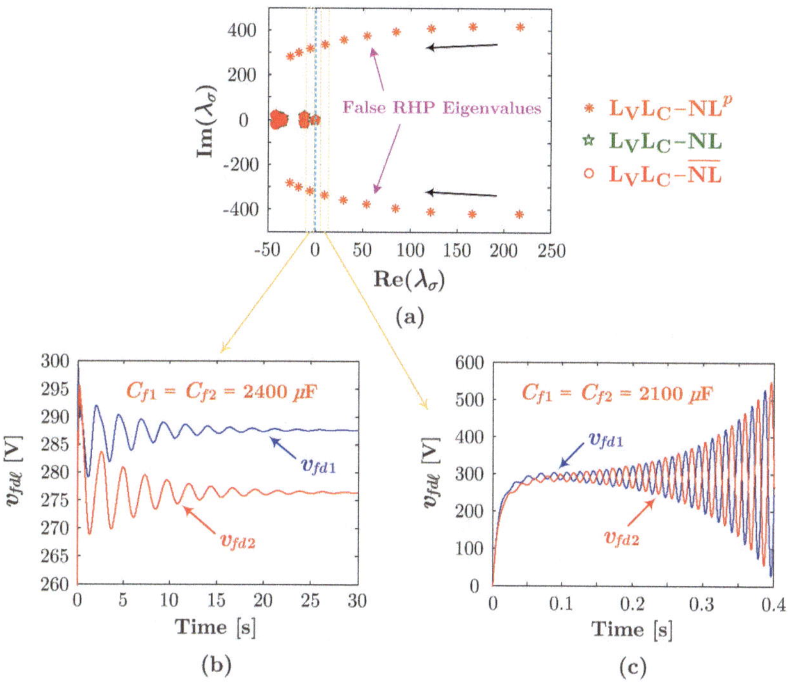

Fig. 6.5 False RHP eigenvalues of the $L_V L_C$–NL^p model. Parameter values are given in Tables 6.1, 6.2, and 6.4 and Case 2 of Table 6.3. (**a**) Root loci of eigenvalues λ_σ of $L_V L_C$–NL, $L_V L_C$–NL^p, and $L_V L_C$–$\overline{NL}$ models, where the black arrow indicates increasing $C_{f\ell}$ direction. Here, $C_{f\ell}$ varies from 300 μF to 3000 μF. Waveforms of $v_{fd\ell}$ (**b**) when $C_{f\ell} = 2400$ μF and (**c**) when $C_{f\ell} = 2100$ μF, based on the $L_V L_C$–NL^p model and Runge–Kutta numerical method with initial values of all state variables set to zero

behavior, which also shows the feasibility to omit the dynamics of the power network and impedance load.

6.2.2.5 Extension to Complex Power Network

For the islanded microgrid shown in Fig. 6.1a, it is beneficial to omit the dynamics of the power network and impedance load in the microgrid model. Denote $Z_{l\ell} = r_{l\ell} + j\omega_1 L_{l\ell}$, $Z_{b\ell} = R_{b\ell} + j\omega_1 L_{b\ell}$, $Z_{cp} = r_{cp} + j\omega_1 L_{cp}$, $Y_{l\ell} = 1/Z_{l\ell}$, $Y_{b\ell} = 1/Z_{b\ell}$, and $Y_{cp} = 1/Z_{cp}$. Then, using an admittance model [105], we get

$$\begin{bmatrix} I_{l1} \\ I_{l2} \\ \vdots \\ I_{lN} \end{bmatrix} = (\mathbf{I} + \mathbf{Y_{BC}} \mathbf{Z_L})^{-1} \mathbf{Y_{BC}} \begin{bmatrix} V_{f1} \\ V_{f2} \\ \vdots \\ V_{fN} \end{bmatrix} \tag{6.18}$$

where $\boldsymbol{I}_{l\ell} = i_{lD\ell} + ji_{lQ\ell}$, $\boldsymbol{V}_{f\ell} = v_{fD\ell} + jv_{fQ\ell}$,

$$\mathbf{Z_L} = \text{diag}\{Z_{l\ell}\}, \quad \Lambda_Y = Y_{b2} + Y_{c1} + Y_{c2},$$

$$\mathbf{Z_{BC}} = \begin{bmatrix} Y_{b1} + Y_{c1} & -Y_{c1} & 0 & \cdots & 0 \\ -Y_{c1} & \Lambda_Y & -Y_{c2} & \cdots & 0 \\ \vdots & \vdots & \vdots & \ddots & \vdots \\ 0 & 0 & 0 & \cdots & Y_{bN} + Y_{c(N-1)} \end{bmatrix}.$$

Hence, we can modify (6.17) as

$$\mathbf{i_{lDQ}} = (\mathbf{I} + \mathbf{D_N B_L})^{-1} \mathbf{D_N v_{fDQ}} \tag{6.19}$$

where $\mathbf{B_L} = \text{diag}\{\mathbf{B_{l\ell}}\}$, $\boldsymbol{\Lambda_D} = \mathbf{D_{b2}} + \mathbf{D_{c1}} + \mathbf{D_{c2}}$, $\mathbf{D_N} =$

$$\begin{bmatrix} \mathbf{D_{b1}} + \mathbf{D_{c1}} & -\mathbf{D_{c1}} & 0 & \cdots & 0 \\ -\mathbf{D_{c1}} & \boldsymbol{\Lambda_D} & -\mathbf{D_{c2}} & \cdots & 0 \\ \vdots & \vdots & \vdots & \ddots & \vdots \\ 0 & 0 & 0 & \cdots & \mathbf{D_{bN}} + \mathbf{D_{c(N-1)}} \end{bmatrix},$$

$$\mathbf{D_{b\ell}} = \frac{1}{R_{b\ell}^2 + (\omega_1 L_{b\ell})^2} \begin{bmatrix} R_{b\ell} & \omega_1 L_{b\ell} \\ -\omega_1 L_{b\ell} & R_{b\ell} \end{bmatrix},$$

$$\mathbf{D_{c\rho}} = \frac{1}{r_{c\rho}^2 + (\omega_1 L_{c\rho})^2} \begin{bmatrix} r_{c\rho} & \omega_1 L_{c\rho} \\ -\omega_1 L_{c\rho} & r_{c\rho} \end{bmatrix}.$$

6.2.3 *Reduced-Order Models for Grid-Forming Converters*

In this subsection, the dynamics of the power network and impedance load are omitted in the microgrid model. Here, we consider in particular four models, namely, $\mathrm{L_V L_C}$–$\overline{\mathrm{NL}}$, $\mathrm{L_V}\overline{\mathrm{L_C}}$–$\overline{\mathrm{NL}}$, $\mathrm{L_V^p}\overline{\mathrm{L_C}}$–$\overline{\mathrm{NL}}$, and $\overline{\mathrm{L_V L_C}}$–$\overline{\mathrm{NL}}$ models.

6.2.3.1 $\mathrm{L_V}\overline{\mathrm{L_C}}$–$\overline{\mathrm{NL}}$ Model

In this model, the dynamics of the current loop are omitted. Hence, $i_{fd\ell} = i^*_{fd\ell}$ and $i_{fq\ell} = i^*_{fq\ell}$. Then, the structure in Fig. 6.2 can be simplified, as shown in Fig. 6.6. The voltage loop is described by

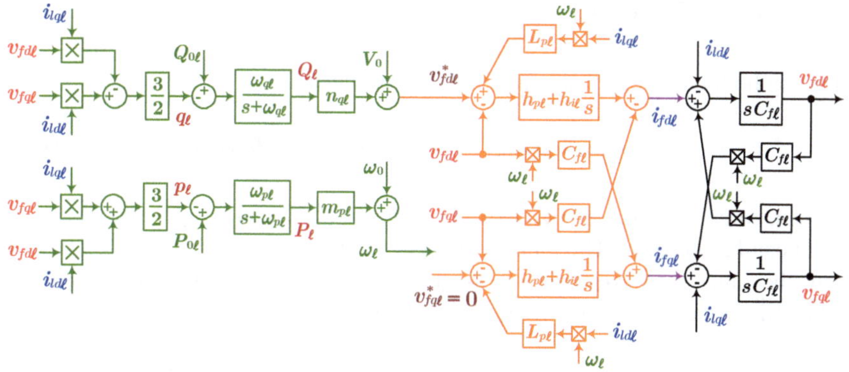

Fig. 6.6 Block diagram of the ℓth GFMC in the $\mathrm{L_V\overline{L_C}}$–$\overline{\mathrm{NL}}$ model

$$
\begin{aligned}
\frac{\mathrm{d}}{\mathrm{d}t}\begin{bmatrix} v_{fd\ell} \\ v_{fq\ell} \end{bmatrix} = \frac{1}{C_{f\ell}}\Bigg(h_{p\ell}\left(\begin{bmatrix} v^*_{fd\ell} \\ v^*_{fq\ell} \end{bmatrix} - \begin{bmatrix} v_{fd\ell} \\ v_{fq\ell} \end{bmatrix}\right) + h_{i\ell}\begin{bmatrix} x_{vd\ell} \\ x_{vq\ell} \end{bmatrix} \\
- \begin{bmatrix} 1 & -h_{p\ell}\omega_\ell L_{p\ell} \\ h_{p\ell}\omega_\ell L_{p\ell} & 1 \end{bmatrix}\begin{bmatrix} i_{ld\ell} \\ i_{lq\ell} \end{bmatrix}\Bigg).
\end{aligned}
\tag{6.20}
$$

Combining (6.1) to (6.4), (6.11) to (6.13), (6.17), and (6.20), the $\mathrm{L_V\overline{L_C}}$–$\overline{\mathrm{NL}}$ model is derived. The number of state equations is $N_4 = 7N - 1$. The state vector becomes $\mathbf{x} = [\mathbf{x}_{\mathbf{GFMC1}} \cdots \mathbf{x}_{\mathbf{GFMC}N}\ \delta_{\mathbf{GFMC}}]^T$, where $\mathbf{x}_{\mathbf{GFMC}\ell} = [\mathbf{S}_\ell\ \mathbf{x}_{\mathbf{vdq}\ell}\ \mathbf{v}_{\mathbf{fdq}\ell}]^T$. The reduced-order system is thus represented by (6.15) with $k = 4$.

6.2.3.2 $\mathrm{L_V^p\overline{L_C}}$–$\overline{\mathrm{NL}}$ Model

It has been shown that the voltage loop representation can be simplified without sacrificing the salient dynamics [113], i.e.,

$$
\frac{1}{h_{i\ell}}\frac{\mathrm{d}}{\mathrm{d}t}\begin{bmatrix} i_{ld\ell} \\ i_{lq\ell} \end{bmatrix} = \begin{bmatrix} v^*_{fd\ell} \\ v^*_{fq\ell} \end{bmatrix} - \omega_\ell L_{p\ell}\begin{bmatrix} 0 & -1 \\ 1 & 0 \end{bmatrix}\begin{bmatrix} i_{ld\ell} \\ i_{lq\ell} \end{bmatrix} - \begin{bmatrix} v_{fd\ell} \\ v_{fq\ell} \end{bmatrix}. \tag{6.21}
$$

Combining (6.1) to (6.3), (6.11) to (6.13), (6.17) and (6.21), the $\mathrm{L_V^p\overline{L_C}}$–$\overline{\mathrm{NL}}$ model can be obtained. The number of state equations is $N_5 = 5N - 1$, and the state vector is $\mathbf{x} = \left[\mathbf{x}_{\mathbf{GFMC1}} \cdots \mathbf{x}_{\mathbf{GFMC}N}\ \delta_{\mathbf{GFMC}}\ \mathbf{i}_{\mathbf{IDQ1}} \cdots \mathbf{i}_{\mathbf{IDQ}N}\right]^T$, where $\mathbf{x}_{\mathbf{GFMC}\ell} = \mathbf{S}_\ell$. The reduced-order model is represented by (6.15) with $k = 5$.

6.2.3.3 $\overline{\mathrm{L_V L_C}}$–$\overline{\mathrm{NL}}$ Model

Omitting both the voltage and current loops, we have

$$v_{fd\ell} = v_{fd\ell}^{*} + \omega_{\ell} L_{p\ell} i_{lq\ell}, \quad v_{fq\ell} = v_{fq\ell}^{*} - \omega_{\ell} L_{p\ell} i_{ld\ell}. \tag{6.22}$$

Then, combining (6.1) to (6.3), (6.11) to (6.13), (6.17), and (6.22), the $\overline{\mathrm{L_V L_C}}$–$\overline{\mathrm{NL}}$ model can be obtained. The number of state equations is $N_6 = 3N - 1$. The state vector becomes $\mathbf{x} = [\mathbf{x}_{\mathbf{GFMC1}} \cdots \mathbf{x}_{\mathbf{GFMC}N}\ \delta_{\mathbf{GFMC}}]^T$, where $\mathbf{x}_{\mathbf{GFMC}\ell} = \mathbf{S}_{\ell}$. The $\overline{\mathrm{L_V L_C}}$–$\overline{\mathrm{NL}}$ model is represented by (6.15) with $k = 6$.

6.2.3.4 Comparison of the Models in Detecting Hopf Bifurcation

The islanded microgrid shown in Fig. 6.1b with default parameters given in Tables 6.1, 6.2, and 6.4 and Case 2 of Table 6.3 is adopted in our study. Here, the $\mathrm{L_V L_C}$–$\overline{\mathrm{NL}}$, $\mathrm{L_V}\overline{\mathrm{L_C}}$–$\overline{\mathrm{NL}}$, $\mathrm{L_V^{\mathit{p}}}\overline{\mathrm{L_C}}$–$\overline{\mathrm{NL}}$, and $\overline{\mathrm{L_V L_C}}$–$\overline{\mathrm{NL}}$ models are compared. By adjusting $m_{p\ell}$, the root locus of eigenvalues λ_{σ} can be derived, as shown in Fig. 6.7 for different combinations of angular frequencies. In Fig. 6.7, the complex eigenvalues of the $\mathrm{L_V L_C}$–$\overline{\mathrm{NL}}$ and $\mathrm{L_V}\overline{\mathrm{L_C}}$–$\overline{\mathrm{NL}}$ models almost overlap regardless of how $m_{p\ell}$ varies. Besides, the complex eigenvalues of the $\overline{\mathrm{L_V L_C}}$–$\overline{\mathrm{NL}}$ model are quite distinct from those of the $\mathrm{L_V L_C}$–$\overline{\mathrm{NL}}$, $\mathrm{L_V}\overline{\mathrm{L_C}}$–$\overline{\mathrm{NL}}$, and $\mathrm{L_V^{\mathit{p}}}\overline{\mathrm{L_C}}$–$\overline{\mathrm{NL}}$ models. As shown in Fig. 6.7, the eigenvalues of the $\mathrm{L_V^{\mathit{p}}}\overline{\mathrm{L_C}}$–$\overline{\mathrm{NL}}$ and $\mathrm{L_V L_C}$–$\overline{\mathrm{NL}}$ models differ significantly, especially when $\omega_{p\ell}$ is large and the complex eigenvalues are far from the imaginary axis. The above conclusion is still valid for Case 1 of Table 6.3, where the root loci of λ_{σ} are similar to Fig. 6.7.

More quantitative results can be derived in terms of the *relative error*, which is defined as [113]

$$e_{\zeta} = \left| \frac{\lambda_{\zeta} - \lambda_{\mathrm{L_V L_C-NL}}}{\lambda_{\mathrm{L_V L_C-NL}}} \right| \times 100\% \tag{6.23}$$

where $\lambda_{\mathrm{L_V L_C-NL}}$ is the eigenvalue of the $\mathrm{L_V L_C}$–NL model at the Hopf bifurcation point (i.e., $\mathrm{Im}(\lambda_{\mathrm{L_V L_C-NL}}) > 0$ and $\mathrm{Re}(\lambda_{\mathrm{L_V L_C-NL}}) = 0$), and λ_{ζ} is the corresponding eigenvalue of the reduced-order model. Besides, the *relative computation time* is defined as

$$d_{\zeta} = \frac{t_{\zeta}}{t_{\mathrm{L_V L_C-NL}}} \times 100\% \tag{6.24}$$

where $t_{\mathrm{L_V L_C-NL}}$ is the computation time using the $\mathrm{L_V L_C}$–NL model to calculate the eigenvalues, and t_{ζ} is the corresponding computation time of the reduced-order model. Results of e_{ζ} and d_{ζ} are summarized in Tables 6.5 and 6.6, respectively.

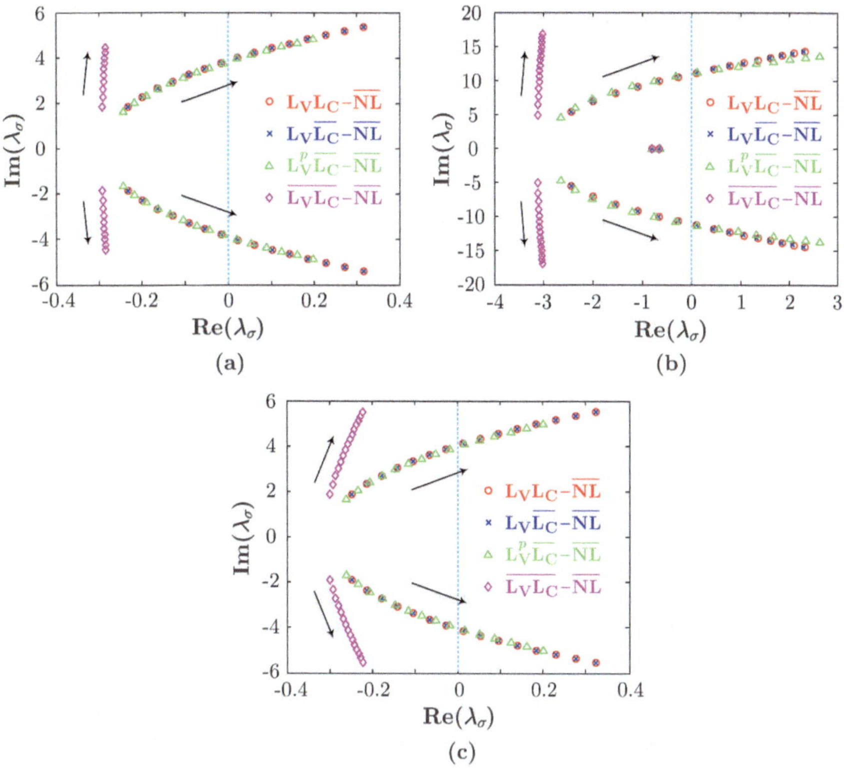

Fig. 6.7 Root loci of eigenvalues λ_σ of L_VL_C–$\overline{NL}$, $L_V\overline{L_C}$–$\overline{NL}$, $L_V^p\overline{L_C}$–$\overline{NL}$, and $\overline{L_VL_C}$–$\overline{NL}$ models. Parameter values are given in Tables 6.1, 6.2, and 6.4 and Case 2 of Table 6.3. The black arrow indicates increasing $m_{p\ell}$ direction, and $m_{p1} = 0.5m_{p2}$. Here, m_{p1} varies from 0.01 p.u. to 0.08 p.u. (**a**) $\omega_{p\ell} = \omega_{q\ell} = 0.1$ p.u.; (**b**) $\omega_{p\ell} = 1$ p.u. and $\omega_{q\ell} = 0.1$ p.u.; (**c**) $\omega_{p\ell} = 0.1$ p.u. and $\omega_{q\ell} = 1$ p.u.

Table 6.5 Relative error of eigenvalues at the Hopf bifurcation point

			Relative error e_ζ [%]			
$\omega_{p\ell}$ [p.u.]	$\omega_{q\ell}$ [p.u.]	$f_{cv\ell}$ [Hz]	L_VL_C–$\overline{NL}$	$L_V\overline{L_C}$–$\overline{NL}$	$L_V^p\overline{L_C}$–$\overline{NL}$	$\overline{L_VL_C}$–$\overline{NL}$
0.1	0.1	59.3	0.222	0.226	11.408	10.680
0.1	0.1	120	0.286	0.288	11.101	7.513
1	0.1	59.3	0.873	0.953	6.808	44.117
1	0.1	120	1.171	1.202	7.468	28.442
0.1	1	59.3	0.216	0.220	11.247	9.666
0.1	1	120	0.263	0.264	10.925	6.590

From these results, the $L_V\overline{L_C}$–$\overline{NL}$ model is the best reduced-order model in practice, as its precision is relatively high, and its computation cost is relatively low.

Finally, the bifurcation diagrams based on the L_VL_C–NL and $L_V\overline{L_C}$–$\overline{NL}$ models are derived, as shown in Fig. 6.8, indicating the Hopf bifurcation boundaries where

Table 6.6 Relative computation time to derive the eigenvalues

$\mathrm{L_VL_C}$–$\overline{\mathrm{NL}}$	$\mathrm{L_V}\overline{\mathrm{L_C}}$–$\overline{\mathrm{NL}}$	$\mathrm{L_V^p}\overline{\mathrm{L_C}}$–$\overline{\mathrm{NL}}$	$\overline{\mathrm{L_VL_C}}$–$\overline{\mathrm{NL}}$
81.08%	36.15%	26.56%	25.28%

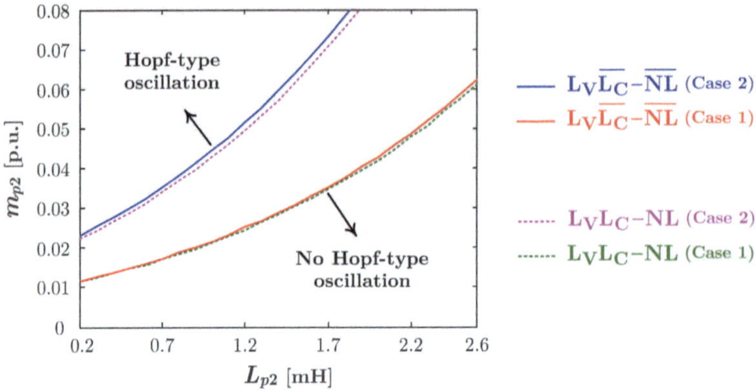

Fig. 6.8 Hopf bifurcation diagrams on the L_{p2}–m_{p2} parameter plane. Solid curves are derived from the $\mathrm{L_V}\overline{\mathrm{L_C}}$–$\overline{\mathrm{NL}}$ model, while dashed curves are derived from the $\mathrm{L_VL_C}$–NL model. System parameters are given in Tables 6.1 to 6.4 with $m_{p1} = 0.5m_{p2}$ and $L_{p1} = 1.5L_{p2}$

the operating point becomes unstable. From Fig. 6.8, the solid curves are very close to the corresponding dashed curves, especially for Case 1, which verifies the effectiveness of the proposed reduced-order model.

6.3 Saddle-Node and Homoclinic Bifurcations

In our recent work [55], the saddle-node bifurcation and homoclinic bifurcation in a GFMC connected to an infinite bus are studied by a third-order model, where the dynamics of the voltage loop, current loop, and power network are omitted. According to the aforedescribed definition, such a model is effectively an $\overline{\mathrm{L_VL_C}}$–$\overline{\mathrm{NL}}$ model. In this section, the effectiveness of the $\overline{\mathrm{L_VL_C}}$–$\overline{\mathrm{NL}}$ model for describing the islanded microgrid will be examined. An $\overline{\mathrm{L_VL_C}}$–$\overline{\mathrm{NL}}$ model will be rebuilt here in terms of power flows. Specifically, to facilitate comparison, four different forms of the $\overline{\mathrm{L_VL_C}}$–$\overline{\mathrm{NL}}$ models will be discussed, which are named as M1, M2, M3, and M4. For M1, the $\overline{\mathrm{L_VL_C}}$–$\overline{\mathrm{NL}}$ model is considered in full. For M2, the frequency variation in the admittance matrix is omitted. M3 and M4 are second-order models based on M2. For M3, $v^*_{fd\ell}$ is set as the SEP value derived from M2, while for M4, $v^*_{fd\ell}$ is set as the average value of the corresponding $v^*_{fd\ell}$-nullcline of M2.

6.3.1 Power-Flow-Based Model

According to Sect. 6.2, the $\overline{\mathrm{L_V L_C}}$–$\overline{\mathrm{NL}}$ model is built in the dq-frame. However, as only the dynamics of the power loop is persevered in the $\overline{\mathrm{L_V L_C}}$–$\overline{\mathrm{NL}}$ model, the dq-frame is unnecessary. Thus, we can rebuild the $\overline{\mathrm{L_V L_C}}$–$\overline{\mathrm{NL}}$ model as [105]

$$\begin{bmatrix} I_{l1} \\ I_{l2} \\ \vdots \\ I_{lN} \end{bmatrix} = \mathbf{Y_{NL1}} \begin{bmatrix} \boldsymbol{V_{f1}^*} \\ \boldsymbol{V_{f2}^*} \\ \vdots \\ \boldsymbol{V_{fN}^*} \end{bmatrix} \tag{6.25}$$

where $\boldsymbol{V_{f\ell}^*} = v_{fD\ell}^* + jv_{fQ\ell}^*$. Besides, for the microgrid shown in Fig. 6.1a, $\mathbf{Y_{NL1}} = (\mathbf{I}+\mathbf{Y_{BC}}(\mathbf{Z_L}+\mathbf{Z_P}))^{-1}\mathbf{Z_{BC}}$, where $\mathbf{Z_P} = \mathrm{diag}\{j\omega_\ell L_{p\ell}\}$. Moreover, for the microgrid shown in Fig. 6.1b, we have $\mathbf{Y_{NL1}} = (\mathbf{Z_B} + \mathbf{Z_L} + \mathbf{Z_P})^{-1}$, where all elements of $\mathbf{Z_B}$ are $Z_b = R_b + j\omega_1 L_b$. For brevity, the elements of $\mathbf{Y_{NL1}}$ are denoted as $\{\mathbf{Y_{NL1}}\}_{\ell k} = G_{\ell k}^1 + jB_{\ell k}^1$, where $\ell, k = 1, 2, \cdots, N$. Then, for the ℓth GFMC, the output power is $p_\ell + jq_\ell = 1.5\boldsymbol{V_{f\ell}}\bar{\boldsymbol{I}}_{l\ell}$, where $\bar{\boldsymbol{I}}_{l\ell} = i_{lD\ell} - ji_{lQ\ell}$. From (6.13) and (6.25), we get

$$p_\ell = \frac{3}{2}\sum_{k=1}^{N} v_{fd\ell}^* v_{fdk}^* \Big(G_{\ell k}^1 \cos(\delta_\ell - \delta_k) + B_{\ell k}^1 \sin(\delta_\ell - \delta_k) \Big) \tag{6.26}$$

$$q_\ell = \frac{3}{2}\sum_{k=1}^{N} v_{fd\ell}^* v_{fdk}^* \Big(G_{\ell k}^1 \sin(\delta_\ell - \delta_k) - B_{\ell k}^1 \cos(\delta_\ell - \delta_k) \Big) + q_{E\ell} \tag{6.27}$$

where

$$q_{E\ell} = -\frac{3}{2}\omega_\ell L_{p\ell}\Bigg(\Big(\sum_{k=1}^{N} v_{fdk}^* \left(G_{\ell k}^1 \cos\delta_k - B_{\ell k}^1 \sin\delta_k \right) \Big)^2 + \Big(\sum_{k=1}^{N} v_{fdk}^* \left(G_{\ell k}^1 \sin\delta_k + B_{\ell k}^1 \cos\delta_k \right) \Big)^2 \Bigg).$$

The above equation highlights the difference between virtual inductors and real inductors. We note that $q_{E\ell}$ is only dependent on the virtual inductor $L_{p\ell}$. Thus, the virtual inductors may help reduce the output reactive power of the GFMC. An example is provided in Fig. 6.9, where $L_{p1} + L_{l1} = 6$ mH and $L_{p2} + L_{l2} = 4$ mH. Compared with the real inductors $L_{l\ell}$, the virtual inductors $L_{p\ell}$ can more effectively increase $v_{fd\ell}^*$ and p_ℓ and decrease q_ℓ.

Combining (6.2) to (6.3), (6.11), (6.26), and (6.27), the M1 version of the $\overline{\mathrm{L_V L_C}}$–$\overline{\mathrm{NL}}$ model is derived, and the order is $N_7 = 3N-1$. Based on M1, $\mathbf{Y_{NL1}}$ in (6.25) can

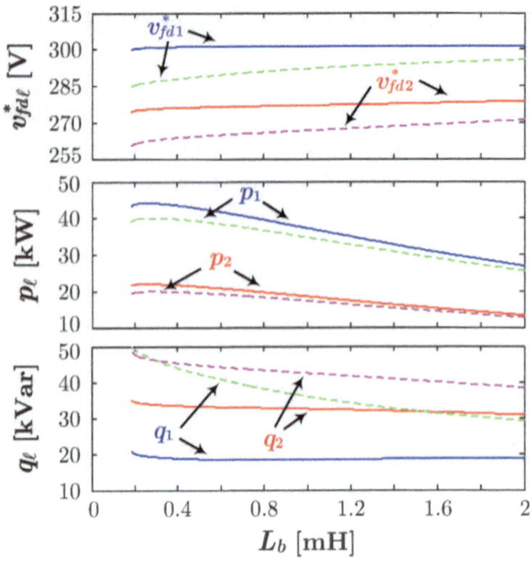

Fig. 6.9 Values of $v^*_{fd\ell}$, p_ℓ, and q_ℓ at the SEP, based on the $\overline{\text{L}_\text{V}\text{L}_\text{C}}$–$\overline{\text{NL}}$ (M2) model. Here, $L_{p1} = 3$ mH, $L_{p2} = 2$ mH, $L_{l1} = 3$ mH, and $L_{l2} = 2$ mH for the solid curves, while $L_{p1} = 0$, $L_{p2} = 0$, $L_{l1} = 6$ mH, and $L_{l2} = 4$ mH for the dashed curves. Other parameters are given in Tables 6.1, 6.2, and 6.4 and Case 2 of Table 6.3, and $R_b = 0.5\ \Omega$

be modified as $\mathbf{Y_{NL2}} = \mathbf{Y_{NL1}}\big|_{\omega_\ell=\omega_0}$, giving M2. The element of $\mathbf{Y_{NL2}}$ is denoted as $\{\mathbf{Y_{NL2}}\}_{\ell k} = G^2_{\ell k} + jB^2_{\ell k}$, where $\ell, k = 1, 2, \cdots, N$. Then, $G^1_{\ell k}$ and $B^1_{\ell k}$ in (6.26) to (6.27) can be replaced by $G^2_{\ell k}$ and $B^2_{\ell k}$, respectively.

6.3.2 *Bifurcation Analysis*

For the islanded microgrid shown in Fig. 6.1b, using the default parameters given in Tables 6.1, 6.2, and 6.4 and setting $\omega_{p1} = \omega_{p2}$, $m_{p1} = 0.5m_{p2}$, and $P_{01} = 2P_{02}$, M2 can be rewritten as a fourth-order system, i.e.,

$$\begin{cases} \dot{\delta}_2 &= y_2 \\ \dot{y}_2 &= \omega_{p2}\Big(- y_2 + m_{p2}\big(0.5p_1 - p_2\big)\Big) \\ \dot{v}^*_{fd1} &= \omega_{q1}\left(V_0 + n_{q1}Q_{01} - v^*_{fd1} - n_{q1}q_1\right) \\ \dot{v}^*_{fd2} &= \omega_{q2}\left(V_0 + n_{q2}Q_{02} - v^*_{fd2} - n_{q2}q_2\right). \end{cases} \tag{6.28}$$

Using this set of equations, the bifurcation diagrams regarding the saddle-node bifurcation can be plotted, as shown in Fig. 6.10. The bifurcation diagrams from the $\text{L}_\text{V}\overline{\text{L}_\text{C}}$–$\overline{\text{NL}}$ and $\text{L}_\text{V}\text{L}_\text{C}$–NL model are also provided for comparison. From Fig. 6.10, the diagrams based on M2 are very close to those based on the $\text{L}_\text{V}\overline{\text{L}_\text{C}}$–$\overline{\text{NL}}$ and $\text{L}_\text{V}\text{L}_\text{C}$–NL model, implying that the M2 model is sufficient for capturing the saddle-node bifurcation.

The attractors and some typical trajectories of (6.28) are plotted in the (δ_2, y_2, v^*_{fd1})-space, as shown in Fig. 6.11a. We can see that the SEP and the SPO are

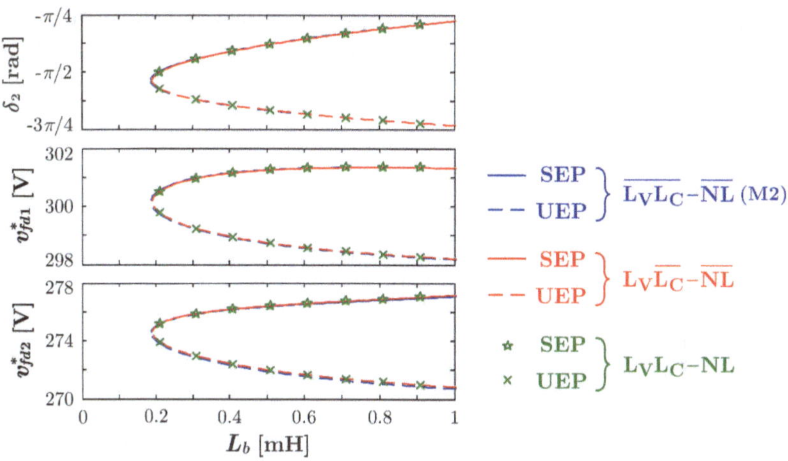

Fig. 6.10 Bifurcation diagrams showing EQPs. Solid curves denote SEP, and dashed curves denote UEP. System parameters are given in Tables 6.1, 6.2, and 6.4 and Case 2 of Table 6.3, and $R_b = 0.5\ \Omega$

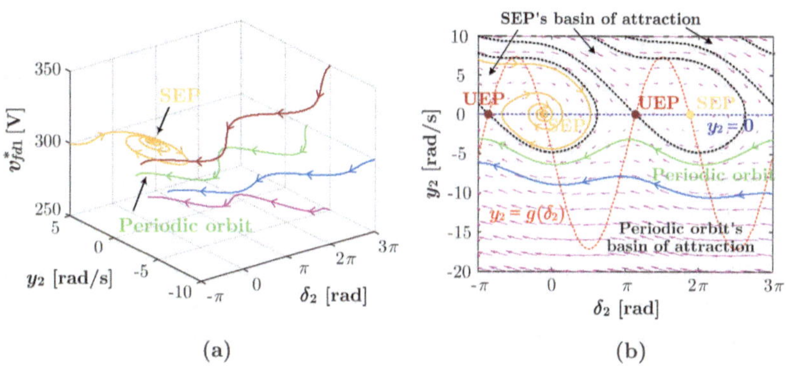

Fig. 6.11 (**a**) Attractors and trajectories of M2 model (i.e., (6.28)) in dim-3 space. The green solid curve is the SPO, and other curves are the typical trajectories. (**b**) Phase portraits from the M3 model (i.e., (6.29)). The vector field is drawn in purple. The red dashed curve is the y_2-nullcline, and the blue dashed line is the δ_2-nullcline. The black dashed curve is the boundary of the SEP's basin of attraction and the SPO's basin of attraction. The green solid curve is the SPO. The yellow and blue solid curves are the typical trajectories. System parameters are given in Tables 6.1, 6.2, and 6.4

almost in the plane $v^*_{fd1} = v^*_{fd1,\mathrm{SEP}} = 309$ V. Besides, all trajectories seems to converge to this plane first. Here, we note that the brown, blue, and purple solid curves all eventually converge to the SPO. Therefore, we may use a second-order subsystem to describe the essential dynamics of (6.28). According to (6.28), the nullclines can be derived. Among them, the y_2-nullcline is $y_2 = h_0(v^*_{fd1}, v^*_{fd2}, \delta_2)$, and the $v^*_{fd\ell}$-nullcline is $v^*_{fd\ell} = h_\ell(\delta_2)$, which is an implicit function. Similar to the third-order system analyzed in a previous work [55], the fourth-order system here can be decomposed into two second-order subsystems, and the subsystem involving $\dot{\delta}_2$ and $\dot{y}_2$ determines the dynamics of (6.28). Such a subsystem has two different forms, i.e., M3 and M4. For M3,

$$\begin{cases} \dot{\delta}_2 = y_2 \\ \dot{y}_2 = \omega_{p2}\left(-y_2 + h_0(v^*_{fd1,\mathrm{SEP}}, v^*_{fd2,\mathrm{SEP}}, \delta_2)\right) \end{cases} \tag{6.29}$$

where $v^*_{fd\ell,\mathrm{SEP}}$ is the SEP value derived from M2. For M4,

$$\begin{cases} \dot{\delta}_2 = y_2 \\ \dot{y}_2 = \omega_{p2}\left(-y_2 + h_0\big(v^*_{fd1,\mathrm{AVG}}, v^*_{fd2,\mathrm{AVG}}, \delta_2\big)\right) \end{cases} \tag{6.30}$$

where $v^*_{fd\ell,\mathrm{AVG}} = \mathrm{avg}\{h_\ell(\delta_2)\}$ is the average value of the $v^*_{fd\ell}$-nullcline.

For M3, the phase portraits of (6.29) are given in Fig. 6.11b, where the y_2-nullcline is

$$y_2 \triangleq g(\delta_2) = h_0(v^*_{fd1,\mathrm{SEP}}, v^*_{fd2,\mathrm{SEP}}, \delta_2).$$

As shown in Fig. 6.11b, the SEP and the SPO coexist, and the phase space is divided into two basins of attraction. Besides, for the system operating at the SEP, some large enough disturbances (e.g., transient faults) may push the trajectory out of its original basin of attraction and induce a sustained low-frequency oscillation. As it is straightforward to derive (6.29), M3 may be a useful model to describe the salient dynamics of the islanded microgrid.

The bifurcation diagrams based on different models are derived, as shown in Fig. 6.12, exposing the homoclinic bifurcation boundaries where the SPO passes the UEP. Above the boundary curve, there are two attractors, namely, the SEP and the SPO, whereas below the curve, only the SEP exists. Some full-circuit cycle-by-cycle simulation results are also provided in Fig. 6.12 for comparison. From Fig. 6.12, the bifurcation diagrams based on M1 are similar to those based on M2, reflecting the weak influence of the frequency variation in $\mathbf{Y}_{\mathbf{NL1}}$. The full-circuit simulation results are highly consistent with the bifurcation diagrams based on the $\mathrm{L_V}\overline{\mathrm{L_C}}$–$\overline{\mathrm{NL}}$ model, but quite different from those based on the $\overline{\mathrm{L_V L_C}}$–$\overline{\mathrm{NL}}$ model, especially for large $\omega_{p\ell}$ or large R_b. Therefore, the $\mathrm{L_V}\overline{\mathrm{L_C}}$–$\overline{\mathrm{NL}}$ model is the most precise reduced-order model, while the $\overline{\mathrm{L_V L_C}}$–$\overline{\mathrm{NL}}$ model (especially M3) can be applied when the islanded microgrid has high inertia and heavy load (i.e., small $\omega_{p\ell}$ and R_b). Besides, from the bifurcation diagrams based on the $\mathrm{L_V}\overline{\mathrm{L_C}}$–$\overline{\mathrm{NL}}$ model, the dynamical response of the voltage loops has a significant impact on the homoclinic bifurcation. If the voltage loops have fast dynamical response (i.e., large $f_{cv\ell}$), the single-attractor area becomes large. Comparing Fig. 6.12a with Fig. 6.12b, the single-attractor area will shrink as R_b gets smaller.

In summary, M2 of the $\overline{\mathrm{L_V L_C}}$–$\overline{\mathrm{NL}}$ model can be applied for the saddle-node bifurcation analysis, whereas the $\mathrm{L_V}\overline{\mathrm{L_C}}$–$\overline{\mathrm{NL}}$ model should be applied for the

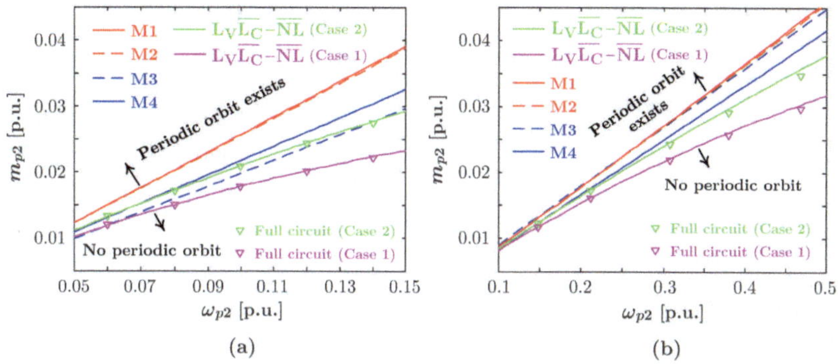

Fig. 6.12 Homoclinic bifurcation diagrams on the ω_{p2}–m_{p2} parameter plane, with $m_{p1} = 0.5m_{p2}$ and $\omega_{p1} = \omega_{p2}$. **(a)** $R_b = 2.1\ \Omega$; **(b)** $R_b = 0.5\ \Omega$. All other parameters follow the default set shown in Tables 6.1 to 6.4

homoclinic bifurcation analysis. The behavior of an islanded microgrid is thus markedly different from the GFMC connected to an infinite bus [55]. Specifically, the voltage loops cannot be neglected. Although M3 of the $\overline{L_V L_C}$–$\overline{NL}$ model is not precise, it can be quickly derived to generate intuitive observations of the salient nonlinear dynamics.

6.4 Case Study

In this section, the slow-scale stability of an islanded microgrid with a complex power network will be analyzed. Then, exact full-circuit cycle-by-cycle simulations based on MATLAB/Simulink R2021b are performed for verification. The topology of the microgrid is shown in Fig. 6.1a, where three GFMCs are contained, i.e., $N = 3$. In this section, the default parameters given in Case 2 of Table 6.3, and Tables 6.1, 6.2, and 6.7 are adopted, unless otherwise specified. Without loss of generality, the admittance matrix is asymmetric here.

6.4.1 Stability Evaluation and Enhancement

Based on the $L_V\overline{L_C}$–$\overline{NL}$ model, the bifurcation diagrams corresponding to Hopf bifurcation are given in Fig. 6.13, where the red and blue solid curves correspond to Cases 1 and 2 of Table 6.3, respectively. For each case, the system is stable below the solid curve and exhibits a Hopf-type instability above the solid curve. As shown in Fig. 6.13, the full-circuit simulation results are perfectly consistent with the analytical results. Also, we may avoid the Hopf-type low-frequency oscillation by reducing $m_{p\ell}$, increasing $L_{p\ell}$, or increasing $h_{p\ell}$ and $h_{i\ell}$ (i.e., increasing $f_{cv\ell}$).

Table 6.7 Circuit parameters of the complex power network

Parameter	Value	Parameter	Value
L_{l1}	3 mH	r_{l1}	0.8 Ω
L_{l2}	2 mH	r_{l2}	0.5 Ω
L_{l3}	2 mH	r_{l3}	0.5 Ω
L_{cp}	2 mH	r_{lp}	0.5 Ω
$R_{b\ell}$	4.2 Ω	$L_{b\ell}$	2 mH

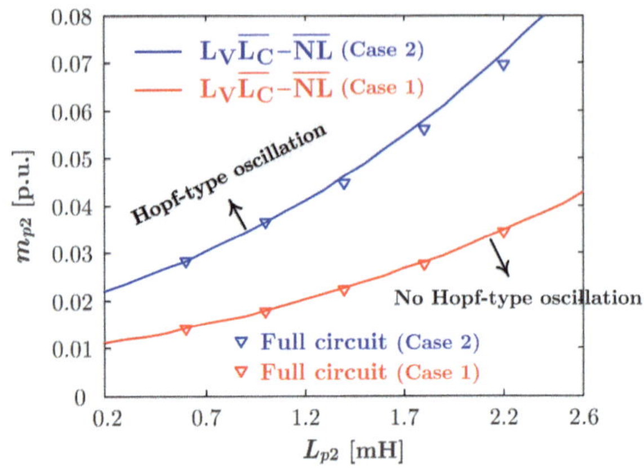

Fig. 6.13 Hopf bifurcation diagrams on the L_{p2}–m_{p2} parameter plane. Solid curves separate unstable (above) and stable (below) regions, with $m_{p1} = 0.5m_{p2} = 0.5m_{p3}$ and $L_{p1} = 1.5L_{p2} = 1.5L_{p3}$. Other parameters follow the default set

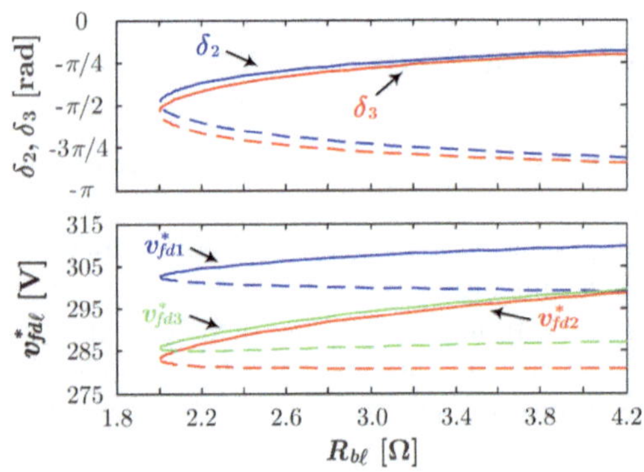

Fig. 6.14 Saddle-node bifurcation diagrams on selected parameter planes. Solid curves denote SEP, and dashed curves denote UEP. $R_{b1} = R_{b2} = R_{b3}$. Other parameters follow the default set

Based on M2 of the $\overline{L_V L_C}$–$\overline{NL}$ model, the bifurcation diagrams corresponding to saddle-node bifurcation are given in Fig. 6.14. The system exhibits saddle-node bifurcation when $R_{b\ell} \approx 2$ Ω. Two EQPs exist for $R_{b\ell} > 2$ Ω and none for $R_{b\ell} < 2$ Ω. For $R_{b\ell} > 2$ Ω, the SEP gets closer to the UEP as $R_{b\ell}$ decreases. Thus, we may increase $R_{b\ell}$ to enhance stability.

Based on the $L_V\overline{L_C}$–$\overline{NL}$ model, the bifurcation diagrams corresponding to homoclinic bifurcation are given in Fig. 6.15, where the red and blue curves correspond to Cases 1 and 2 of Table 6.3, respectively. Homoclinic bifurcation

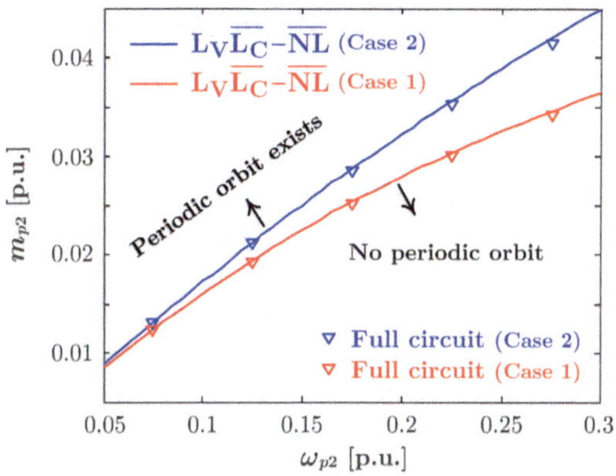

Fig. 6.15 Homoclinic bifurcation diagrams on the ω_{p2}–m_{p2} parameter plane, with $m_{p1} = 0.5\,m_{p2} = 0.5\,m_{p3}$ and $\omega_{p1} = \omega_{p2} = \omega_{p3}$. Other parameters follow the default set

Table 6.8 Stability enhancement trends

	$f_{cv\ell}$	$P_{0\ell}$	$m_{p\ell}$	$\omega_{p\ell}$	$Q_{0\ell}$	V_0	$n_{q\ell}$	$\omega_{q\ell}$	$L_{p\ell}$	$R_{b\ell}$	$L_{b\ell}$	$L_{l\ell}$	$r_{l\ell}$	$L_{c\rho}$	$r_{c\rho}$
Hopf	↑	–	↓	↓↑	↓	↓	↓	↓↑	↑	↓	↑	↑	↓↑	↑	↓↑
Saddle-node	–	–	–	–	↓	↑	↑	–	↓	↑	↑	↓	↓	↓	↓
Homoclinic	↑	–	↓	↑	↓	↓	↑	–	↓	↑	↑	↓	↓	–	↓

Note: All parameters increase or decrease at the same proportion based on their default values, where $\ell = 1, 2, 3$, and $\rho = 1, 2$. "↑" ("↓") denotes improvement in system stability by increasing (decreasing) the corresponding parameter. "↓↑" denotes improvement in stability by decreasing the corresponding parameter when its value is small and by increasing the parameter when its value is large. "–" denotes weak or negligible effect on stability by increasing the corresponding parameter

occurs on the solid curve. Above the curve, an SPO and an SEP coexist, and below the curve, only the SEP exists. In Fig. 6.15, full-circuit simulation results are also provided, which are perfectly consistent with the analytical bifurcation diagrams. From Fig. 6.15, we may eliminate the SPO and enhance stability by reducing $m_{p\ell}$, increasing $\omega_{p\ell}$, or increasing $h_{p\ell}$ and $h_{i\ell}$ (i.e., increasing $f_{cv\ell}$).

Using the same procedure, we can obtain bifurcation diagrams on other parameter planes or spaces, which provide clear guidance on the ways to enhance system's stability, as summarized in Table 6.8. For the Hopf and homoclinic bifurcations, the most practical and effective way to broaden the stability region is to increase $f_{cv\ell}$ (i.e., increase $h_{p\ell}$ and $h_{i\ell}$) or decrease $m_{p\ell}$, without changing the power supply or modifying the circuit parameters. Moreover, harmonic instability may occur if $h_{p\ell}$ or $k_{p\ell}$ is too large [116].

From Table 6.8, we observe that for the Hopf bifurcation, the system stability can be enhanced by increasing $L_{p\ell}$, or $L_{l\ell}$, or $L_{c\rho}$, whereas for the saddle-node and homoclinic bifurcations, the system stability can be enhanced by decreasing $L_{p\ell}$, or $L_{l\ell}$, or $L_{c\rho}$.

6.4.2 Simulation Verification

Exact cycle-by-cycle simulations represent a viable form of verification for power converters, as the primitive switching behavior of power converters is fully considered, and the actual system dynamics are simulated strictly. In the simulations, the step size is set to 2 μs, the relative and absolute tolerances are set to 10^{-5}, and the digital algorithms, computational delay, and modulator delay (i.e., zero-order-hold delay) are also considered. All the controllers and filters are discretized through the Tustin method [117]. These digital algorithms are implemented in the MATLAB function blocks, using the persistent variables to simulate the registers of the digital chips. Here, all digital algorithms are carried out every control period, which simulates the actual behavior in practical control systems.

6.4.2.1 Comparison

First, the transient responses from the $\mathrm{L_V}\overline{\mathrm{L_C}}$–$\overline{\mathrm{NL}}$ model and full-circuit simulations are compared, as shown in Fig. 6.16. For Fig. 6.16a, $L_{p2} = 2$ mH when $t < 5$ s, and $L_{p2} = 0.5$ mH when $t > 5$ s. According to Fig. 6.13, the system exhibits a Hopf-type oscillation after $t = 5$ s, which is verified by the full-circuit simulation given in Fig. 6.16a. For Fig. 6.16b, $R_{b\ell} = 40\ \Omega$ when $t < 5$ s, and $R_{b\ell} = 2.5\ \Omega$ when $t > 5$ s. Sustained oscillation emerges after $t = 5$ s, which means that under such a large load change, the trajectory is pushed out of the SEP's basin of attraction and attracted to the SPO. Furthermore, it is also observed that significant voltage and current fluctuation may emerge after such a load change. In a word, Figs. 6.16a and b verify the effectiveness of the $\mathrm{L_V}\overline{\mathrm{L_C}}$–$\overline{\mathrm{NL}}$ model.

Second, the transient responses from the $\overline{\mathrm{L_V L_C}}$–$\overline{\mathrm{NL}}$ and $\mathrm{L_V}\overline{\mathrm{L_C}}$–$\overline{\mathrm{NL}}$ models are compared, as shown in Fig. 6.17. It is found that for the $\overline{\mathrm{L_V L_C}}$–$\overline{\mathrm{NL}}$ model, the

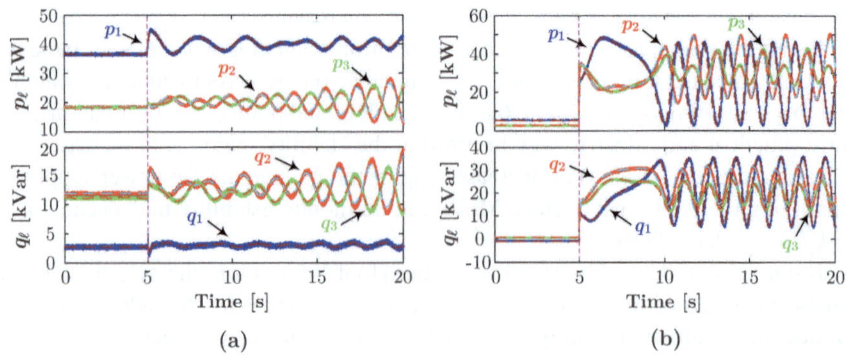

Fig. 6.16 Transient responses from the $\mathrm{L_V}\overline{\mathrm{L_C}}$–$\overline{\mathrm{NL}}$ model (dashed curves) and full-circuit simulations (solid curves). (**a**) $L_{p1} = 3$ mH and $L_{p2} = L_{p3} = 2$ mH when $t < 5$ s, while $L_{p1} = 0.75$ mH and $L_{p2} = L_{p3} = 0.5$ mH when $t > 5$ s; (**b**) $R_{b\ell} = 40\ \Omega$ when $t < 5$ s, and $R_{b\ell} = 2.5\ \Omega$ when $t > 5$ s. Other parameters follow their default values

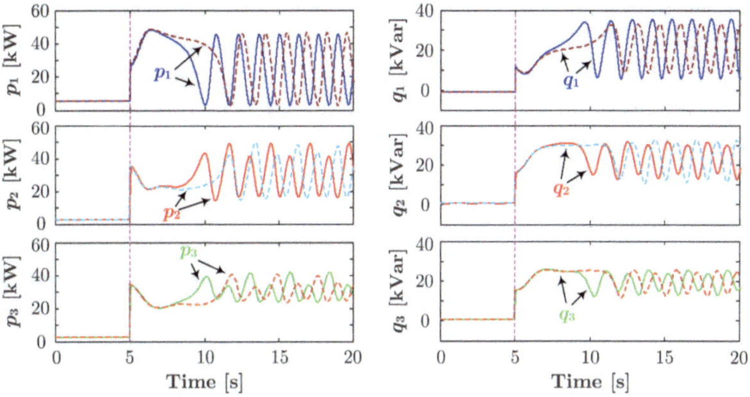

Fig. 6.17 Transient responses from the $\overline{L_V}\overline{L_C}$–$\overline{NL}$ model (dashed curves) and the $L_V\overline{L_C}$–$\overline{NL}$ model (solid curves). Here, $R_{b\ell}$ = 40 Ω when t < 5 s, and $R_{b\ell}$ = 2.5 Ω when t > 5 s. Other parameters follow their default values

sustained oscillation still emerges after the large load change. Besides, the transient responses of the aforementioned two models are quite different. Such differences show that the power loop is the cause for the sustained oscillation, while the voltage loop has an auxiliary but significant impact on shaping the sustained oscillation.

6.4.2.2 Hopf Bifurcation

The influence of the voltage loops on the Hopf bifurcation is demonstrated in Fig. 6.18. Initially, the parameters in Case 2 of Table 6.3 are adopted, and the system operates normally. After $t = 10$ s, the parameters in Case 1 of Table 6.3 are adopted. Then, low-frequency oscillation emerges, confirming the bifurcation diagrams shown in Fig. 6.13.

6.4.2.3 Saddle-Node Bifurcation

As the loads change gradually, i.e., $R_{b\ell}$ = 4.2 Ω when $t = 0 \sim 10$ s, $R_{b\ell}$ = 2.5 Ω when $t = 10 \sim 30$ s, $R_{b\ell}$ = 2.05 Ω when $t = 30 \sim 50$ s, and $R_{b\ell}$ = 1.95 Ω when $t = 50 \sim 70$ s, the system operates normally before $t = 50$ s and begins to oscillate after $t = 50$ s, as shown in Fig. 6.19a. This confirms the bifurcation diagrams shown in Fig. 6.14, where $R_{b\ell} \approx 2$ Ω marks the onset of saddle-node bifurcation. After the onset of saddle-node bifurcation, there is no SEP, and only an SPO exists, which manifests itself as sustained low-frequency oscillation. Comparing Fig. 6.16b with Fig. 6.19a, we see that there is no sustained oscillation when $R_{b\ell}$ switches to 2.5 Ω from 4.2 Ω, whereas sustained oscillation emerges when $R_{b\ell}$ switches to 2.5 Ω from 40 Ω. This verifies the previous analysis that a large load change may push

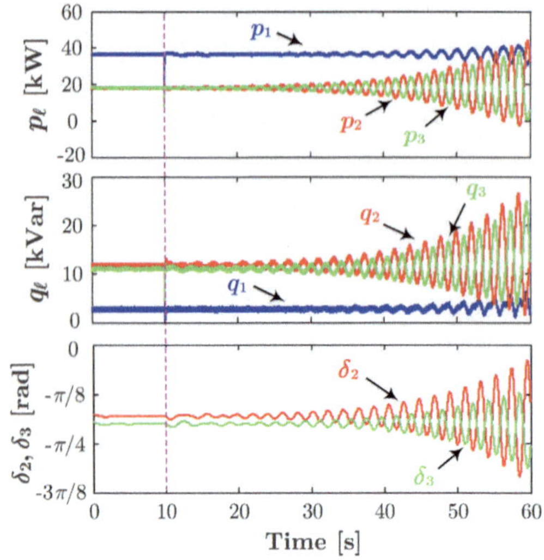

Fig. 6.18 Simulated waveforms of p_ℓ, q_ℓ, δ_2, and δ_3 showing emergence of oscillation when parameters are changed. Case 2 of Table 6.3 is adopted when $t < 10$ s, and Case 1 of Table 6.3 is adopted when $t > 10$ s. Other parameters follow their default values

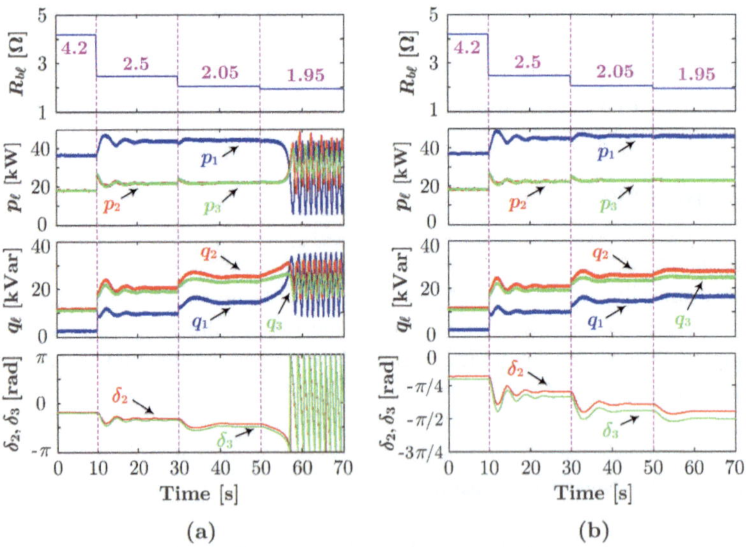

Fig. 6.19 Simulated waveforms of p_ℓ, q_ℓ, δ_2, and δ_3 when the loads $R_{b\ell}$ change gradually. (**a**) $L_{p1} = 3$ mH, and $L_{p2} = L_{p3} = 2$ mH. (**b**) $L_{p1} = 2.7$ mH, and $L_{p2} = L_{p3} = 1.8$ mH. Other parameters follow their default values

the trajectory from one basin of attraction to another. Furthermore, for the same scenario, if $L_{p\ell}$ is decreased, there is no sustained oscillation when $R_{b\ell} = 1.95$ Ω, as shown in Fig. 6.19b. This verifies the stability enhancement method given in Table 6.8.

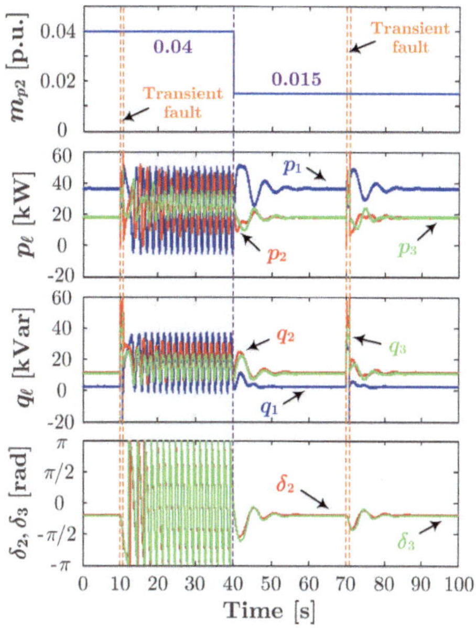

Fig. 6.20 Simulated waveforms of p_ℓ, q_ℓ, δ_2, and δ_3. $m_{p2} = m_{p3} = 2m_{p1} = 0.04$ p.u. when $t < 40$ s, and $m_{p2} = m_{p3} = 2m_{p1} = 0.015$ p.u. when $t > 40$ s. A symmetrical three-phase-to-ground fault occurs when $t = 10 \sim 10.8$ s and $t = 70 \sim 70.8$ s, and the fault is cleared at other times. Other parameters follow their default values

6.4.2.4 Homoclinic Bifurcation

The influence of the transient faults and the homoclinic bifurcation are demonstrated in Fig. 6.20. Initially, $m_{p2} = m_{p3} = 2m_{p1} = 0.04$ p.u., and the system operates with an SEP. Then, a symmetrical three-phase-to-ground fault is applied to the second bus with duration of 0.8 s, i.e., $v_{b2} \approx 0$. Under this transient fault, the system exhibits sustained oscillation. At $t = 40$ s, the parameters are adjusted as $m_{p2} = m_{p3} = 2m_{p1} = 0.015$ p.u., and then the sustained oscillations subside. According to Fig. 6.15, there is no SPO in this case, which is consistent with the simulation results. Finally, another transient fault takes place when $t = 70 \sim 70.8$ s. However, the system recovers to its normal working state automatically, which implies the absence of a periodic-orbit attractor. Also, for the Hopf-type oscillation, δ_ℓ oscillates near its original value at the SEP, as shown in Fig. 6.18. However, for the aforedescribed sustained oscillation, δ_ℓ varies from $-\pi$ to π.

6.5 Recapitulation

In this chapter, the nonlinear dynamics of an islanded microgrid comprising a number of GFMCs is studied. Several slow-scale reduced-order models are proposed for addressing different bifurcation scenarios, as shown in Fig. 6.21.

1. For the Hopf and homoclinic bifurcations, the influence of the voltage loops in the converters must be taken into consideration, while the dynamics of

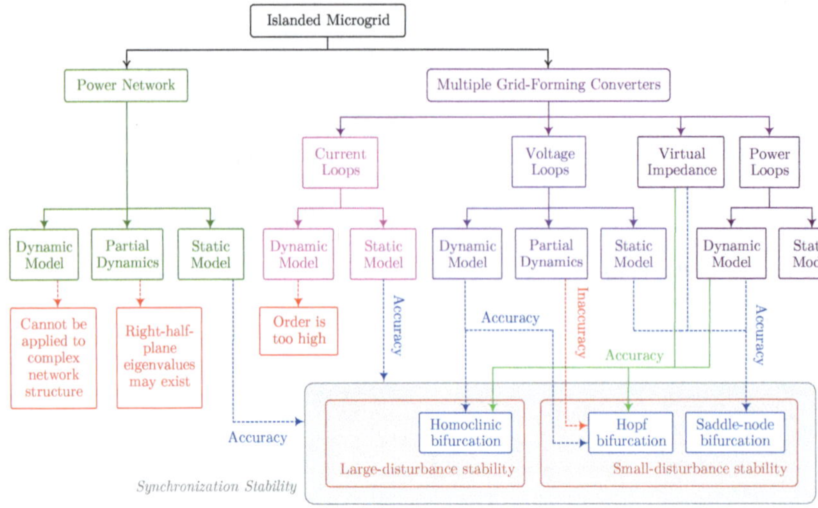

Fig. 6.21 Reduced-order models of the islanded microgrid containing a number of GFMCs

the current loops and power network can be omitted. Furthermore, the Hopf bifurcation behavior essentially arises from the stability of voltage loops and only considering the dynamics of power loops is unable to predict this behavior. On the contrary, the homoclinic bifurcation behavior is mainly determined by the power loops, and the voltage loops have auxiliary but significant impacts on shaping the SPOs and bifurcation diagrams (i.e., the exact bifurcation onset point).

2. For the saddle-node bifurcation, the dynamics of the voltage and current loops can be omitted. On this basis, power-flow-based models are proposed. It is shown that the influence of the frequency variation on the admittance matrix is so weak that power-flow-based models can be significantly simplified. Such models are capable of capturing the saddle-node bifurcation behavior.

This chapter emphasizes the importance of the voltage loops and virtual inductors in the GFMCs, which are responsible for the distinct dynamical behavior of an islanded microgrid. Using these results, we can establish the model of a given islanded microgrid quickly. For example, when studying the Hopf or homoclinic bifurcation behavior, the modeling procedure are as follows:

- The model of each GFMC is built in the corresponding dq-frame, where only the dynamics of voltage and power loops are considered.
- One particular dq-frame is chosen as the common coordinate frame. The power angle between each dq-frame and this common coordinate frame reflects the synchronization status among the GFMCs, i.e., constant values indicate the synchronization stability, while the oscillating or diverging values indicate the synchronization instability.

- The model of the power network and the impedance loads is established in the selected common coordinate frame. Moreover, the transformation matrices associated with the aforementioned power angles are used to connect all coordinate frames. Then, the model of the islanded microgrid is established.

On this basis, we can further reduce the microgrid model by using the methods of model aggregation [118, 119] and/or the techniques of truncation and residualization (i.e., removing the fast-mode state variables related to the eigenvalues far from the imaginary axis) [120]. Notably, for the later techniques, the truncated and residual state variables usually lack clear physical meaning. Therefore, such techniques cannot be applied to the initial stage of modeling. On the other hand, the Kron reduction technique can be adopted to simplify the structure of the power network [118] and therefore can be used prior to the aforementioned modeling procedure.

Chapter 7
Sub-Synchronous Oscillations Under Transient Disturbances in Islanded Microgrids

Due to the interaction between power converters and the grid, a power system may lose stability, giving rise to harmonic oscillations (usually at hundreds of Hz), or sub-/super-synchronous oscillations (at near or far below 50 Hz) [121]. For the latter case, it is also referred to as rotor angle instability or synchronization instability. This instability can be either *small-disturbance synchronization* type or *transient synchronization* type [1–3, 39, 41]. The former is associated with the status change of the EQP, typically the SEP being lost with a UEP emerged, leading to sub-/super-synchronous oscillation. From the nonlinear dynamical viewpoint, a Hopf bifurcation occurs in this case [1, 5]. Some examples can be found in Chaps. 2 and 6. For the transient synchronization instability, the initially stable system may fail to maintain its synchronization under a transient disturbance (e.g., a transient grid voltage dip) [41]. Upon removal of the disturbance, the system may collapse due to the diverging state variables [56, 90, 91, 122, 123], or exhibit a sustained sub-/super-synchronous oscillation, which differs fundamentally from the aforementioned Hopf-type oscillation. In Chaps. 3, 4, and 6, this new type of oscillation has been found to associate with a periodic orbit, which is remote from, but coexisting with, the SEP. Also, this periodic orbit has been found to emerge or subside via a homoclinic bifurcation. Thus, such periodic orbit is a Shil'nikov-type attractor.

The aforementioned Hopf-type and Shil'nikov-type periodic orbits can all be regarded as standard self-excited attractors [69, 70, 72, 124]. The self-excited attractor can be observed naturally by putting the initial conditions in the neighborhood of UEPs [5, 125–127]. On the other hand, another type of attractors, namely, hidden attractors, may coexist with the SEP, and each hidden attractor has its own basin of attraction which does not cover the neighborhood of any EQPs. Hence, the hidden attractors cannot be observed naturally. However, due to the transient disturbance, the system initially operating at the SEP may be pushed out of its original basin of attraction and driven to one of the hidden attractors, inducing sustained oscillations. In the previous chapters, the physical origin of sustained sub-

J. Yang, C. K. Tse, *Complex Behavior of Grid-Connected Power Electronics Systems*, https://doi.org/10.1007/978-3-031-69340-3_7

synchronous oscillation is found to be a self-excited periodic orbit. However, the sub-synchronous oscillations associated with various types of hidden attractors have not been discovered.

This chapter studies an islanded microgrid consisting of GFMCs. For this system, various types of hidden attractors are identified, and their characteristics, typical bifurcation behaviors, and basins of attraction are explored, following with the development of transient stability enhancement methods.

7.1 Microgrid Model

Figure 7.1 shows an N-bus islanded microgrid containing N GFMCs. The output filter of GFMC-ℓ comprises $L_{f\ell}$ and $C_{f\ell}$, where $\ell = 1, 2, \cdots, N$. Each GFMC connects to a local bus through a transmission line, which is modeled as a series connection of $L_{l\ell}$ and $r_{l\ell}$. Each bus has a constant impedance load, i.e., $Z_{b\ell} = R_{b\ell} + j\omega_0 L_{b\ell}$. Bus-$\rho$ connects to Bus-$(\rho + 1)$ through a transmission line, which is represented by $L_{c\rho}$ and $r_{c\rho}$, where $\rho = 1, 2, \cdots, N - 1$. Here, p_ℓ and q_ℓ are the output active and reactive power of GFMC-ℓ, respectively. Besides, $v_{f\ell}$ is the voltage of $C_{f\ell}$, $v_{fm\ell}$ is the magnitude of $v_{f\ell}$, ω_ℓ is the output frequency, θ_ℓ is the phase angle, $v_{f\ell,\text{ref}}$ is the reference of $v_{f\ell}$, and $v_{fm\ell,\text{ref}}$ is the reference of $v_{fm\ell}$. Furthermore, $P_{0\ell}$, $Q_{0\ell}$, ω_0, and V_0 are the nominal values of p_ℓ, q_ℓ, ω_ℓ, and $v_{fm\ell,\text{ref}}$, respectively. Moreover, $\omega_{p\ell}$ and $\omega_{q\ell}$ are the cutoff frequencies of the low-pass filters, while $m_{p\ell}$ and $n_{q\ell}$ are the droop coefficients. We first examine a 5-bus microgrid and extend the study to the 30-bus microgrid. The default parameters are summarized in Tables 7.1 and 7.2.

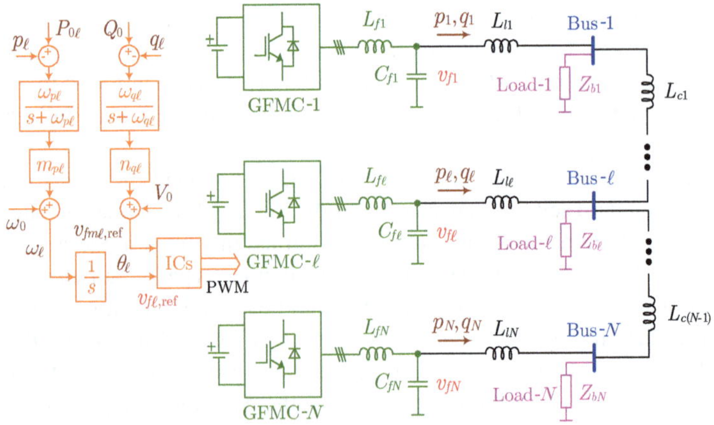

Fig. 7.1 Islanded microgrid. The ESR of $L_{f\ell}$, $L_{l\ell}$, and $L_{c\rho}$ are $r_{f\ell}$, $r_{l\ell}$ and $r_{c\rho}$, respectively, where $\ell = 1, 2, \cdots, N$ and $\rho = 1, 2, \cdots, N - 1$. "ICs" denotes voltage and current inner-loop controllers

Table 7.1 Default circuit parameters of the islanded microgrid

Parameter	ℓ or ρ	Value	Parameter	ℓ or ρ	Value
$L_{f\ell}$	$1, 2, \cdots, N$	1.8 mH	$r_{f\ell}$	$1, 2, \cdots, N$	0.1 Ω
$C_{f\ell}$	$1, 2, \cdots, N$	300 μF			
L_{cp}	$1, 2, \cdots, N-1$	2 mH	r_{cp}	$1, 2, \cdots, N-1$	0.5 Ω
$L_{l\ell}$	$1, 2, 3, 4, 6, 7, 8, 9, \cdots, N-1$	3 mH	$r_{l\ell}$	$1, 2, 3, 4, 6, 7, 8, 9, \cdots, N-1$	0.8 Ω
	$5, 10, \cdots, N$	3.8 mH		$5, 10, \cdots, N$	1 Ω
$L_{b\ell}$	$1, 3, 6, 8, \cdots, N-2$	3 mH	$R_{b\ell}$	$1, 3, 6, 8, \cdots, N-2$	6 Ω
	$2, 4, 7, 9, \cdots, N-1$	2 mH		$2, 4, 7, 9, \cdots, N-1$	4.2 Ω
	$5, 10, \cdots, N$	1.5 mH		$5, 10, \cdots, N$	3 Ω
L_{gn}	/	0.2 mH	r_{gn}	/	0.1 Ω

Table 7.2 Default control parameters of GFMCs

Parameter	ℓ	Value	p.u.
$\omega_{p\ell}$	$1, 2, \cdots, N$	$2\pi \times 0.1$	0.1
$\omega_{q\ell}$	$1, 2, \cdots, N$	$2\pi \times 0.1$	0.1
$P_{0\ell}$	$1, 4, 6, 9, \cdots, N-1$	50 kW	2
	$2, 3, 7, 8, \cdots, N-2$	25 kW	1
	$5, 10, \cdots, N$	100 kW	4
$m_{p\ell}$	$1, 4, 6, 9, \cdots, N-1$	$0.02\,\omega_0/P_m$	0.02
	$2, 3, 7, 8, \cdots, N-2$	$0.04\,\omega_0/P_m$	0.04
	$5, 10, \cdots, N$	$0.01\,\omega_0/P_m$	0.01
$n_{q\ell}$	$1, 4, 6, 9, \cdots, N-1$	$0.025\,V_0/Q_m$	0.025
	$2, 3, 7, 8, \cdots, N-2$	$0.05\,V_0/Q_m$	0.05
	$5, 10, \cdots, N$	$0.0125\,V_0/Q_m$	0.0125

Note: $P_m = 50$ kW, $Q_m = 15$ kVar, $\omega_0 = 100\pi$, $Q_0 = 0$, and $V_0 = 311$ V (i.e., $V_0 = 1$ p.u.). Besides, $m_{p\ell} P_{0\ell} = m_{p1} P_{01}$ for power sharing

In GFMCs, the dynamics of the voltage loop and current loop are much faster than the dynamics of the power loop. We set $v_{f\ell} \approx v_{f\ell,\text{ref}}$ and $v_{fm\ell} \approx v_{fm\ell,\text{ref}}$. The phase difference between GFMC-ℓ and GFMC-1 is denoted as δ_ℓ, i.e., $\delta_\ell = \theta_\ell - \theta_1$. From Fig. 7.1, the output voltage of GFMC-ℓ is

$$\boldsymbol{V_{f\ell}} \triangleq v_{fm\ell} \angle \delta_\ell = v_{fm\ell} \cos \delta_\ell + j v_{fm\ell} \sin \delta_\ell$$

Denoting the output current of GFMC-ℓ as $\boldsymbol{I_{l\ell}}$ and using an admittance model [105], we get

$$\begin{bmatrix} \boldsymbol{I}_{l1} \\ \boldsymbol{I}_{l2} \\ \vdots \\ \boldsymbol{I}_{lN} \end{bmatrix} = \mathbf{Y_{NL}} \begin{bmatrix} \boldsymbol{V}_{f1} \\ \boldsymbol{V}_{f2} \\ \vdots \\ \boldsymbol{V}_{fN} \end{bmatrix} \tag{7.1}$$

where $\mathbf{Y_{NL}} = (\mathbf{I} + \mathbf{Y_{BC}}\mathbf{Z_L})^{-1}\mathbf{Y_{BC}}$, $\mathbf{Z_L} = \mathrm{diag}\{Z_{l\ell}\} = \mathrm{diag}\{r_{l\ell} + j\omega_0 L_{l\ell}\}$,

$$\mathbf{Y_{BC}} = \begin{bmatrix} Y_{b1} + Y_{c1} & -Y_{c1} & 0 & \cdots & 0 \\ -Y_{c1} & Y_{b2} + Y_{c1} + Y_{c2} & -Y_{c2} & \cdots & 0 \\ \vdots & \vdots & \vdots & \ddots & \vdots \\ 0 & 0 & 0 & \cdots & Y_{bN} + Y_{c(N-1)} \end{bmatrix},$$

$$Y_{b\ell} = \frac{1}{Z_{b\ell}}, \quad Y_{c\rho} = \frac{1}{Z_{c\rho}} = \frac{1}{r_{c\rho} + j\omega_0 L_{c\rho}}.$$

Here, the elements of $\mathbf{Y_{NL}}$ are denoted as $\{\mathbf{Y_{NL}}\}_{\ell k} = G_{\ell k} + jB_{\ell k}$, where $\ell, k = 1, 2, \cdots, N$. As $p_\ell + jq_\ell = 1.5\boldsymbol{V}_{f\ell}\bar{\boldsymbol{I}}_{l\ell}$,

$$p_\ell = \frac{3}{2}\sum_{k=1}^{N} v_{fm\ell} v_{fmk}\Big(G_{\ell k}\cos(\delta_\ell - \delta_k) + B_{\ell k}\sin(\delta_\ell - \delta_k)\Big) \tag{7.2}$$

$$q_\ell = \frac{3}{2}\sum_{k=1}^{N} v_{fm\ell} v_{fmk}\Big(G_{\ell k}\sin(\delta_\ell - \delta_k) - B_{\ell k}\cos(\delta_\ell - \delta_k)\Big). \tag{7.3}$$

According to Fig. 7.1, we have

$$\begin{cases} \dot{\delta}_\ell &= x_\ell - x_1 \\ \dot{x}_\ell &= \omega_{p\ell}\Big(- x_\ell + m_{p\ell}\big(P_{0\ell} - p_\ell\big)\Big) \\ \dot{v}_{fm\ell} &= \omega_{q\ell}\Big(V_0 - v_{fm\ell} + n_{q\ell}\big(Q_0 - q_\ell\big)\Big). \end{cases} \tag{7.4}$$

In particular, $\dot{\delta}_1 = 0$ and $\delta_1 = 0$. Here, we consider a special case where $\omega_{p1} = \omega_{p2} = \cdots = \omega_{pN}$, i.e.,

$$\begin{cases} \dot{\delta}_\ell &= y_\ell \\ \dot{y}_\ell &= \omega_{p1}\Big(- y_\ell + m_{p1}\big(p_1 - \mu_\ell p_\ell\big)\Big) \\ \dot{v}_{fm\ell} &= \omega_{q\ell}\Big(V_0 - v_{fm\ell} + n_{q\ell}\big(Q_0 - q_\ell\big)\Big) \end{cases} \tag{7.5}$$

where $\mu_\ell = m_{p\ell}/m_{p1}$, and $y_\ell = x_\ell - x_1$. Here, $m_{p\ell}P_{0\ell} = m_{p1}P_{01}$ for power sharing. As $\dot{\delta}_1 = 0$ and $\dot{y}_1 = 0$, the state order of (7.5) for an N-bus microgrid is $M = 3N - 2$.

7.2 Sub-Synchronous Oscillation and Hidden Attractors

For the system described by (7.5), the SEP coexists with multiple invisible attractors (e.g., periodic, quasi-periodic, and chaotic attractors). Most of these invisible attractors are hidden attractors, which will be explained later (see Fig. 7.15). The distinction between the invisible attractor and hidden attractor can be found in Fig. 1.33. This section presents some typical examples of hidden attractors.

7.2.1 Hidden Periodic Orbit and Hidden Quasi-Periodic Orbit

7.2.1.1 Characteristics

In this subsection, the specific hidden attractor is found by setting the initial conditions as $\delta_\ell(0) = 0$, $y_2(0) = -10$, $y_\ell(0) = 0$ for $\ell \neq 2$, and $v_{fm\ell}(0) = V_0$. When m_{p4} is adjusted from 0.026 to 0.018 p.u., this attractor may be a hidden periodic or quasi-periodic orbit. Here, we use the steady-state data of δ_2, δ_3, and δ_5 and plot the attractor in (δ_2, δ_5)-torus and (δ_3, δ_5)-torus.

The relationship between the torus and the 3-dim space is presented as

$$\begin{cases} u_x = (R_x + R_z \cos\delta_5)\cos\delta_k \\ u_y = (R_y + R_z \cos\delta_5)\sin\delta_k \\ u_z = R_z \sin\delta_5 \end{cases} \tag{7.6}$$

where $k = 2$ or 3. Also, we set $R_x = R_y = 5$ and $R_z = 2$ for brevity of presentation.

Four typical attractors are shown in Fig. 7.2. These attractors share a common characteristic, i.e., δ_2 and δ_5 vary from $-\pi$ to π, and other angles oscillate slightly. In addition, the interaction between δ_2 and δ_5 is different among these attractors. For $m_{p4} = 0.025$ p.u., δ_2 and $-\delta_5$ are phase-locked, as shown in Fig. 7.3a. Therefore, the attractor is a circle, i.e., blue curve in Fig. 7.2a and red curve in Fig. 7.2b. This attractor is a periodic orbit. For $m_{p4} = 0.022$ p.u., δ_2 and $-\delta_5$ are still phase-locked, but the synchronization is not as stable as when $m_{p4} = 0.025$ p.u. Hence, the original circle becomes a collection of many nearby and dense circles, resembling a circle girdle, as shown in Fig. 7.2c, d. This attractor is still a periodic orbit. For $m_{p4} = 0.02$ p.u., δ_2 and δ_5 are uncoupled, as shown in Fig. 7.3b. Here, the oscillation frequency of δ_2 and δ_5 are denoted as f_{osc2} and f_{osc5}, respectively. In this case, the ratio of f_{osc2} to f_{osc5} is constant but irrational. Thus, the attractor is dense on the surface of (δ_2, δ_5)-torus, i.e., the solution trajectory will visit every point on that surface eventually, as shown in Fig. 7.2e, f [5]. This attractor is a quasi-periodic orbit. For $m_{p4} = 0.0194$ p.u., δ_2 and δ_5 are uncoupled as well; however, the ratio of f_{osc2} to f_{osc5} is a rational constant, i.e., $f_{\text{osc2}}/f_{\text{osc5}} \approx 1.25$. Therefore, the attractor becomes a simple orbit on the tori, as shown in Fig. 7.2g, h. This attractor is strictly a periodic orbit.

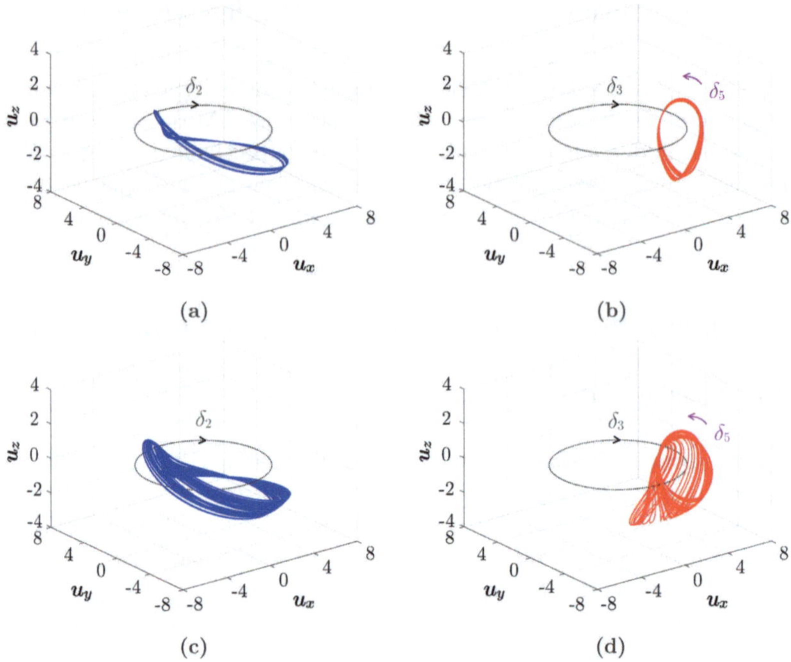

Fig. 7.2 Type-II hidden periodic orbit in (**a**) (δ_2, δ_5)-torus and (**b**) (δ_3, δ_5)-torus for $m_{p4} = 0.025$ p.u. Type-II hidden periodic orbit in (**c**) (δ_2, δ_5)-torus and (**d**) (δ_3, δ_5)-torus for $m_{p4} = 0.022$ p.u. Here, $\delta_\ell(0) = 0$, $y_2(0) = -10$, $y_\ell(0) = 0$ for $\ell \neq 2$, and $v_{fm\ell}(0) = V_0$; Hidden quasi-periodic orbit in (**e**) (δ_2, δ_5)-torus and (**f**) (δ_3, δ_5)-torus for $m_{p4} = 0.02$ p.u. and Type-III hidden periodic orbit in (**g**) (δ_2, δ_5)-torus and (**h**) (δ_3, δ_5)-torus for $m_{p4} = 0.0194$ p.u. Here, $\delta_\ell(0) = 0$, $y_2(0) = -10$, $y_\ell(0) = 0$ for $\ell \neq 2$, and $v_{fm\ell}(0) = V_0$

Depending on δ_ℓ, the hidden periodic orbits belong to three possible categories:

(i) Type-I: Only one angle varies from $-\pi$ to π, and other angles oscillate slightly.
(ii) Type-II: Two (or more) angles vary from $-\pi$ to π, and these angles are phase-locked, as shown in Fig. 7.2a, c.
(iii) Type-III: Two uncoupled angles vary from $-\pi$ to π, and the ratio of their frequencies is a rational constant, as shown in Fig. 7.2g.

7.2.1.2 Bifurcation

Based on (7.5), we get

$$\begin{cases} \dot{\phi}_{25} = z_{25} \\ \dot{z}_{25} = \omega_{p1}(-z_{25} + g(\phi_{25})) \end{cases} \tag{7.7}$$

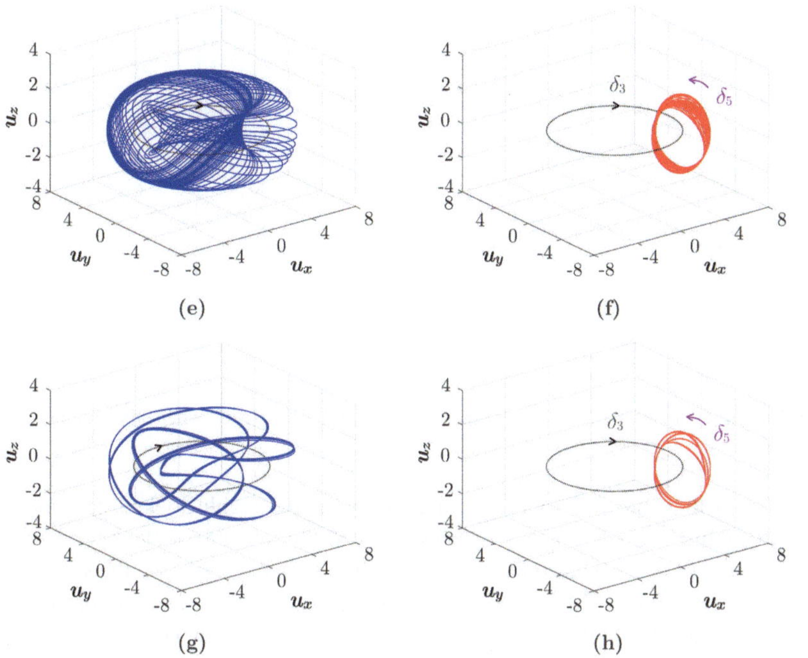

Fig. 7.2 (continued)

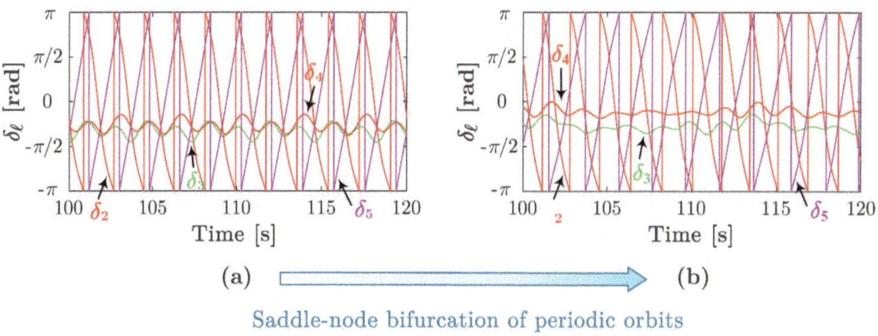

Fig. 7.3 Waveforms of δ_2, δ_3, δ_4, and δ_5 for (**a**) the Type-II hidden periodic orbit when $m_{p4} = 0.025$ p.u. and (**b**) the hidden quasi-periodic orbit when $m_{p4} = 0.02$ p.u.

where $\phi_{25} = \delta_2 + \delta_5$, $z_{25} = y_2 + y_5$, and $g(\phi_{25}) = m_{p1}(2p_1 - \mu_2 p_2 - \mu_5 p_5)$. Also, the choice of m_{p4} determines δ_3, δ_4, and $v_{fm\ell}$ and hence also affects $g(\phi_{25})$.

We begin with some qualitative analysis. If $g(\phi_{25}) = 0$ has two fixed points for $\phi_{25} \in (-\pi, \pi)$, the system (7.7) has two EQPs, i.e., SEP and UEP, as shown in Fig. 7.4a. Normally, the trajectory will be attracted to the SEP. Thus, $z_{25} = 0$, and $\phi_{25} = \delta_2 - (-\delta_5) =$ constant eventually. In this case, δ_2 and $-\delta_5$ are phase-locked.

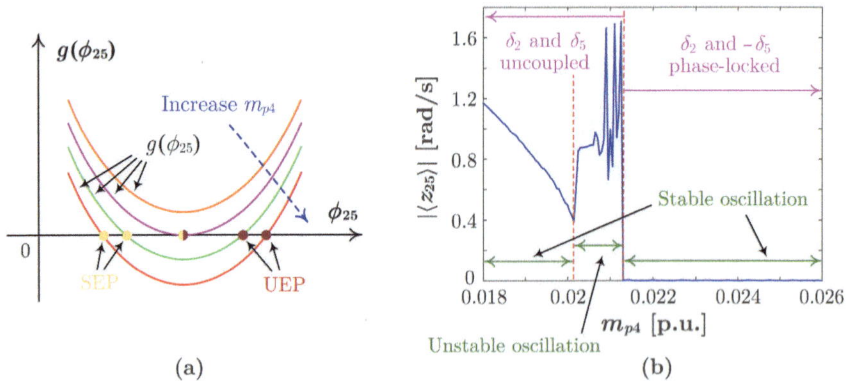

Fig. 7.4 (**a**) Generation mechanism of the SNPO. (**b**) Value of $|\langle z_{25}\rangle|$ when adjusting m_{p4}

It is noted that the SEP of (7.7) corresponds to the stable locked periodic orbit of (7.5), while the UEP of (7.7) corresponds to the unstable locked periodic orbit of (7.5). If $g(\phi_{25}) = 0$ has only one fixed point for $\phi_{25} \in (-\pi, \pi)$, the system (7.7) has one EQP. Here, a SNPO occurs, where the two locked periodic orbits merge and disappear [128]. If $g(\phi_{25}) = 0$ has no fixed point for $\phi_{25} \in (-\pi, \pi)$, the system (7.7) has no EQP, where δ_2 and δ_5 become uncoupled. In this case, $|z_{25}| \gg 0$, and $\phi_{25} \neq$ constant.

Next, the bifurcation behavior is quantitatively observed through z_{25}. As z_{25} may contain ripples, its long-term average value is adopted, i.e.,

$$|\langle z_{25}\rangle| \triangleq \left|\frac{1}{T_{p2} - T_{p1}} \int_{T_{p1}}^{T_{p2}} (y_2 + y_5)\mathrm{d}t\right|. \tag{7.8}$$

Here, we set $T_{p1} = 200$ s and $T_{p2} = 600$ s to permit precise identification of the real system behavior near the SNPO point.

As shown in Fig. 7.4b, $|\langle z_{25}\rangle| \approx 0$ when $m_{p4} = 0.0213 \backsim 0.026$ p.u., implying that δ_2 and $-\delta_5$ are phase-locked, and a Type-II hidden periodic orbit exists. Then, when $m_{p4} = 0.02015 \backsim 0.0213$ p.u., the m_{p4}-$|\langle z_{25}\rangle|$ curve appears irregular. It is observed that δ_2 and $-\delta_5$ are phase-locked initially; however, the locked condition is broken soon, leading finally to another hidden attractor after a short transient chaotic oscillation. The explanation for this strange phenomenon is as follows. As the SNPO occurs, $\min |g(\phi_{25})|$ is just a little larger than zero, and the trajectory may spend a long time to pass the extinct locked periodic orbit. Besides, as there is no stable hidden quasi-periodic orbit (with oscillating δ_2 and δ_5) yet the trajectory diverges to other hidden attractors eventually. Finally, when $m_{p4} = 0.018 \backsim 0.02015$ p.u., $|\langle z_{25}\rangle| \gg 0$, and the m_{p4}-$|\langle z_{25}\rangle|$ curve is a smooth and continuous curve. In this case, the stable hidden quasi-periodic orbit or Type-III hidden periodic orbit (with oscillating δ_2 and δ_5) exists.

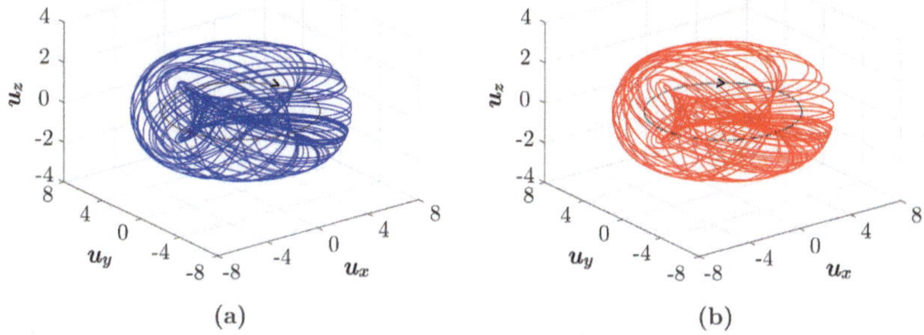

Fig. 7.5 Hidden chaotic attractor in (**a**) (δ_2, δ_5)-torus and (**b**) (δ_3, δ_5)-torus for $m_{p5} = 0.01$ p.u. Here, $\delta_\ell(0) = 0$, $y_2(0) = -5$, $y_\ell(0) = 0$ for $\ell \neq 2$, and $v_{fm\ell}(0) = V_0$

7.2.2 *Hidden Chaotic Attractor*

7.2.2.1 Characteristics

The initial conditions are set as $\delta_\ell(0) = 0$, $y_2(0) = -5$, $y_\ell(0) = 0$ for $\ell \neq 2$, and $v_{fm\ell}(0) = V_0$. The value of m_{p5} is varied, and other parameters adopt their default values. The hidden chaotic attractor when $m_{p5} = 0.01$ p.u. is plotted in (δ_2, δ_5)-torus and (δ_3, δ_5)-torus, as shown in Fig. 7.5. The trajectory in the tori is irregular comparing with the previous quasi-periodic orbit shown in Fig. 7.2e. Furthermore, as shown in Fig. 7.6a, δ_2, δ_3, δ_4, and δ_5 all vary from $-\pi$ to π, implying chaotic motion. Moreover, not all angles vary from $-\pi$ to π in a hidden chaotic attractor. An example is given in Fig. 7.6b, where $m_{p5} = 0.025$ p.u., and only δ_2 and δ_3 vary from $-\pi$ to π. It is clear that the variation frequencies of δ_2 and δ_3 are not constant, but change very quickly. The ratio of these frequencies varies as well.

The hidden chaotic attractor when $m_{p5} = 0.01$ p.u. is replotted in (δ_2, y_2, v_{fm2})-space for further investigation, as shown in Fig. 7.7a. Here, we select a 2-dim cross section where $\delta_2 = 0$ (Poincaré section) and record the value of v_{fm2} when the attractor trajectory crosses this section. Hence, by varying m_{p5} and recording the corresponding v_{fm2}, we can derive the bifurcation diagram on the m_{p5}-v_{fm2} plane, as shown in Fig. 7.7b. From Fig. 7.7b, we observe countless distinct intersection points when $m_{p5} = 0.01 \backsim 0.0136$ p.u., and $m_{p5} = 0.0232 \backsim 0.0262$ p.u., implying chaotic motion. For most other m_{p5} values, there is only one or two distinct intersection points, showing quasi-periodic or periodic behavior. Thus, the chaotic attractor emerges via quasi-periodicity as m_{p5} is varied near 0.0136 p.u. or 0.0232 p.u.

7.2.2.2 Lyapunov Exponents

The state equations given in (7.5) can be rewritten as

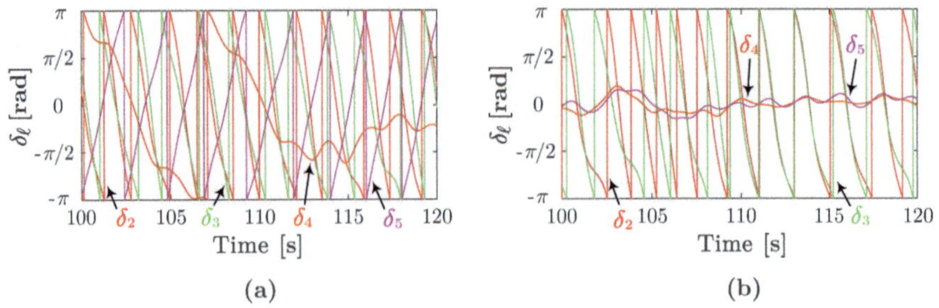

Fig. 7.6 Waveforms of δ_2, δ_3, δ_4, and δ_5 for hidden chaotic attractors for (**a**) $m_{p5} = 0.01$ p.u. and (**b**) $m_{p5} = 0.025$ p.u.

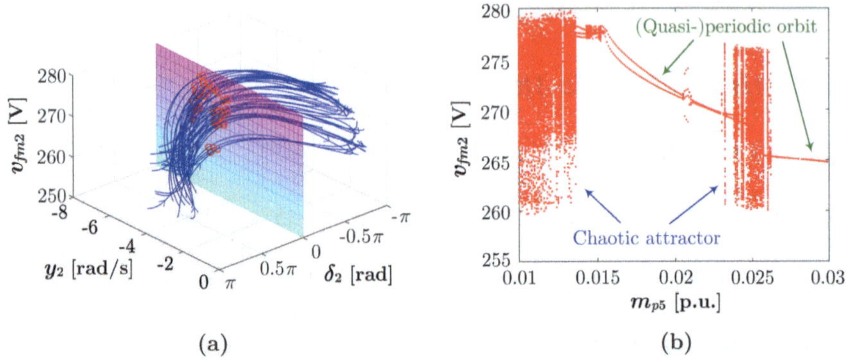

Fig. 7.7 (**a**) Hidden chaotic attractor in (δ_2, y_2, v_{fm2})-space when $m_{p5} = 0.01$ p.u., where $\delta_2 = 0$ is the selected 2-dim cross section. (**b**) Codimension-1 bifurcation diagram on m_{p5}-v_{fm2} plane. Here, v_{fm2} is the value of the intersection point of the attractor trajectory and the selected 2-dim cross section

$$\frac{\mathrm{d}\mathbf{u_o}}{\mathrm{d}t} = \mathbf{F}(\mathbf{u_o}) \tag{7.9}$$

where $\mathbf{u_o} = \left[v_{fm1}\ \delta_2\ y_2\ v_{fm2} \cdots \delta_N\ y_N\ v_{fmN}\right]^T \triangleq \left[u_{01}\ u_{02} \cdots u_{0M}\right]^T \in \mathbb{R}^M$, and

$$\mathbf{F}(\mathbf{u_o}) = \left[f_1(\mathbf{u_o})\ f_2(\mathbf{u_o}) \cdots f_M(\mathbf{u_o})\right]^T \in \mathbb{R}^M.$$

The variational equations are [89, 129]

$$\frac{\mathrm{d}\mathbf{E_d}}{\mathrm{d}t} = D\mathbf{F}(\mathbf{u_o})\mathbf{E_d} \tag{7.10}$$

where $\mathbf{E_d} = \left[\boldsymbol{\varepsilon}_{\mathbf{d1}}\ \boldsymbol{\varepsilon}_{\mathbf{d2}} \cdots \boldsymbol{\varepsilon}_{\mathbf{dM}}\right] \in \mathbb{R}^M \times \mathbb{R}^M$ measures the growth of infinitesimal displacement in M different orthogonal directions, and $D\mathbf{F}(\mathbf{u_o})$ is the matrix of first partial derivates of $\mathbf{F}(\mathbf{u_o})$, i.e.,

$$DF(\mathbf{u_o}) = \left[\frac{\partial f_\sigma}{\partial u_{ok}}\right]\Bigg|_{\sigma,k=1,2,\cdots,M} \in \mathbb{R}^M \times \mathbb{R}^M.$$

Thus, $\mathbf{E_d}$ can be found by combining (7.9) and (7.10) using the Runge–Kutta numerical method.

Here, the initial condition of $\mathbf{E_d}$ is set as $\mathbf{E_d}|_{t=0} = \mathbf{I}$, where $\mathbf{I}$ is the identity matrix. After each numerical step, the Gram–Schmidt orthogonalization technique is applied to make each $\boldsymbol{\varepsilon}_{\mathbf{d}\sigma}$ orthogonal with others. For example, at $t = t_f$, $\mathbf{w}_\sigma \triangleq \boldsymbol{\varepsilon}_{\mathbf{d}\sigma}|_{t=t_f^-}$, where $\boldsymbol{\varepsilon}_{\mathbf{d}\sigma}|_{t=t_f^-}$ is derived from (7.9) and (7.10) in the last round. Then,

$$\mathbf{v}_\sigma = \mathbf{w}_\sigma - \sum_{k=1}^{\sigma-1}\left(\frac{\mathbf{w}_\sigma \cdot \mathbf{v}_k}{\|\mathbf{v}_k\|^2}\right)\mathbf{v}_k \tag{7.11}$$

where $\mathbf{w}_\sigma \cdot \mathbf{v}_k$ represents the scalar product of $\mathbf{w}_\sigma$ and $\mathbf{v}_k$ [89]. Hence, $\boldsymbol{\varepsilon}_{\mathbf{d}\sigma}|_{t=t_f^+} = \mathbf{v}_\sigma$ becomes the initial condition of $\mathbf{E_d}$ to begin the new round of numerical calculation of (7.9) and (7.10).

After calculation time T_{L}, we get

$$\mathbf{E_d}|_{t=T_{\mathrm{L}}} = \mathbf{Q_d}\mathbf{R_d} \tag{7.12}$$

using the QR decomposition technique. Here, $\mathbf{Q_d}$ is an orthogonal matrix, and $\mathbf{R_d}$ is an upper triangular matrix. The σth row σth column diagonal element of $\mathbf{R_d}$ is denoted as γ_σ [129]. The σth largest Lyapunov exponent is [89, 129]

$$\lambda_{\mathrm{L}\sigma} = \frac{\ln|\gamma_\sigma|}{T_{\mathrm{L}}}. \tag{7.13}$$

Furthermore, in order to reduce the accumulated error during the numerical calculation, the calculation time T_{L} is divided to N_{L} equal intervals. After each interval, γ_σ is recorded as γ_σ^ζ (where $\zeta = 1, 2, \cdots, N_{\mathrm{L}}$), and $\mathbf{E_d}$ is reassigned as $\mathbf{I}$ to begin the new interval of numerical calculation. Then, the σth largest Lyapunov exponent is updated as

$$\lambda_{\mathrm{L}\sigma} = \frac{1}{T_{\mathrm{L}}}\sum_{\zeta=1}^{N_{\mathrm{L}}} \ln|\gamma_\sigma^\zeta|. \tag{7.14}$$

Here, we set $T_{\mathrm{L}} = 1500\,\mathrm{s}$ and $N_{\mathrm{L}} = 5$. The numerical results of $\lambda_{\mathrm{L}\sigma}$ with varying m_{p5} are shown in Fig. 7.8, with $\lambda_{\mathrm{L}1}$ being the largest Lyapunov exponent. Therefore, $\lambda_{\mathrm{L}1} > 0$ implies chaotic behavior, whereas $\lambda_{\mathrm{L}1} \approx 0$ implies quasi-periodic or periodic behavior. From Fig. 7.8, $\lambda_{\mathrm{L}1} = 0.2191$ for $m_{p5} = 0.01$ p.u., while $\lambda_{\mathrm{L}1} = 0.5022$ for $m_{p5} = 0.025$ p.u., proving quantitatively that the attractors shown in Figs. 7.5, 7.6, and 7.7 are chaotic attractors. Besides, when $m_{p5} = 0.0117 \sim 0.0136$ p.u., there are two positive Lyapunov exponents (i.e.,

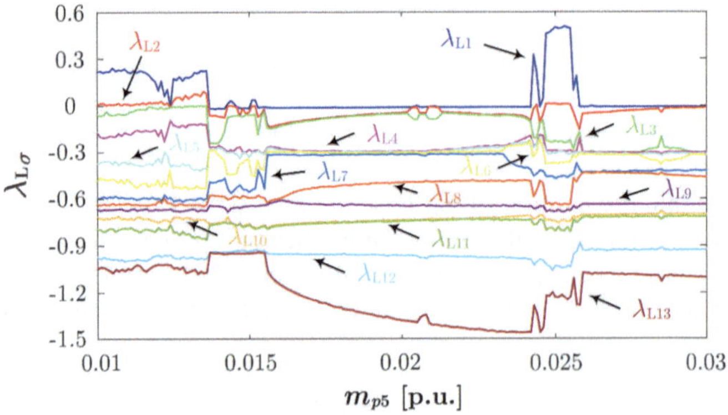

Fig. 7.8 Lyapunov exponents $\lambda_{L\sigma}$ under varying m_{p5}

Table 7.3 Lyapunov exponents and Lyapunov dimension

m_{p5}	λ_{L1}	λ_{L2}	λ_{L3}	D_L	Attractor
0.01 p.u.	0.2191	≈ 0	−0.0558	3.9926	Chaos
0.0117 p.u.	0.1973	0.0240	≈ 0	4.1597	Hyperchaos
0.0135 p.u.	0.2230	0.0945	≈ 0	4.4978	Hyperchaos
0.025 p.u.	0.5022	≈ 0	−0.2259	3.9966	Chaos

$\lambda_{L1} > 0$, and $\lambda_{L2} > 0$), implying the existence of hyperchaotic attractors [124], as shown in Table 7.3. It is noted that the general chaotic attractors have at least one positive Lyapunov exponent, while the hyperchaotic attractors have at least two positive Lyapunov exponents, showing that the hyperchaotic attractors have multiple expanding directions. Moreover, to reflect the fractal geometry of these hidden chaotic attractors, the Lyapunov dimension (i.e., fractal dimension) can be derived [130], i.e.,

$$D_L = \sigma_m + \frac{1}{|\lambda_{L(\sigma_m+1)}|} \sum_{k=1}^{\sigma_m} \lambda_{Lk} \tag{7.15}$$

where σ_m is the maximum integer with $\lambda_{L1} + \lambda_{L2} + \cdots + \lambda_{L\sigma_m} \geqslant 0$. From Table 7.3, $D_L \approx 4$ for $m_{p5} = 0.01$ p.u. and 0.025 p.u., while $D_L > 4$ for $m_{p5} = 0.0117$ p.u. and 0.0135 p.u., showing that the hidden hyperchaotic attractors may have higher fractal dimension.

7.2.3 Frequency Spectrum

For the default parameters, there are four invisible attractors, namely, periodic orbit (with oscillating δ_5), hidden quasi-periodic orbit (with oscillating δ_3 and δ_5),

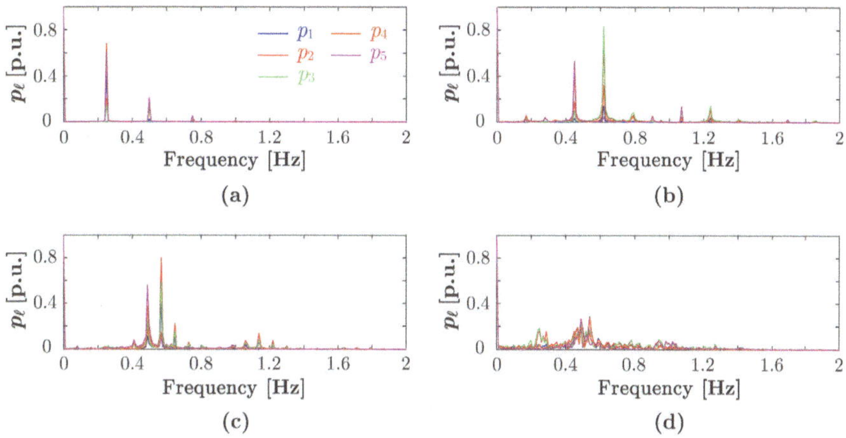

Fig. 7.9 Frequency spectra of sub-synchronous oscillations, which correspond to (**a**) periodic orbit (with oscillating δ_5), (**b**) hidden quasi-periodic orbit (with oscillating δ_3 and δ_5), (**c**) hidden quasi-periodic orbit (with oscillating δ_2 and δ_5), and (**d**) hidden chaotic attractor (with oscillating δ_2, δ_3, δ_4, and δ_5). Here, the p.u. value of p_ℓ is calculated by taking the DC component of the corresponding p_ℓ as the basis

hidden quasi-periodic orbit (with oscillating δ_2 and δ_5), and hidden chaotic attractor (with oscillating δ_2, δ_3, δ_4, and δ_5). Here, the oscillating δ_ℓ means that δ_ℓ varies from $-\pi$ to π. These attractors manifest themselves as sustained sub-synchronous oscillations, and their frequency spectra are shown in Fig. 7.9. For the periodic orbit, the most significant peak appears at 0.25 Hz. For the hidden quasi-periodic orbit (with oscillating δ_3 and δ_5), the most significant peaks appear at 0.45 Hz and 0.62 Hz. For the hidden quasi-periodic orbit (with oscillating δ_2 and δ_5), the most significant peaks appear at 0.49 Hz and 0.57 Hz. For the hidden chaotic attractor, the magnitude is small, and the spectrum is relatively flat. From Fig. 7.9, the frequency range of the aforementioned four attractors is $0 \sim 2$ Hz.

Besides, when V_0 increases, the frequency range may be enlarged, as shown in Fig. 7.10, where the frequency range is $0 \sim 4$ Hz for $V_0 = 2$ p.u., $0 \sim 7$ Hz for $V_0 = 3$ p.u., $0 \sim 10$ Hz for $V_0 = 4$ p.u., and $0 \sim 14$ Hz for $V_0 = 5$ p.u. Similarly, it is found that the frequency range may be broadened if $m_{p\ell}$ or $\omega_{p\ell}$ is uniformly increased, or $n_{q\ell}$ is uniformly decreased.

7.3 Basins of Attraction

As will be shown, several invisible attractors may coexist with the SEP. An example is given in Fig. 7.11 for $m_{p5} = 0.01$ p.u. The attractors are projected in (y_2, y_3, y_5)-space. As shown in Fig. 7.11, the invisible attractors are all far from the SEP, and none of them surrounds the SEP. Moreover, each of them occupies a specific

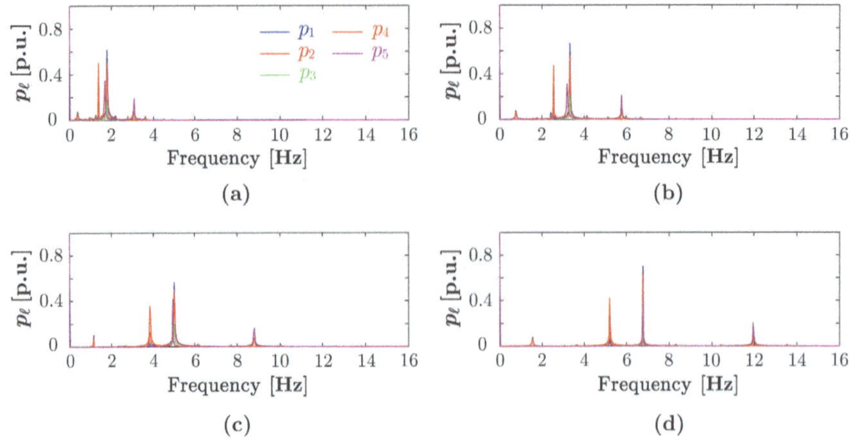

Fig. 7.10 Frequency spectra of sub-synchronous oscillations, which correspond to (**a**) $V_0 = 2$ p.u., (**b**) $V_0 = 3$ p.u., (**c**) $V_0 = 4$ p.u., and (**d**) $V_0 = 5$ p.u. Here, the p.u. value of p_ℓ is calculated by taking the DC component of the corresponding p_ℓ as the basis. The initial conditions are $\delta_\ell = 0$, $y_2 = -20$, $y_3 = -20$, $y_\ell = 0$ for $\ell \neq 2, 3$, and $v_{fm\ell} = V_0$

Fig. 7.11 SEP and several invisible attractors in (y_2, y_3, y_5)-space. Among them, two quasi-periodic orbits and the chaotic attractor belong to the hidden attractor. All parameters adopt their default values

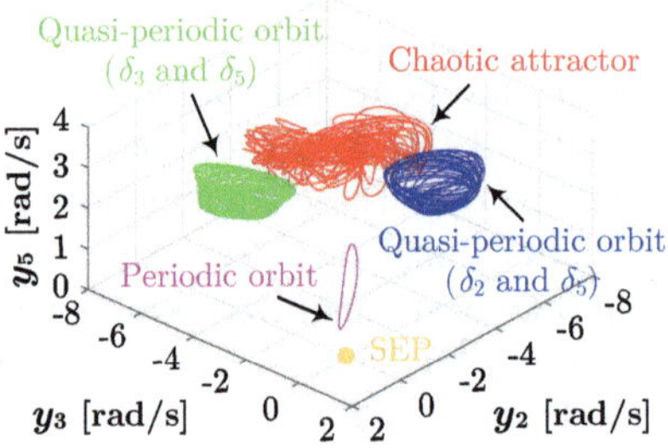

space. The whole phase space is separated into several parts. Each part is a basin of attraction of a specific attractor. For ease of observation and analysis, the basins of attraction are projected onto the (y_2, y_3)-plane and (y_4, y_5)-plane, as shown in Fig. 7.12.

First, the changes of invisible attractors under varying m_{p5} are studied. As shown in Fig. 7.12a, b for $m_{p5} = 0.01$ p.u., there are five attractors, including the SEP, a periodic orbit, two hidden quasi-periodic orbits, and a hidden chaotic attractor. However, as m_{p5} is increased, all these four types of invisible attractors may subside and be replaced. For example, when $m_{p5} = 0.02$ p.u., the SEP coexists with three hidden attractors, namely, hidden periodic orbit (with oscillating δ_2), hidden periodic orbit (with oscillating δ_3), and hidden quasi-periodic orbit (with oscillating δ_2 and δ_3), as shown in Fig. 7.12c, d. It is noted that for $m_{p5} = 0.01$ p.u., the attractors contain the oscillating δ_5, while for $m_{p5} = 0.02$ p.u., the attractors do not contain the oscillating δ_5. Furthermore, when $m_{p5} = 0.025$ p.u., the hidden quasi-periodic orbit becomes a hidden chaotic attractor with the same oscillating angles (i.e., δ_2 and δ_3), and other attractors remain unchanged, as shown in Fig. 7.12e, f.

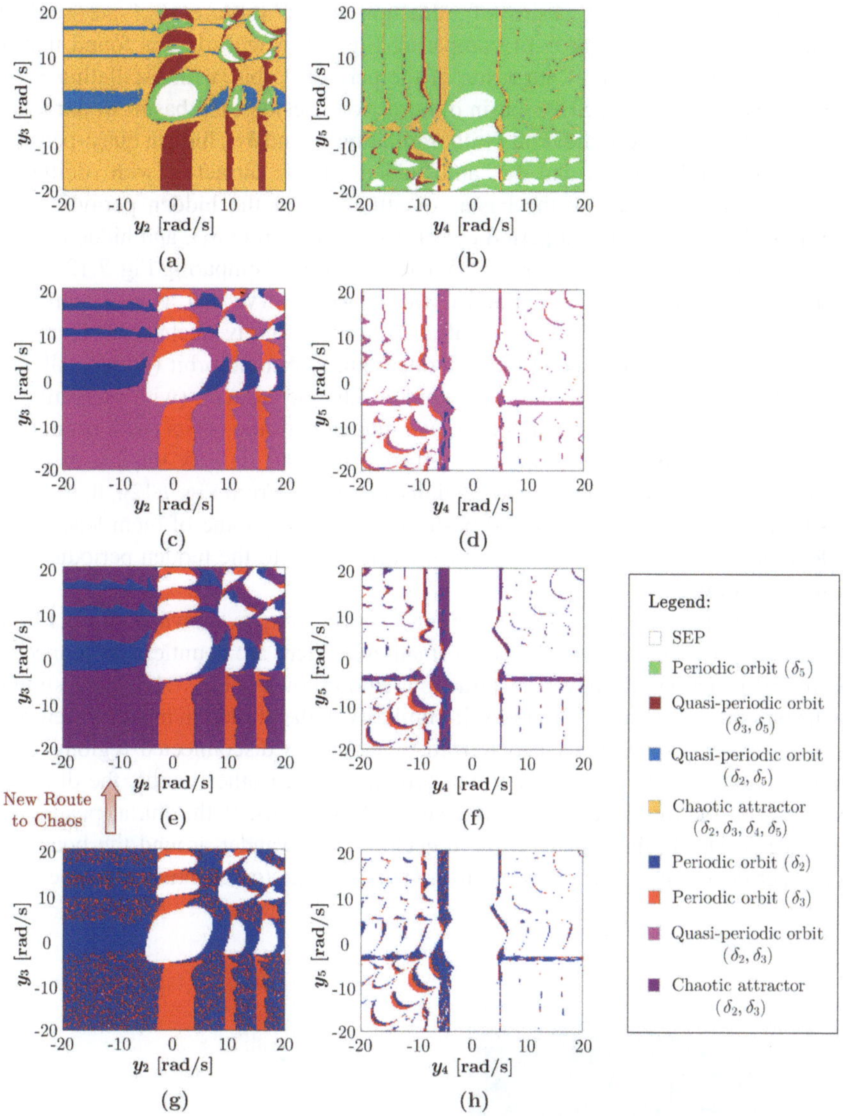

Fig. 7.12 Basins of attraction projected onto **(a)** (y_2, y_3)-plane and **(b)** (y_4, y_5)-plane for $m_{p5} = 0.01$ p.u.; basins of attraction projected onto **(c)** (y_2, y_3)-plane and **(d)** (y_4, y_5)-plane for $m_{p5} = 0.02$ p.u.; basins of attraction projected onto **(e)** (y_2, y_3)-plane and **(f)** (y_4, y_5)-plane for $m_{p5} = 0.025$ p.u.; and basins of attraction projected onto **(g)** (y_2, y_3)-plane and **(h)** (y_4, y_5)-plane for $m_{p5} = 0.03$ p.u.

Here, the chaotic attractor emerges via the quasi-periodicity [5]. Then, when $m_{p5} = 0.03$ p.u., the hidden chaotic attractor is disrupted, and only the SEP and two hidden periodic orbits exist, as shown in Fig. 7.12g, h.

Next, the changes of basins of attraction under varying m_{p5} are studied. Comparing Fig. 7.12a with c, or comparing Fig. 7.12b with d, it is found that as m_{p5} increases, the basin of attraction of the periodic orbit (with oscillating δ_5) is almost absorbed into the SEP's basin of attraction. Besides, the basins of attraction of the hidden quasi-periodic orbit (with oscillating δ_2 and δ_5), hidden quasi-periodic orbit (with oscillating δ_3 and δ_5), and hidden chaotic attractor (with oscillating δ_2, δ_3, δ_4 and δ_5) become the basins of attraction of the hidden periodic orbit (with oscillating δ_2), hidden periodic orbit (with oscillating δ_3), and hidden quasi-periodic orbit (with oscillating δ_2 and δ_3), respectively. Comparing Fig. 7.12c and e with g, or comparing Fig. 7.12d and f with h, we observe that as m_{p5} increases, the location and shape of all basins of attraction are nearly unchanged. Here, the basin of attraction may belong to the hidden quasi-periodic orbit (with oscillating δ_2 and δ_3) when $m_{p5} = 0.02$ p.u., or the hidden chaotic attractor (with oscillating δ_2 and δ_3) when $m_{p5} = 0.025$ p.u., or the two hidden periodic orbits (with oscillating δ_2 and δ_3 respectively) when $m_{p5} = 0.03$ p.u. In particular, for the last case, the attraction area is completely chaotic and disconnected. From Fig. 7.12g, h, this area is split into countless disconnected basins of attraction. Some of them lead to the hidden periodic orbit with oscillating δ_2, while others to the hidden periodic orbit with oscillating δ_3.

Furthermore, we enlarge Fig. 7.12g for a closer look, as shown in Fig. 7.13a. It is surprising to see that even in such a small area, there are countless disconnected basins of attraction. Moreover, an attraction region shown in Fig. 7.13a looks similar to the one shown in Fig. 7.12g. The self-similarity of the attraction regions resembles *fractal geometry,* which is characterized by countless disconnected regions, each upon enlargement being similar to the original one. In other words, the different basins of attraction have fractal boundaries [130]. Thus, if the initial position is near the *fractal* basin boundary, the trajectory may wander around the boundary for a period of time and converge to a specific attractor eventually. In the time

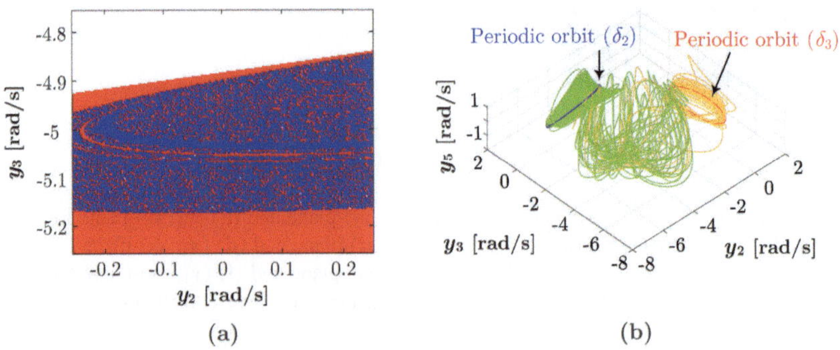

Fig. 7.13 (**a**) Enlargement of Fig. 7.12g. (**b**) Two different hidden periodic orbits (blue and red solid curves) and two typical trajectories (green and yellow solid curves) for $m_{p5} = 0.03$ p.u. Here, $y_3(0) = -4.96$ for the green trajectory, while $y_3(0) = -4.95$ for the yellow trajectory. For both trajectories, $\delta_\ell(0) = 0$, $y_\ell(0) = 0$ for $\ell \neq 3$, and $v_{fm\ell}(0) = V_0$

domain, the trajectory exhibits transient chaos before it settles to an attractor. On this basis, if we adjust the initial position slightly, a different final attractor results, e.g., the hidden periodic orbit with oscillating δ_2 for the previous case, but the hidden periodic orbit with oscillating δ_3 for this case. This phenomenon is termed *final-state sensitivity* [130]. An example is shown in Fig. 7.13b. The initial conditions of the two trajectories are almost the same, except $y_3(0) = -4.96$ for the green trajectory and $y_3(0) = -4.95$ for the yellow trajectory. However, the final states of these trajectories are different, and the transient is chaotic motion, especially for the green trajectory.

It is worth noting that the final-state sensitivity may exist for normal basin boundaries. Therefore, such a phenomenon can also be observed when $m_{p5} =$ 0.01 p.u. or 0.02 p.u. or 0.025 p.u. However, as the fractal characteristic of the basin boundaries is not obvious, it may be hard to observe transient chaos in these cases. On the other hand, when adjusting m_{p5} from 0.03 to 0.025 p.u., the hidden chaotic attractor emerges due to the fractal basin boundaries, i.e., the trajectory wanders around the basin boundaries permanently, evolving into a hidden chaotic attractor.

7.4 Transient Stability Enhancement

As multiple attractors exist in the phase space, the system originally operating at the SEP may be pushed to one of the invisible attractors under a large transient disturbance. In this case, sub-synchronous oscillation will emerge, and the GFMCs will easily desynchronize. Hence, this transient instability should be avoided. Basically, there are three ways to enhance transient synchronization stability.

First, we can adjust the control or circuit parameters to trigger a homoclinic bifurcation to eliminate the Shil'nikov-type invisible attractors (and, meanwhile, may trigger other bifurcations to remove the hidden attractors). In Chap. 3, it has been found that the 2-dim/3-dim periodic orbit (which is a Shil'nikov-type attractor) may pass the UEP and disappear at the onset of homoclinic bifurcation. Hence, for the islanded microgrid, the Shil'nikov-type attractors may also be close to the UEP and subside when a homoclinic bifurcation occurs. In this section, we will consider adjusting $m_{p\ell}$ and/or $\omega_{p\ell}$ to eliminate the attractors.

Second, we may try to avoid any significantly large transient disturbance to ensure that the system does not escape the SEP's basin of attraction. For the islanded microgrid, the disturbance may be a transient SCF, a phase disturbance, a large load change, or a malfunction of some GFMCs. We will explore the behavior of the islanded microgrid when one GFMC is permanently or temporarily shut down in this section.

Third, if the sub-synchronous oscillations have already emerged under a transient disturbance, we may trip off one or more specific GFMCs to eliminate such oscillations. According to Sect. 7.2, for Type-I (hidden) periodic orbit, only one angle δ_ℓ varies from $-\pi$ to π, implying that the sub-synchronous oscillations mainly arise from the serious interaction between GFMC-ℓ and the microgrid. Besides, for other hidden attractors, more angles may vary from $-\pi$ to π, implying that the oscillations are generated from the interaction between the corresponding GFMCs and the microgrid. Therefore, if these GFMCs are tripped off, the sub-synchronous oscillations will subside. This method will be studied in Sect. 7.5.

7.4.1 Homoclinic Bifurcation

7.4.1.1 Uniform Adjustment

As $m_{p\ell}$ uniformly decreases, the invisible attractors may change their internal structures, and all subside eventually, as shown in Fig. 7.14. From Figs. 7.11 and 7.14, as $m_{p\ell}$ decreases, the hidden quasi-periodic orbits and hidden chaotic attractor may become the Type-II hidden periodic orbit, approach the periodic orbit (with oscillating δ_5), and disappear at last. Besides, as $m_{p\ell}$ decreases, the periodic orbit (with oscillating δ_5) approaches the UEP and may become a collection of many nearby and dense circles, as shown in Fig. 7.14f. Then, if we keep decreasing $m_{p\ell}$, the aforementioned attractor will subside eventually, which is the onset result of a homoclinic bifurcation (see Chaps. 3 and 4).

Specifically, the UEP is a saddle, which has only one positive real eigenvalue (i.e., the real part of other eigenvalues is negative). Therefore, the UEP has two unstable manifolds, corresponding to the aforementioned positive eigenvalue [54]. For the unstable manifold, its initial condition is the UEP, while its final state is the SEP, or some other specific attractor, or the infinity. As shown in Fig. 7.15, one unstable manifold converges to the periodic orbit (with oscillating δ_5), and the other converges to the SEP. This also confirms that the aforementioned periodic orbit generates or subsides via a homoclinic bifurcation. The mechanism in detail can be found in Chap. 3. Therefore, this periodic orbit belongs to the Shil'nikov-type invisible attractor, as indicated in Sect. 1.4.3.3. It is also noteworthy that for other invisible attractors, they do not generate from the interaction of the UEP's manifolds. In other words, they do not follow Shil'nikov's theorem [69, 131], and they may generate and vanish through other types of bifurcations (other than Hopf bifurcation). Thus, they belong to the hidden attractor.

On this basis, we can derive the boundaries separating the SEP regime and multi-attractor regime in the parameter space. Here, $m_{p\ell}$ and $\omega_{p\ell}$ are uniformly varied, i.e., $0.5\,m_{p1} = 0.25\,m_{p2} = 0.25\,m_{p3} = 0.5\,m_{p4} = m_{p5}$, and $\omega_{p1} = \omega_{p2} = \cdots = \omega_{p5}$. For every $(\omega_{p\ell}, m_{p\ell})$, various initial conditions are used to test whether the trajectories converge to the SEP. The homoclinic bifurcation may occur, and no invisible attractor may exist when all trajectories converge to the SEP. The initial conditions are set as $\delta_\ell(0) = 0$, $y_\ell(0) \in [-20\ -10\ 0\ 10\ 20]$, and $v_{fm\ell}(0) = V_0$. Then, a bifurcation diagram displayed on $\omega_{p\ell}$-m_{p5} plane can be derived, as given in Fig. 7.16. Here, the SEP coexists with the invisible attractors above the solid curve, and only the SEP exists below the solid curve. From Fig. 7.16, we can uniformly

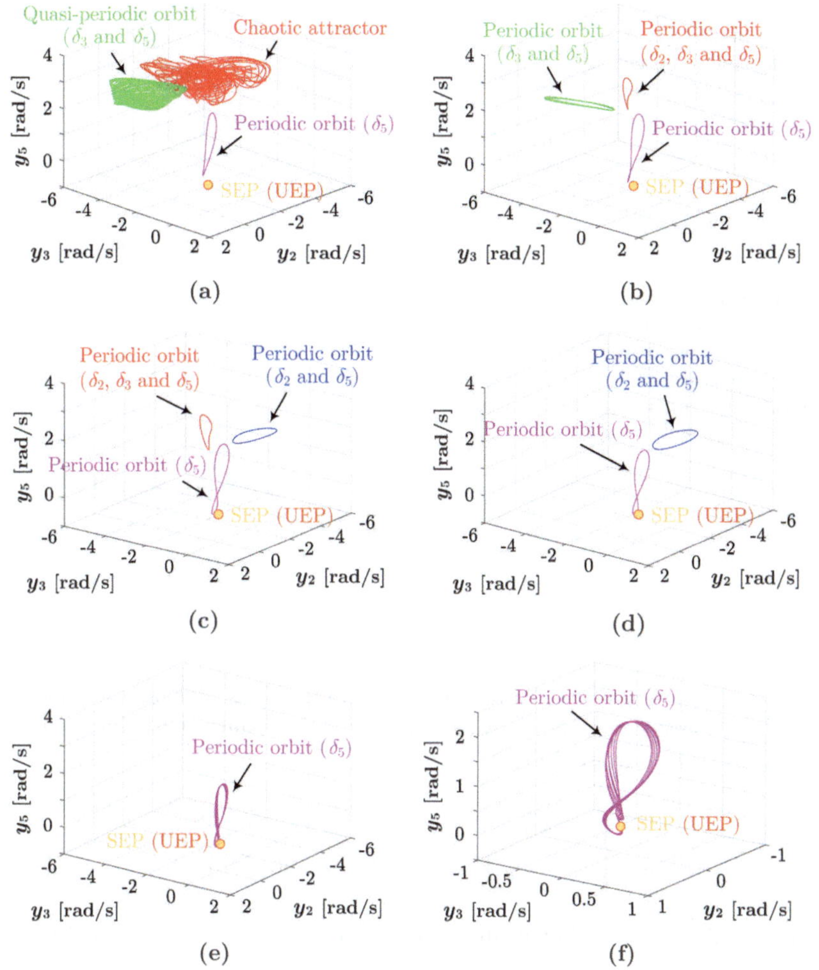

Fig. 7.14 SEP, UEP, and several hidden attractors in (y_2, y_3, y_5)-space for **(a)** $m_{p5} = 0.009$ p.u., **(b)** $m_{p5} = 0.008$ p.u., **(c)** $m_{p5} = 0.007$ p.u., **(d)** $m_{p5} = 0.00688$ p.u., and **(e, f)** $m_{p5} = 0.0067$ p.u. Other parameters adopt their default values with $0.5\,m_{p1} = 0.25\,m_{p2} = 0.25\,m_{p3} = 0.5\,m_{p4} = m_{p5}$

decrease $m_{p\ell}$, and/or increase $\omega_{p\ell}$ to eliminate all invisible attractors to enhance the system transient stability.

7.4.1.2 Individual Adjustment

From (7.5), we can derive the variable values of SEP, which are denoted as $\delta_{2,\text{SEP}}$, $\cdots$, $\delta_{N,\text{SEP}}$, $y_{2,\text{SEP}}$, $\cdots$, $y_{N,\text{SEP}}$, $v_{fm1,\text{SEP}}$, $v_{fm2,\text{SEP}}$, $\cdots$, $v_{fmN,\text{SEP}}$. In particular,

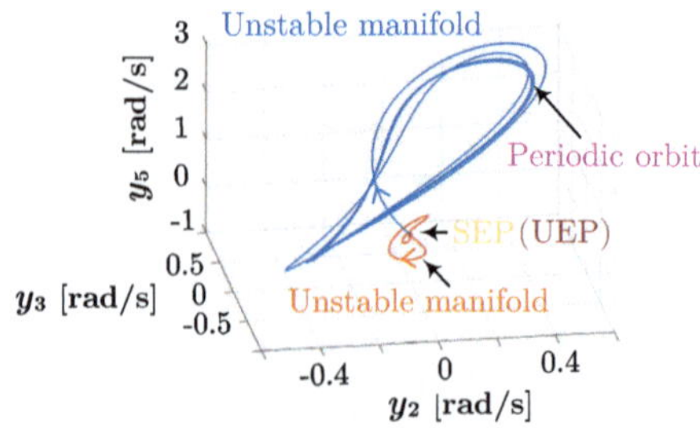

Fig. 7.15 Unstable manifolds of the UEP. All parameters follow the default set

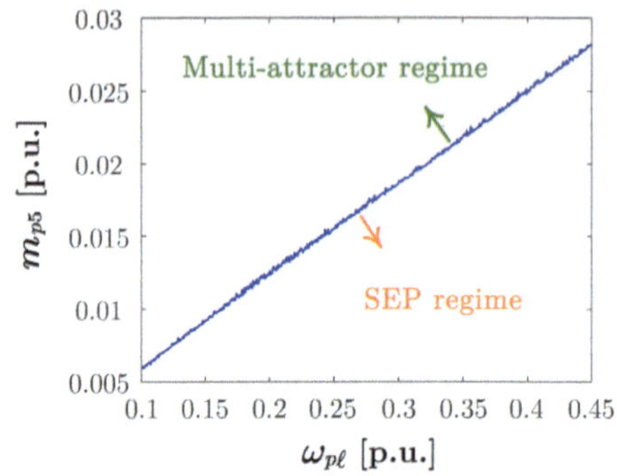

Fig. 7.16 Codimension-2 bifurcation diagram of (7.5) on the $\omega_{p\ell}$-m_{p5} plane, where $0.5\,m_{p1} = 0.25\,m_{p2} = 0.25\,m_{p3} = 0.5\,m_{p4} = m_{p5}$, and $\omega_{p1} = \omega_{p2} = \cdots = \omega_{p5}$

$y_{2,\mathrm{SEP}} = \cdots = y_{N,\mathrm{SEP}} = 0$. At first, we neglect the voltage dynamics, divide (7.5) into $(N-1)$ subsystems, and study the behavior of these subsystems separately. The subsystems derived from (7.5) are

$$\begin{cases} \dot{\delta}_\ell = y_\ell \\ \dot{y}_\ell = \omega_{p1}\Big(- y_\ell + m_{p1} p_1 - m_{p\ell} p_\ell \Big) \end{cases} \tag{7.16}$$

where $\ell = 2, 3, \cdots, N$. In the ℓth subsystem, $\delta_k = \delta_{k,\mathrm{SEP}}$ for $k \neq \ell$, and $v_{fmk} = v_{fmk,\mathrm{SEP}}$.

From (7.16), we find that for some $m_{p\ell}$, an SEP and a periodic orbit coexist (i.e., before the onset of homoclinic bifurcation), while for other $m_{p\ell}$, only an SEP exists (i.e., after the onset of homoclinic bifurcation). From Fig. 7.17a, when $m_{p2} = 0.0076 \backsim 0.0244$ p.u., or $m_{p3} = 0.0054 \backsim 0.0203$ p.u., or $m_{p4} = 0.0061 \backsim 0.0185$ p.u., or $m_{p5} = 0.0085 \backsim 0.0254$ p.u., there is only one attractor in the phase space, i.e., the SEP. In this case, δ_ℓ cannot oscillate from $-\pi$ to π sustainably in the corresponding subsystem (7.16). Therefore, if all $m_{p\ell}$ is in the above range, sustained oscillation of all δ_ℓ in the whole system (7.5) can be avoided. Thus, (7.16) offers a simple method to find a suitable value of $m_{p\ell}$ to enhance the transient stability.

Furthermore, we can obtain more precise results directly from (7.5), as shown in Fig. 7.17b, c. For Fig. 7.17b, $m_{p2} = m_{p3}$, $m_{p4} = 0.015$ p.u., and m_{p2} (m_{p3}) and m_{p5} are adjusted. For Fig. 7.17c, $m_{p2} = m_{p3} = 0.015$ p.u., and m_{p4} and m_{p5} are adjusted. In Fig. 7.17b, c, the blue region corresponds to the SEP regime, while the

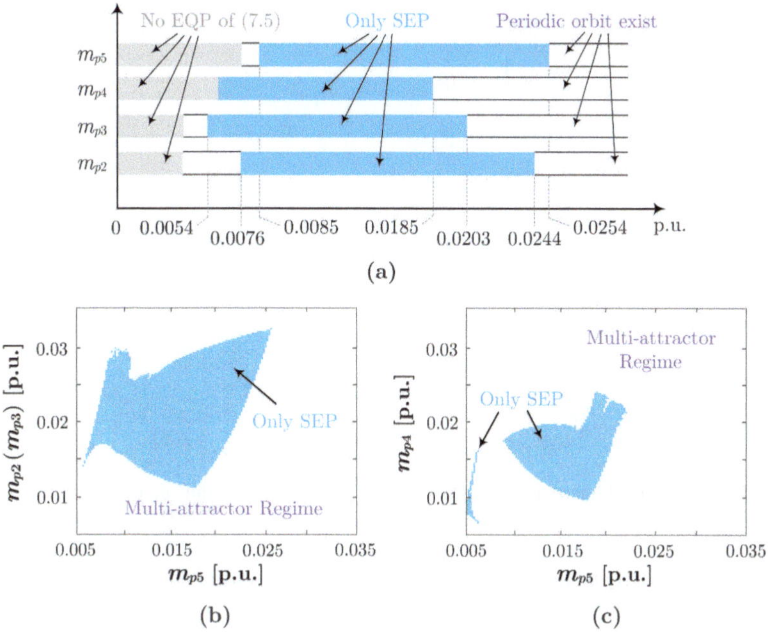

Fig. 7.17 (**a**) Codimension-1 bifurcation diagram of (7.16) for varying $m_{p\ell}$. (**b**) Codimension-2 bifurcation diagram of (7.5) on the m_{p5}-m_{p2} plane, where $m_{p2} = m_{p3}$ and $m_{p4} = 0.015$ p.u. (**c**) Codimension-2 bifurcation diagram of (7.5) on the m_{p5}-m_{p4} plane, where $m_{p2} = m_{p3} = 0.015$ p.u.

white region corresponds to the multi-attractor regime. The variation range of $m_{p\ell}$ of the blue regions in Fig. 7.17b, c basically coincides with the corresponding one in Fig. 7.17a. From the above analysis, the system transient stability can be enhanced by individually adjusting $m_{p\ell}$. It is worth noting that $m_{p\ell}$ is relatively large, and $\omega_{p\ell}$ is relatively small here. This also helps to improve the regulation speed of the power loops and increase the virtual inertia of the islanded microgrid.

7.4.2 Effects of Faulty Grid-Forming Converters

The malfunctioning of GFMCs can be permanent or temporary. For a permanent malfunction, the GFMC is shut down and cannot restart, for example, due to a faulty semiconductor power device. For a temporary malfunction, the GFMC can be restarted and reconnected to the microgrid after removing the fault, e.g., in overvoltage, overcurrent, or overheat protection/alarm. In practice, in order to restart the GFMCs reliably and safely, the manufacturers or operators may set a relatively long time delay to restart the GFMCs even when the temporary fault is removed [132]. If the GFMC is restarted too soon, the same protection may be

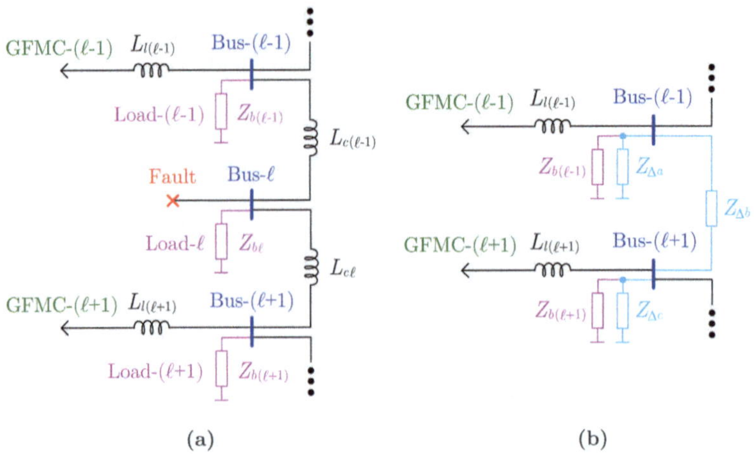

Fig. 7.18 (**a**) Islanded microgrid with faulty GFMC-ℓ and (**b**) the equivalent circuit of (**a**)

triggered repeatedly as the circumstance is unchanged, which may lead to permanent shutdown of the GFMC.

Here, it is assumed that GFMC-ℓ is protectively shut down. With a decreasing number of operating GFMCs, the topology of the islanded microgrid is changed, and the equivalent circuit is shown in Fig. 7.18. Applying the Y-Δ transformation gives

$$Z_{\Delta a} = \frac{Z_{Y\Delta}}{Z_{c\ell}}, \quad Z_{\Delta b} = \frac{Z_{Y\Delta}}{Z_{b\ell}}, \quad Z_{\Delta c} = \frac{Z_{Y\Delta}}{Z_{c(\ell-1)}}, \tag{7.17}$$

where $Z_{Y\Delta} = Z_{c(\ell-1)}Z_{c\ell} + Z_{c(\ell-1)}Z_{b\ell} + Z_{c\ell}Z_{b\ell}$. According to Sect. 7.1, the microgrid model with a faulty GFMC-ℓ can be derived. In this subsection, a 30-bus microgrid is studied.

7.4.2.1 Permanent Fault

Using the above model, the effect of a permanently faulted GFMC is studied. For ease of analysis, the loads are set as $R_{b\ell} = K_{ld} \times$ (Default Value), and $L_{b\ell} = K_{ld} \times$ (Default Value), where $\ell = 1, 2, \cdots, N$. For example, $R_{b1} = 6K_{ld}\ \Omega$, and $L_{b1} = 3K_{ld}$ mH.

For each load level K_{ld}, after the microgrid enters the steady state, a specific GFMC is shut down permanently, and then all other GFMCs either still operate normally or exhibit a sustained oscillation. In Fig. 7.19a, a yellow (red) square corresponds to normal operation (sustained oscillation) of all but the faulted GFMC. Specifically, the faulted GFMC may have a relatively large impact on the microgrid. For example, for $K_{ld} = 1.1$, the malfunctioned GFMC-5 may lead to sustained

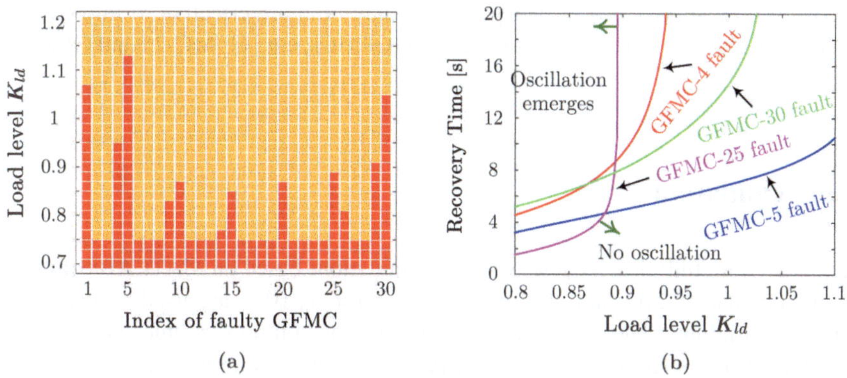

Fig. 7.19 (**a**) Stability status after a faulty GFMC is permanently removed. Yellow (red) square indicates post-fault operation being normal (oscillatory) for all GFMCs other than the faulty one. (**b**) Operation boundaries for temporarily faulted GFMC-4 (red solid curve), GFMC-5 (blue solid curve), GFMC-25 (purple solid curve), and GFMC-30 (green solid curve) for specific recovery time and load condition. $R_{b\ell} = K_{ld} \times$ (Default Value), and $L_{b\ell} = K_{ld} \times$ (Default Value), where $\ell = 1, 2, \cdots, N$

oscillation of the rest of the system. However, no sustained oscillation emerges if any one of other GFMCs is shut down permanently (yellow boxes). Furthermore, for permanently faulted GFMC-5, the rest of the system may oscillate if $K_{ld} \leqslant 1.12$; and for permanently faulted GFMC-4, the rest of the system may oscillate if $K_{ld} \leqslant 0.94$. In other words, protecting the critical GFMCs (e.g., GFMC-5) would enhance system transient stability.

7.4.2.2 Temporary Fault

The islanded microgrid with a temporarily faulted GFMC-ℓ is shown in Fig. 7.18. Here, we investigate the influence of temporarily faulted GFMCs on the microgrid. In Fig. 7.19b, four curves, corresponding to faulted GFMC-4, GFMC-5, GFMC-25, and GFMC-30, represent boundaries of normal operation. For each case, the region below the curve corresponds to normal operation after the GFMC restarts. Above the curve, moreover, the sustained oscillation emerges after the GFMC restarts. From Fig. 7.19b, a long recovery time may lead to sustained oscillation. Besides, for light loads (i.e., large K_{ld}), the recovery time may become insignificant. For example, when $K_{ld} \geqslant 0.9$, the system remains stable regardless of the recovery time of GFMC-25. Such a finding coincides with Fig. 7.19a. However, for heavy loads (i.e., small K_{ld}), the system can be easily destabilized due to a temporarily faulted GFMC-25, as the tolerable recovery time of GFMC-25 is much shorter than other GFMCs.

In summary, we can enhance the system transient stability by accelerating the restarting process of the GFMC, especially for light loads. However, as the tolerable recovery time shortens for a heavy load, it may be more effective to improve the transient stability by protecting some critical GFMCs, e.g., GFMC-25 and GFMC-5 here.

7.5 Validation

In this section, full-circuit cycle-by-cycle simulations based on MATLAB/Simulink with step size of 2 μs and absolute/relative error of 10^{-5} are used to verify the analytical findings. The switching behavior and the control dynamics are properly considered in the simulations. Default parameters given in Tables 7.1 and 7.2 are adopted unless otherwise specified.

Specifically, there are two different transient SCFs denoted as SCF1 and SCF2. For SCF1, the fault occurs in Bus-ℓ. As SCF1 is far from the GFMCs, the fault current may be relatively small and the fault-clearing time may be relatively long. For SCF2, the fault occurs in the output of GFMC-ℓ, and thus, the fault-clearing time may be relatively short. In practice, the secondary control and communication network may be adopted for the microgrid. In this case, a cyberattack may perturb ω_0, thereby perturbing θ_ℓ and δ_ℓ [13, 133]. Therefore, we also focus on the possible influence of the phase disturbance on the islanded microgrid in this section.

7.5.1 *Hidden Attractors in 5-Bus Microgrid*

7.5.1.1 Influence of SCF1

The influence of SCF1 on the 5-bus microgrid is described in Fig. 7.20. At first, the system operates normally. Then, an SCF1 with duration Δt_{gf} is applied, where Bus-3 is connected to the ground through $r_{gn} + j\omega_0 L_{gn}$. The values of r_{gn} and L_{gn} are given in Table 7.1.

As shown in Fig. 7.20a, with $\Delta t_{gf} = 0.5$ s, after the transient fault, only δ_5 varies from $-\pi$ to π, and sustained oscillation occurs in p_ℓ and q_ℓ, which exhibits a significant peak in the frequency spectrum. Hence, this sub-synchronous oscillation is induced by the periodic orbit with oscillating δ_5.

From Fig. 7.20b, with $\Delta t_{gf} = 0.8$ s, after the transient fault, δ_3 and δ_5 vary from $-\pi$ to π, and sustained rhythmic oscillation occurs in p_ℓ and q_ℓ, which exhibits two significant peaks in the frequency spectrum. We observe that δ_3 and δ_5 are uncoupled, and the ratio of their frequencies is irrational, which implies the existence of the hidden quasi-periodic orbit with oscillating δ_3 and δ_5.

Also, as shown in Fig. 7.20c, with $\Delta t_{gf} = 1$ s, after the transient fault, all angles vary from $-\pi$ to π, and sustained irregular oscillation occurs in p_ℓ and q_ℓ. For p_ℓ, the frequency spectrum is relatively flat, and there is no obvious peak. Therefore, this sub-synchronous oscillation is induced by a hidden chaotic attractor.

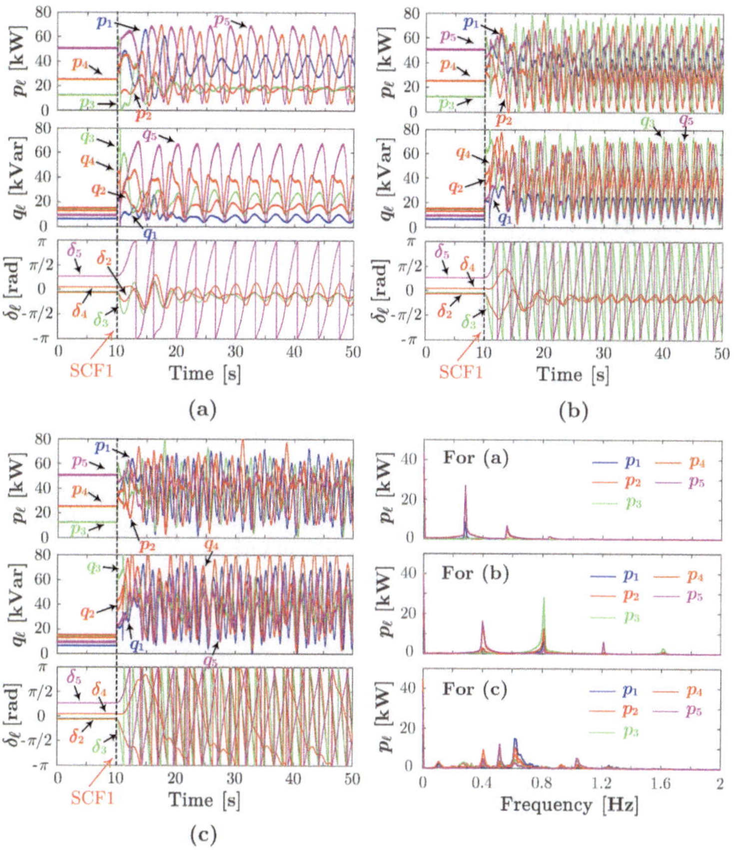

Fig. 7.20 Simulation waveforms of p_ℓ, q_ℓ, and δ_ℓ and frequency spectra of p_ℓ. The SCF1 occurs in Bus-3 (**a**) when $t = 10 \backsim 10.5$ s ($\Delta t_{gf} = 0.5$ s), (**b**) when $t = 10 \backsim 10.8$ s ($\Delta t_{gf} = 0.8$ s), and (**c**) when $t = 10 \backsim 11$ s ($\Delta t_{gf} = 1$ s). This fault is cleared at other times. All parameters adopt their default values

These simulations clearly show that transient disturbances may cause the system to leave the SEP and enter a basin of attraction of some other attractor, eventually inducing a sustained oscillation. Besides, it is noted that the slight difference in the frequency spectra of Figs. 7.9 and 7.20 may be caused by the voltage and current inner loops of GFMCs.

7.5.1.2 Influence of SCF2

The influence of SCF2 is demonstrated in Fig. 7.21. From Fig. 7.21a, after the transient fault is applied with $\Delta t_{gf} = 0.18$ s, only δ_5 varies from $-\pi$ to π, and sustained oscillation occurs in p_ℓ and q_ℓ, implying the existence of a periodic orbit.

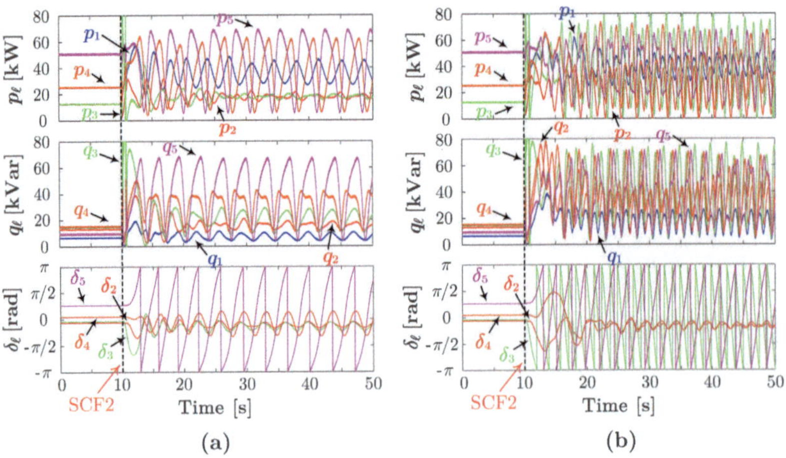

Fig. 7.21 Simulation waveforms of p_ℓ, q_ℓ, and δ_ℓ. The SCF2 occurs in the output of GFMC-3 (**a**) when $t = 10 \backsim 10.18$ s ($\Delta t_{gf} = 0.18$ s) and (**b**) when $t = 10 \backsim 10.2$ s ($\Delta t_{gf} = 0.2$ s). This fault is cleared at other times. All parameters adopt their default values

From Fig. 7.21b, after the transient fault is applied with $\Delta t_{gf} = 0.2$ s, δ_3 and δ_5 vary from $-\pi$ to π, and sustained rhythmic oscillation occurs in p_ℓ and q_ℓ, implying the existence of a quasi-periodic orbit.

7.5.1.3 Transient Stability Enhancement

A transient stability enhancement by triggering the homoclinic bifurcation is described in Fig. 7.22. The microgrid operates normally before $t = 10$ s, but exhibits sustained chaotic oscillation when $t = 10 \sim 40$ s due to SCF1 or a phase disturbance. Then, the control parameters are adjusted after $t = 40$ s, i.e., m_{p2} and m_{p3} become 0.015 p.u. from 0.04 p.u., m_{p4} becomes 0.015 p.u. from 0.02 p.u., and m_{p5} becomes 0.02 p.u. from 0.01 p.u. It is observed that all sustained sub-synchronous oscillations subside after adjusting $m_{p\ell}$. When $t = 80$ s, the same transient SCF is applied in Fig. 7.22a, and the same phase disturbance is applied in Fig. 7.22b. However, there is no sustained sub-synchronous oscillation, and the GFMCs remain synchronized with each other after these disturbances, implying that all invisible attractors subside after adjusting the parameters.

7.5.2 Impact on Synchronous Generator

In this subsection, the 6-bus microgrid shown in Fig. 7.1 with GFMC-6 replaced by a 2-pole synchronous generator is studied. The model of the synchronous generator

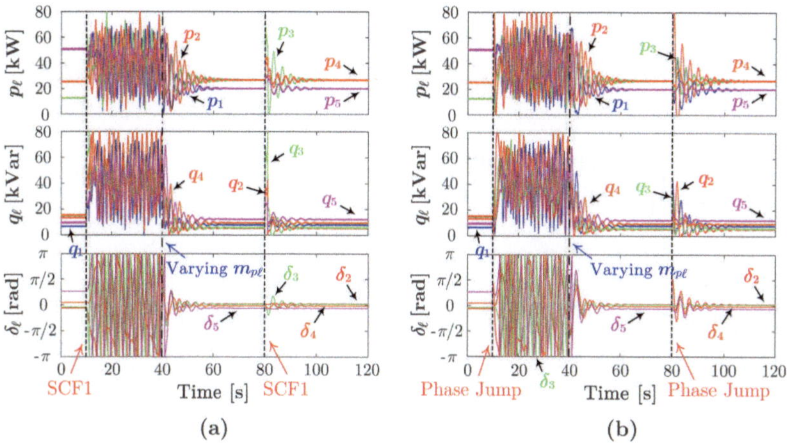

Fig. 7.22 Simulation waveforms of p_ℓ, q_ℓ, and δ_ℓ. (**a**) The SCF1 occurs in Bus-3 when $t = 10 \backsim 11$ s and $t = 80 \backsim 81$ s ($\Delta t_{gf} = 1$ s), and (**b**) a phase disturbance with $\Delta\delta_2 = 0.5\pi$ is applied to GFMC-2 when $t = 10$ s and $t = 80$ s. All $m_{p\ell}$ follow their default values before $t = 40$ s, while $m_{p2} = m_{p3} = m_{p4} = 0.015$ p.u. and $m_{p1} = m_{p5} = 0.02$ p.u. after $t = 40$ s. Other parameters adopt their default values

Fig. 7.23 Model of the synchronous generator [64]

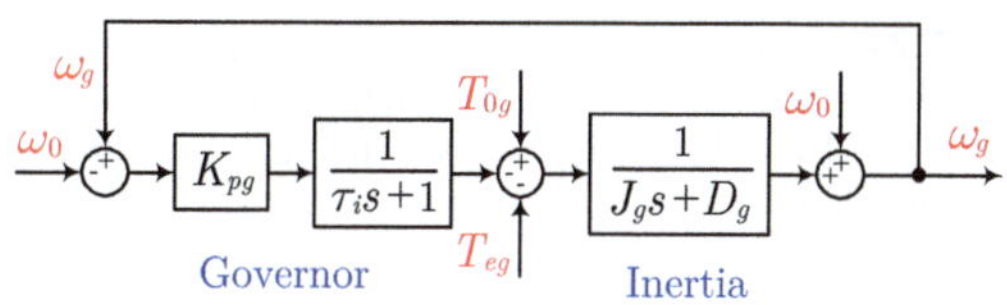

Table 7.4 Default parameters of the synchronous generator [64]

Parameter	Value	Parameter	Value
K_{pg}	16	τ_i	0.6 s
J_g	15.0224	D_g	0.01
ω_0	100π	T_{0g}	127.324

is shown in Fig. 7.23. Here, T_{eg} and ω_g are the electromagnetic torque and rotor angular frequency, respectively. Besides, T_{0g} and ω_0 are the nominal values of T_{eg} and ω_g, respectively. Moreover, K_{pg} and τ_i are the control parameter and time delay constant of the speed governor, respectively, while J_g and D_g are the inertia constant and damping coefficient of the synchronous generator, respectively. The default parameters of the synchronous generator are summarized in Table 7.4.

7.5.2.1 Sub-Synchronous Oscillations Under Transient Faults

The influence of the transient fault is demonstrated in Fig. 7.24.

From Fig. 7.24a, sustained sub-synchronous oscillation occurs in p_ℓ and p_{SG} after application of a phase disturbance, where p_{SG} is the active power of the

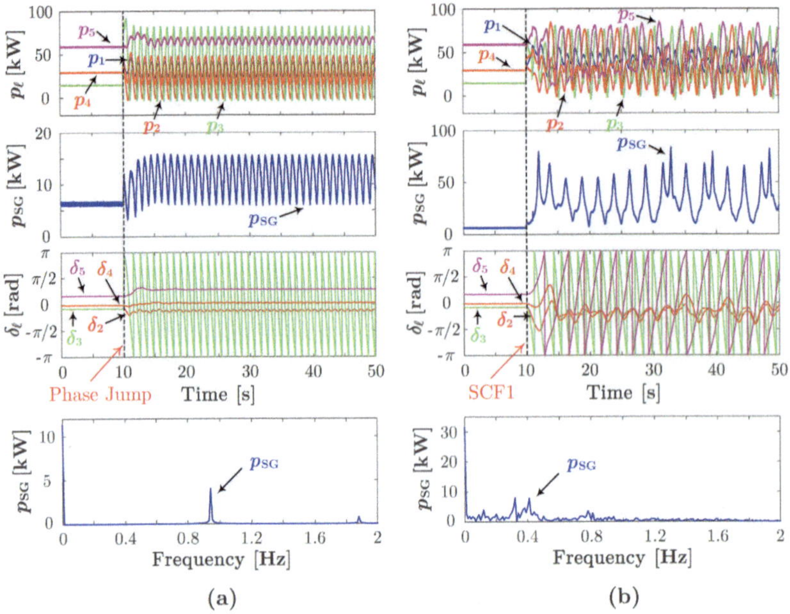

Fig. 7.24 Simulation waveforms of p_ℓ, p_{SG}, and δ_ℓ and frequency spectra of p_{SG}. (**a**) A phase disturbance with $\Delta\delta_3 = \pi$ is applied to GFMC-3 when $t = 10$ s, and (**b**) the SCF1 occurs in Bus-3 when $t = 10 \backsim 11$ s ($\Delta t_{gf} = 1$ s). This fault is cleared at other times. All parameters adopt their default values

synchronous generator. The oscillation magnitude of p_{SG} is much smaller than the oscillation magnitudes of p_2, p_3, and p_4. Besides, only δ_3 varies from $-\pi$ to π, implying that the oscillations mainly arise from the interaction between GFMC-3 and the microgrid. In this case, the oscillation in p_{SG} is passively motivated, and the interaction between the synchronous generator and the microgrid is not strong.

For Fig. 7.24b, sustained irregular oscillation emerges in p_ℓ and p_{SG} even after the SCF1 is removed. From Fig. 7.24b, p_ℓ and p_{SG} oscillate fiercely, and δ_3 and δ_5 vary from $-\pi$ to π. In this case, GFMC-3 and GFMC-5 actively interact with the microgrid, giving rise to the aforementioned oscillations. Unlike the previous scenario, the interaction between the synchronous generator and the microgrid is considerably strong. Besides, according to the frequency spectrum of p_{SG}, there is no obvious frequency peak, and the spectrum from 0 Hz to 1.8 Hz is relatively flat, implying that such oscillations may be induced by a hidden chaotic attractor.

For the synchronous generator, the oscillation in p_{SG} may be transferred to the torque T_{eg} and the rotor angular frequency ω_g. Eventually, the turbine-synchronous-generator rotors may suffer from serious torsional interaction, especially when the frequency of the sub-synchronous oscillation is close to the torsional natural frequency. Long-term torsional interaction is very harmful, which may reinforce the torsional fatigue and damage the turbine-synchronous-generator rotors [40, 134, 135].

7.5.2.2 Mitigation Methods

As analyzed in Sect. 7.4, sub-synchronous oscillations may arise from the interaction between some specific GFMC and the microgrid, where the corresponding angles vary from $-\pi$ to π. Therefore, removing these GFMCs may stop the sub-synchronous oscillations. From Fig. 7.24a, only δ_3 varies from $-\pi$ to π. Hence, GFMC-3 is tripped to eliminate sub-synchronous oscillations, as shown in Fig. 7.25b. From Fig. 7.24b, δ_3 and δ_5 vary from $-\pi$ to π. In this case, only tripping GFMC-3 or GFMC-5 cannot eliminate the sub-synchronous oscillations, as shown in Fig. 7.25b, c. However, when GFMC-3 and GFMC-5 are all tripped off, all sub-synchronous oscillations disappear, as shown in Fig. 7.25c.

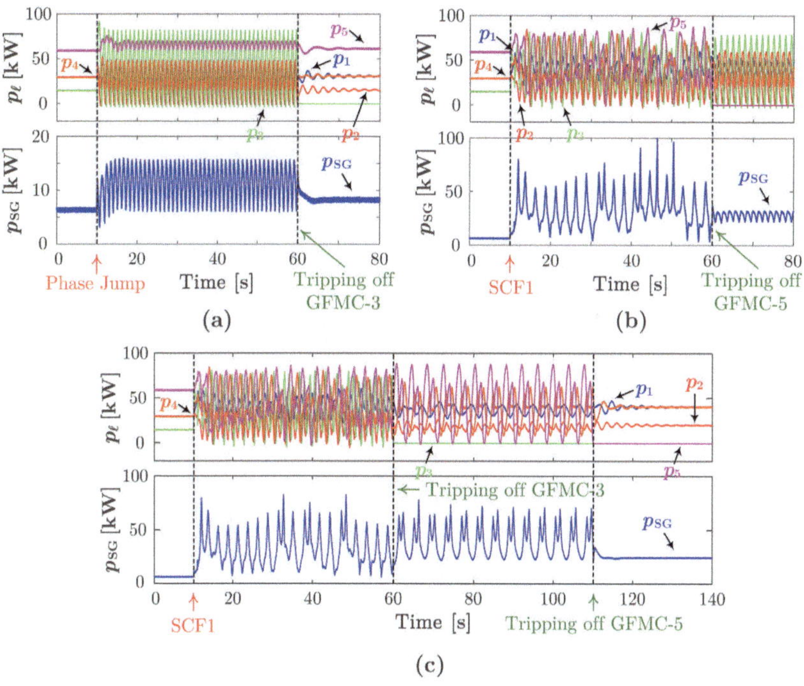

Fig. 7.25 Simulation waveforms of p_ℓ and p_{SG}. (**a**) A phase disturbance with $\Delta\delta_3 = \pi$ is applied to GFMC-3 when $t = 10$ s, and GFMC-3 is tripped off when $t = 60$ s. (**b**) The SCF1 occurs in Bus-3 when $t = 10 \backsim 11$ s, and GFMC-5 is tripped off when $t = 60$ s. (**c**) The SCF1 occurs in Bus-3 when $t = 10 \backsim 11$ s, and then GFMC-3 is tripped off when $t = 60$ s, and GFMC-5 is tripped off when $t = 110$ s. All parameters adopt their default values

7.5.3 Effects of Faulty Grid-Forming Converters

7.5.3.1 Permanent Fault

The effects of a permanent fault of GFMC on the 30-bus microgrid is shown in Fig. 7.26, where the GFMC being permanently shut down is GFMC-5 in Fig. 7.26a and GFMC-4 in Fig. 7.26b. From Fig. 7.26a, after the permanent shutdown of GFMC-5, fierce oscillations appear in p_6, $p_9 \backsim p_{11}$, and $p_{14} \backsim p_{20}$, while weak or unobvious oscillations emerge in other p_ℓ. Furthermore, there is no oscillation in $p_{24} \backsim p_{30}$, implying that the corresponding GFMCs remain in synchronization with the microgrid. From Fig. 7.26b, no oscillation emerges, and some p_ℓ slightly increases after the permanent shutdown of GFMC-4, implying that all GFMCs remain in synchronization with the microgrid. This verifies the analytical findings presented in Fig. 7.19a.

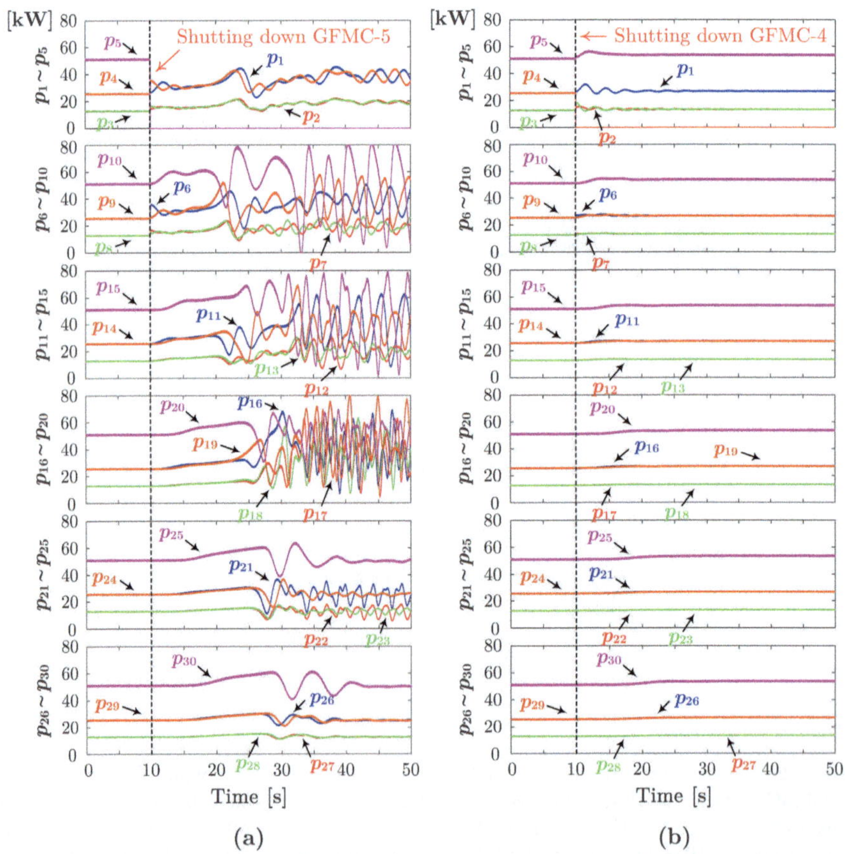

Fig. 7.26 Simulation waveforms of $p_1 \backsim p_{30}$. (**a**) GFMC-5 is permanently shut down after t = 10 s. (**b**) GFMC-4 is permanently shut down after t = 10 s. All parameters follow their default values (i.e., $K_{ld} = 1$)

7.5.3.2 Temporary Fault

The effects of a temporary fault in GFMC-5 on the 30-bus microgrid is shown in Fig. 7.27, where the recovery time is 7 s for Fig. 7.27a and 6 s for Fig. 7.27b. As shown in Fig. 7.27a, significant oscillations occur in $p_{11} \backsim p_{24}$, but oscillations in other p_ℓ are not obvious. It is interesting to find that a faulty GFMC-5 does not cause oscillation in other nearby GFMCs. In other words, the nearby GFMCs maintain their synchronization, but the faraway GFMCs are desynchronized. In Fig. 7.27b, after the GFMC-5 reconnects to the microgrid, the system returns to its normal operation, and all GFMCs synchronize to each other. Figure 7.27a and (b) together verifies the analytical findings presented in Sect. 7.4.2.

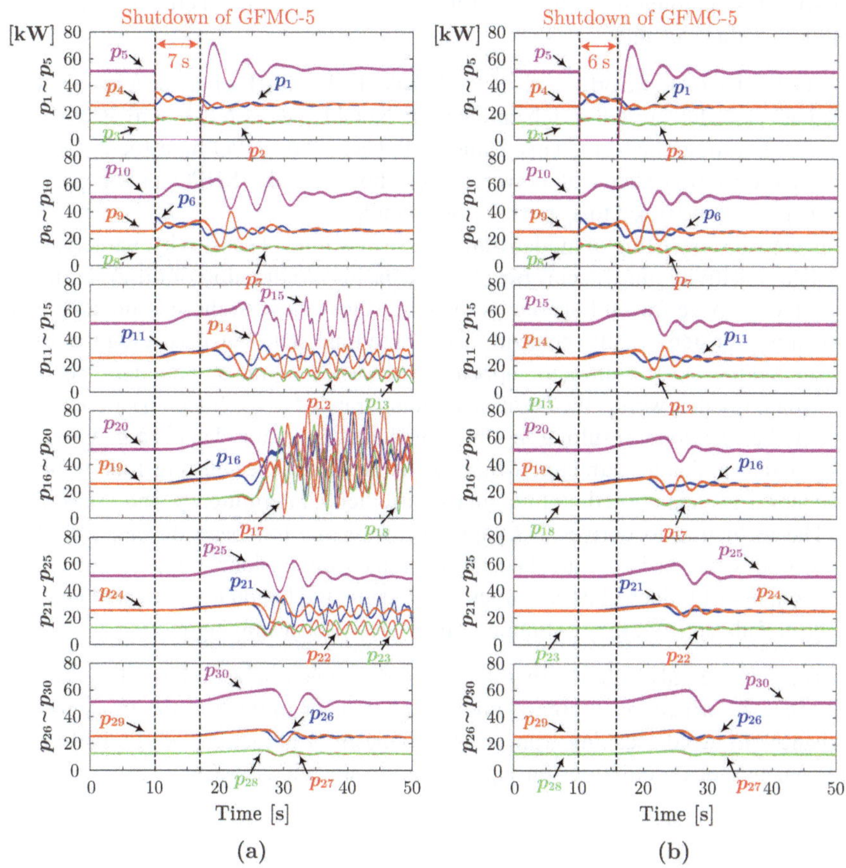

Fig. 7.27 Simulation waveforms of $p_1 \backsim p_{30}$. (**a**) GFMC-5 is shut down when $t = 10 \backsim 17$ s and connected to the microgrid at other times. (**b**) GFMC-5 is shut down when $t = 10 \backsim 16$ s and connected to the microgrid at other times. All parameters follow their default values (i.e., $K_{ld} = 1$)

7.6 Recapitulation

In Chaps. 3, 4, and 6, the physical origin of the sustained sub-synchronous oscillation has been found to be an SPO, which is remote from, but coexisting with, the SEP. This SPO emerges or subsides through a homoclinic bifurcation. Therefore, such an SPO is a typical Shil'nikov-type self-excited attractor (see Fig. 1.33). In this chapter, it is found that due to the interaction between different GFMCs, various new types of attractors may emerge, which does not emerge or vanish via a homoclinic bifurcation or a Hopf bifurcation. Thus, these attractors are hidden attractors. Furthermore, these hidden attractors also manifest themselves as sub-synchronous oscillations in the time domain and coexist with the SEP, thereby significantly influencing the transient synchronization stability of the islanded microgrid. Regarding the behavior of these hidden attractors, there are three major findings in this chapter:

1. The hidden periodic orbit with several phase-locked angles may become a hidden quasi-periodic orbit through an SNPO. Notably, the SNPO here is different from that described in Chaps. 2 and 6. Because the SNPO here focuses on the synchronization status of power angles; however, the SNPO in the previous chapters is restricted to a specific power angle.
2. Due to the coexisting of hidden attractors, a new kind of sensitivity, namely, final-state sensitivity, may occur. Besides, the basins of attraction of different hidden attractors have fractal boundaries. Any trajectory close to these boundaries may exhibit transient chaotic behavior and a high final-state sensitivity. On this basis, the hidden chaotic attractor appears when such a fractal boundary becomes strong. In other words, the duration of being trapped in the fractal basin boundary may be very long, resulting in a hidden chaotic attractor. This is a new route to generate a chaotic attractor, which is distinct from the existing four routes to chaos, i.e., via period doubling, quasi-periodicity, intermittency, and crisis [5].
3. The technique to derive all Lyapunov exponents has been developed, which serves as the basis to calculate the Lyapunov dimension, reflecting the fractal geometry of hidden chaotic attractors. Notably, the Lyapunov exponent here can be calculated from the continuous-time state-space averaging models.

This chapter also shows that the transient synchronization stability can be enhanced by triggering a homoclinic bifurcation, or mitigating the transient disturbances to avoid escaping from the SEP's basin of attraction, or eliminating the sub-synchronous oscillations through removing specific GFMCs.

Chapter 8
Unveiling the Benefits of Hidden Attractors to Transient Synchronization Stability

In Chap. 7, several hidden attractors have been identified in an islanded microgrid comprising a number of GFMCs, and sub-synchronous oscillations have been found to emerge when the trajectory is driven to one of these hidden attractors under a transient disturbance. However, the transient synchronization stability of the islanded microgrid containing multiple GFMCs and GFLCs remains unexplored, e.g., the physical characteristics and bifurcation behavior of hidden attractors under the nonlinear interaction between GFMCs and GFLCs, the influence of varying control parameters of GFMCs and GFLCs on the bifurcation behavior associated with the essential change of SEP's basin of attraction, and the impact of the aforementioned hidden attractors and bifurcation behavior on the microgrid's transient stability. These topics will be studied in this chapter.

8.1 Motivation

8.1.1 Islanded Microgrid

In an islanded microgrid containing multiple GFMCs and GFLCs, there are in general $2N$ GFMCs, N GFLCs, and $3N$ local buses. Each GFMC connects to a local bus through a connecting line, which is modeled as a series connection of L_{lk} and r_{lk}, where $k = 1, 2, \ldots, 2N$. Each GFLC connects to a local bus through a connecting line, which is modeled as a series connection of $L_{s\ell}$ and $r_{s\ell}$, where $\ell = 1, 2, \ldots, N$. Each local bus has a constant impedance load, and Bus-ρ connects to Bus-$(\rho+1)$ through a connecting line, which is represented by $L_{c\rho}$ and $r_{c\rho}$, where $\rho = 1, 2, \ldots, 3N - 1$. Our following study will start from a three-bus islanded microgrid, as shown in Fig. 8.1, where $N = 1$. In this microgrid, there are 2 GFMCs and 1 GFLC, and the load connected to Bus-ζ is $Z_{b\zeta} = R_{b\zeta} + j\omega_0 L_{b\zeta}$, where

J. Yang, C. K. Tse, *Complex Behavior of Grid-Connected Power Electronics Systems*, https://doi.org/10.1007/978-3-031-69340-3_8

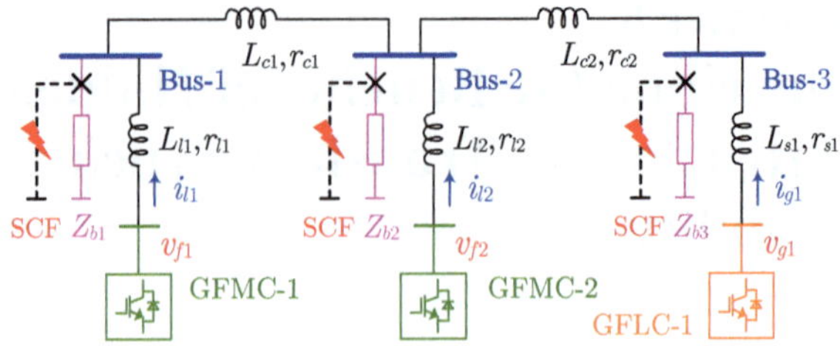

Fig. 8.1 Three-bus islanded microgrid, where $Z_{b\zeta} = R_{b\zeta} + j\omega_0 L_{b\zeta}$, and $\zeta = 1, 2, 3$

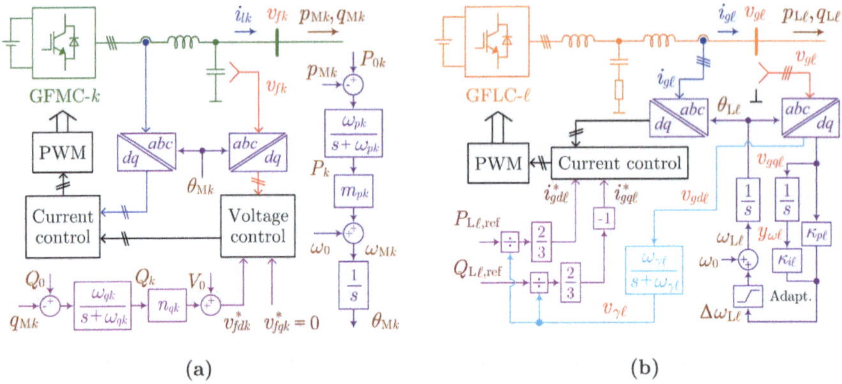

Fig. 8.2 Circuit and control of (**a**) GFMC-k and (**b**) GFLC-ℓ

$\zeta = 1, 2, 3$. In Sect. 8.6, our theoretical findings will be also validated in a 15-bus microgrid, where $N = 5$.

The control strategy of the GFMC is shown in Fig. 8.2a. For GFMC-k, v_{fk} is the AC-side output voltage, i_{lk} is the AC-side output current, p_{Mk} is the instantaneous active power, q_{Mk} is the instantaneous reactive power, ω_{Mk} is the output frequency, and θ_{Mk} is the phase angle. Here, v_{fk} and i_{lk} represent the variables in the abc-frame. For example, v_{fk} denotes v_{fak}, v_{fbk}, and v_{fck}. Likewise, the corresponding variables in the dq-frame are represented as v_{fdk} and v_{fqk}. Besides, v^*_{fdk} and v^*_{fqk} are the dq-frame references of the voltage control. Moreover, P_{0k}, Q_0, ω_0, and V_0 are the nominal values of p_{Mk}, q_{Mk}, ω_{Mk}, and v^*_{fdk}, respectively. Furthermore, ω_{pk} and ω_{qk} are the cutoff frequencies of the low-pass filters, while m_{pk} and n_{qk} are the droop coefficients.

The control strategy of the GFLC is shown in Fig. 8.2b. For GFLC-ℓ, $v_{g\ell}$ is the AC-side output voltage, $i_{g\ell}$ is the AC-side output current, $\omega_{L\ell}$ is the output frequency of PLL, $\theta_{L\ell}$ is the output phase angle of PLL, $y_{\omega\ell}$ is the state variable of PLL, and $\Delta\omega_{L\ell}$ is the output variable of the PLL controller. In the dq-frame, the corresponding variables of $v_{g\ell}$ are represented as $v_{gd\ell}$ and $v_{gq\ell}$, while the corresponding variables of $i_{g\ell}$ are represented as $i_{gd\ell}$ and $i_{gq\ell}$. For the PLL controller, $\kappa_{p\ell}$ is the proportional coefficient, and $\kappa_{i\ell}$ is the integral coefficient. Besides, $i^*_{gd\ell}$ and $i^*_{gq\ell}$ are the dq-frame references of the current control, whereas $P_{L\ell,\mathrm{ref}}$ and $Q_{L\ell,\mathrm{ref}}$ are the injection active and reactive powers of the GFLC,

Table 8.1 Default circuit parameters of the islanded microgrid

Parameter	Index	Value	Parameter	Index	Value
L_{cp}	$1, 2, \ldots,$ $3N-1$	2 mH	r_{cp}	$1, 2, \ldots$ $3N-1$	0.5 Ω
L_{lk}	$1, 2, \ldots, 2N$	3 mH	r_{lk}	$1, 2, \ldots, 2N$	0.8 Ω
$L_{s\ell}$	$1, 2, \ldots, N$	3 mH	$r_{s\ell}$	$1, 2, \ldots, N$	0.8 Ω
$L_{b\zeta}$	1, 3	3 mH	$R_{b\zeta}$	1, 3	6 Ω
	2	2 mH		2	4.2 Ω
L_{gn}	/	0.2 mH	r_{gn}	/	0.1 Ω

Table 8.2 Default control parameters of GFMCs

Parameter	k	Value	p.u.	Parameter	k	Value	p.u.
P_{0k}	Odd	50 kW	2	ω_{pk}	All	$2\pi \times 0.1$	0.1
	Even	25 kW	1	ω_{qk}	All	$2\pi \times 0.1$	0.1
m_{pk}	Odd	$\frac{0.02\omega_0}{P_m}$	0.02	n_{qk}	Odd	$\frac{0.025V_0}{Q_m}$	0.025
	Even	$\frac{0.04\omega_0}{P_m}$	0.04		Even	$\frac{0.05V_0}{Q_m}$	0.05

Note: $P_m = 50\,\text{kW}$, $Q_m = 15\,\text{kVar}$, $\omega_0 = 100\pi$, $Q_0 = 0$, and $V_0 = 311$ V. Besides, $m_{pk}P_{0k} = m_{p1}P_{01}$ for power sharing

Table 8.3 Default control parameters of GFLCs

Parameter	Value	p.u.	Parameter	Value	p.u.
$\kappa_{p\ell}$	0.413	1	$\omega_{\gamma\ell}$	2π	1
$\kappa_{i\ell}$	7.786	1	$\Delta f_{\text{L}\ell,\text{th}}$	2 Hz	/
$P_{\text{L}\ell,\text{ref}}$	11.5 kW	1	$Q_{\text{L}\ell,\text{ref}}$	13.69 kVar	1

respectively. Moreover, $v_{\gamma\ell}$ is found by applying a low-pass filter to $v_{gd\ell}$. The cutoff frequency of this filter is $\omega_{\gamma\ell}$.

The default parameters are summarized in Tables 8.1, 8.2, and 8.3.

8.1.2 Critical Fault Clearing Time

As shown in Fig. 8.1, a temporary SCF may occur in Bus-1, Bus-2, or Bus-3, and the local bus is connected to the ground through $r_{gn} + j\omega_0 L_{gn}$. We derive the critical fault clearing times (CFCTs) for different ω_{pk}, different $Q_{\text{L}\ell,\text{ref}}$, and different buses for which SCFs occur. Here, the post-fault system is stable only if the duration of SCF is smaller than the corresponding CFCT. For the SCF that occurs in Bus-ζ, the CFCT is denoted as CFCT_ζ, where $\zeta = 1, 2, 3$. Then, we have the minimal CFCT, which is denoted as $\min(\text{CFCT}_\zeta)$, as shown in Fig. 8.3. Here, we also find that $\min(\text{CFCT}_\zeta) = \text{CFCT}_2$ in most cases, implying that the SCF that occurs in Bus-2 is the worst.

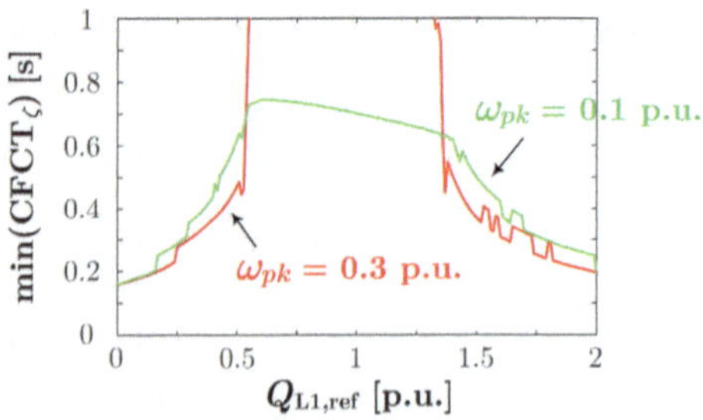

Fig. 8.3 Values of min(CFCT$_\zeta$) versus $Q_{\text{L1,ref}}$ from 0 to 1 p.u. for different ω_{pk}

From Fig. 8.3, there are three operating regions of interest, corresponding to different relative durations of the CFCT:

1. The CFCT is relatively long for medium $Q_{\text{L}\ell,\text{ref}}$, but relatively short otherwise.
2. For medium $Q_{\text{L}\ell,\text{ref}}$, the CFCT when $\omega_{pk} = 0.3$ p.u. is much longer than that when $\omega_{pk} = 0.1$ p.u.
3. For relatively small or large $Q_{\text{L}\ell,\text{ref}}$, the CFCT when $\omega_{pk} = 0.3$ p.u. is shorter than that when $\omega_{pk} = 0.1$ p.u.

The generation mechanisms of these different CFCT durations will be analyzed in the sequel.

8.2 Generalized Saddle-Node Bifurcation

8.2.1 Microgrid Model

8.2.1.1 Grid-Forming and Grid-Following Converters

For GFMC-k, the dynamics of the voltage and current loops are much faster than the dynamics of the power loop. Therefore, we set $v_{fdk} \approx v^*_{fdk}$, and $v_{fqk} \approx v^*_{fqk}$ [52]. The phase difference between GFMC-k and GFMC-1 is denoted as $\delta_{\text{M}k}$, i.e., $\delta_{\text{M}k} = \theta_{\text{M}k} - \theta_{\text{M1}}$. For GFLC-$\ell$, the dynamics of the current loop are much faster than the dynamics of the PLL and the current reference generation. Therefore, we set $i_{gd\ell} \approx i^*_{gd\ell}$, and $i_{gq\ell} \approx i^*_{gq\ell}$ [103]. The phase difference between GFLC-ℓ and GFMC-1 is denoted as $\delta_{\text{L}\ell}$, i.e., $\delta_{\text{L}\ell} = \theta_{\text{L}\ell} - \theta_{\text{M1}}$.

For each converter, the state equations and variables are represented in its dq-frame with angular frequency $\omega_{\text{M}k}$ for GFMCs or $\omega_{\text{L}\ell}$ for GFLCs. The first GFMC's coordinate frame is chosen as the common coordinate frame and denoted as DQ-frame. Therefore, the angle between other GFMC's dq-frame and the DQ-frame is $\delta_{\text{M}k}$, and the angle between the GFLC's dq-frame and the DQ-frame is $\delta_{\text{L}\ell}$. Hence,

$$\frac{\mathrm{d}\delta_{\text{M}k}}{\mathrm{d}t} = \omega_{\text{M}k} - \omega_{\text{M1}}, \quad \frac{\mathrm{d}\delta_{\text{L}\ell}}{\mathrm{d}t} = \omega_{\text{L}\ell} - \omega_{\text{M1}}, \quad \delta_{\text{M1}} = 0. \tag{8.1}$$

According to Fig. 8.2a, for GFMC-k, we have

$$\frac{\mathrm{d}}{\mathrm{d}t}\begin{bmatrix} P_k \\ Q_k \end{bmatrix} = \begin{bmatrix} \omega_{pk} & 0 \\ 0 & \omega_{qk} \end{bmatrix}\left(-\begin{bmatrix} P_k \\ Q_k \end{bmatrix} + \begin{bmatrix} P_{0k} \\ Q_0 \end{bmatrix} - \begin{bmatrix} p_{\mathrm{M}k} \\ q_{\mathrm{M}k} \end{bmatrix}\right), \tag{8.2}$$

$$\omega_{\mathrm{M}k} = \omega_0 + m_{pk}P_k, \quad v^*_{fdk} = V_0 + n_{qk}Q_k, \quad v^*_{fqk} = 0, \tag{8.3}$$

where

$$p_{\mathrm{M}k} = \frac{3}{2}(v_{fdk}i_{ldk} + v_{fqk}i_{lqk}), \quad q_{\mathrm{M}k} = \frac{3}{2}(-v_{fdk}i_{lqk} + v_{fqk}i_{ldk}). \tag{8.4}$$

Moreover, we define

$$y_k \triangleq \omega_{\mathrm{M}k} - \omega_{\mathrm{M}1} = m_{pk}P_k - m_{p1}P_1, \tag{8.5}$$

which reflects the synchronization stability of GFMC-k [53].

According to Fig. 8.2b, for GFLC-ℓ, we have

$$\frac{\mathrm{d}y_{\omega\ell}}{\mathrm{d}t} = v_{gq\ell}, \tag{8.6}$$

$$\frac{\mathrm{d}v_{\gamma\ell}}{\mathrm{d}t} = \omega_{\gamma\ell}(v_{gd\ell} - v_{\gamma\ell}), \tag{8.7}$$

$$i_{gd\ell} = \frac{2P_{\mathrm{L}\ell,\mathrm{ref}}}{3v_{\gamma\ell}}, \quad i_{gq\ell} = -\frac{2Q_{\mathrm{L}\ell,\mathrm{ref}}}{3v_{\gamma\ell}}, \tag{8.8}$$

$$\omega_{\mathrm{L}\ell} = \Delta\omega_{\mathrm{L}\ell} + \omega_0 = \kappa_{p\ell}v_{gq\ell} + \kappa_{i\ell}y_{\omega\ell} + \omega_0, \tag{8.9}$$

where $|\Delta\omega_{\mathrm{L}\ell}|$ is restricted by $2\pi\Delta f_{\mathrm{L}\ell,\mathrm{th}}$.

8.2.1.2 Power Network and Impedance Loads

For the power network and impedance loads, the equations and variables are represented in the DQ-frame, which corresponds to a rather weak and passive slow-scale nonlinear behavior [52]. Therefore, their dynamics can be omitted here. Moreover, their state equations and variables can be represented in the DQ-frame with angular frequency $\omega_{\mathrm{M}1}$. Then, it can be readily shown [52] that

$$\mathbf{i}_{\mathbf{lgDQ}} = (\mathbf{I} + \mathbf{D_N B_{LS}})^{-1}\mathbf{D_N v_{fgDQ}} \triangleq \mathbf{M v_{fgDQ}} \tag{8.10}$$

where

$$\mathbf{v}_{\mathbf{fgDQ}} = \begin{bmatrix} \mathbf{v}_{\mathbf{fDQ1}} \\ \mathbf{v}_{\mathbf{fDQ2}} \\ \mathbf{v}_{\mathbf{gDQ1}} \end{bmatrix}, \quad \mathbf{i}_{\mathbf{lgDQ}} = \begin{bmatrix} \mathbf{i}_{\mathbf{lDQ1}} \\ \mathbf{i}_{\mathbf{lDQ2}} \\ \mathbf{i}_{\mathbf{gDQ1}} \end{bmatrix},$$

$$\mathbf{v}_{\mathbf{fDQ}k} = \left[v_{fDk}\ v_{fQk}\right]^T, \quad \mathbf{v}_{\mathbf{gDQ}\ell} = \left[v_{gD\ell}\ v_{gQ\ell}\right]^T,$$

$$\mathbf{i}_{\mathbf{lDQ}k} \triangleq \left[i_{lDk}\ i_{lQk}\right]^T, \quad \mathbf{i}_{\mathbf{gDQ}\ell} \triangleq \left[i_{gD\ell}\ i_{gQ\ell}\right]^T,$$

$$\mathbf{B}_{\mathbf{LS}} = \mathrm{diag}\{\mathbf{B}_{\mathbf{l1}}, \mathbf{B}_{\mathbf{l2}}, \mathbf{B}_{\mathbf{s1}}\},$$

$$\mathbf{B}_{\mathbf{l}k} = \begin{bmatrix} r_{lk} & -\omega_0 L_{lk} \\ \omega_0 L_{lk} & r_{lk} \end{bmatrix}, \quad \mathbf{B}_{\mathbf{s}\ell} = \begin{bmatrix} r_{s\ell} & -\omega_0 L_{s\ell} \\ \omega_0 L_{s\ell} & r_{s\ell} \end{bmatrix},$$

$$\mathbf{D}_{\mathbf{N}} = \begin{bmatrix} \mathbf{D}_{\mathbf{b1}} + \mathbf{D}_{\mathbf{c1}} & -\mathbf{D}_{\mathbf{c1}} & 0 \\ -\mathbf{D}_{\mathbf{c1}} & \mathbf{D}_{\mathbf{b2}} + \mathbf{D}_{\mathbf{c1}} + \mathbf{D}_{\mathbf{c2}} & -\mathbf{D}_{\mathbf{c2}} \\ 0 & -\mathbf{D}_{\mathbf{c2}} & \mathbf{D}_{\mathbf{b3}} + \mathbf{D}_{\mathbf{c2}} \end{bmatrix},$$

$$\mathbf{D}_{\mathbf{b}\zeta} = \frac{1}{R_{b\zeta}^2 + (\omega_0 L_{b\zeta})^2} \begin{bmatrix} R_{b\zeta} & \omega_0 L_{b\zeta} \\ -\omega_0 L_{b\zeta} & R_{b\zeta} \end{bmatrix},$$

$$\mathbf{D}_{\mathbf{c}\rho} = \frac{1}{r_{c\rho}^2 + (\omega_0 L_{c\rho})^2} \begin{bmatrix} r_{c\rho} & \omega_0 L_{c\rho} \\ -\omega_0 L_{c\rho} & r_{c\rho} \end{bmatrix}.$$

In practice, ω_{M1} is close to ω_0. So, we use ω_0 to replace ω_{M1} in $\mathbf{B}_{\mathbf{l}k}$, $\mathbf{B}_{\mathbf{s}\ell}$, $\mathbf{D}_{\mathbf{b}\zeta}$, and $\mathbf{D}_{\mathbf{c}\rho}$, which significantly reduces the computational cost. Besides, $\mathbf{M}$ is a 6×6 real-value matrix, and $\mathbf{M}[x_{r1} : x_{r2}; x_{c1} : x_{c2}]$ represents the submatrix of $\mathbf{M}$, where the elements of $\mathbf{M}$ in the selected rows (from x_{r1} to x_{r2}) and columns (from x_{c1} to x_{c2}) form the submatrix. From (8.10), we get

$$\begin{bmatrix} \mathbf{v}_{\mathbf{gDQ1}} \\ \mathbf{v}_{\mathbf{gDQ2}} \end{bmatrix} = \mathbf{M}_{\mathbf{c}}^{-1} \times \left(\begin{bmatrix} \mathbf{i}_{\mathbf{gDQ1}} \\ \mathbf{i}_{\mathbf{gDQ2}} \end{bmatrix} - \mathbf{M}_{\mathbf{b}} \begin{bmatrix} \mathbf{v}_{\mathbf{fDQ1}} \\ \mathbf{v}_{\mathbf{fDQ2}} \end{bmatrix} \right), \tag{8.11}$$

$$\begin{bmatrix} \mathbf{i}_{\mathbf{lDQ1}} \\ \mathbf{i}_{\mathbf{lDQ2}} \end{bmatrix} = \mathbf{M}_{\mathbf{a}} \mathbf{v}_{\mathbf{fgDQ}} \tag{8.12}$$

where $\mathbf{M}_{\mathbf{a}} = \mathbf{M}[1:4; 1:6]$, $\mathbf{M}_{\mathbf{b}} = \mathbf{M}[5:6; 1:4]$, and $\mathbf{M}_{\mathbf{c}} = \mathbf{M}[5:6; 5:6]$.

8.2.1.3 Transformation Between Different Frames

The transformation between the local dq-frame and the DQ-frame can be derived as

$$\begin{bmatrix} v_{fDk} \\ v_{fQk} \end{bmatrix} = \mathbf{T}(\delta_{\mathrm{M}k}) \begin{bmatrix} v_{fdk} \\ v_{fqk} \end{bmatrix}, \quad \begin{bmatrix} i_{ldk} \\ i_{lqk} \end{bmatrix} = \mathbf{T}^{-1}(\delta_{\mathrm{M}k}) \begin{bmatrix} i_{lDk} \\ i_{lQk} \end{bmatrix}, \tag{8.13}$$

$$\begin{bmatrix} i_{gD\ell} \\ i_{gQ\ell} \end{bmatrix} = \mathbf{T}(\delta_{\mathrm{L}\ell}) \begin{bmatrix} i_{gd\ell} \\ i_{gq\ell} \end{bmatrix}, \quad \begin{bmatrix} v_{gd\ell} \\ v_{gq\ell} \end{bmatrix} = \mathbf{T}^{-1}(\delta_{\mathrm{L}\ell}) \begin{bmatrix} v_{gD\ell} \\ v_{gQ\ell} \end{bmatrix} \tag{8.14}$$

where

$$\mathbf{T}(\delta_x) = \begin{bmatrix} \cos\delta_x & -\sin\delta_x \\ \sin\delta_x & \cos\delta_x \end{bmatrix}, \quad \mathbf{T}^{-1}(\delta_x) = \begin{bmatrix} \cos\delta_x & \sin\delta_x \\ -\sin\delta_x & \cos\delta_x \end{bmatrix}.$$

Here, $\delta_x = \delta_{\mathrm{M}k}$ or $\delta_{\mathrm{L}\ell}$. Combining (8.1) to (8.14), we get the model of the islanded microgrid, which can be represented as

$$\frac{\mathrm{d}\mathbf{x}}{\mathrm{d}t} = \mathbf{f}(\mathbf{x}) = \left[f_1(x_1, \cdots, x_{N_s}) \ \cdots \ f_{N_s}(x_1, \cdots, x_{N_s}) \right]^T \tag{8.15}$$

where $\mathbf{x}$ is the vector of state variables, and $\mathbf{f}(\mathbf{x})$ is the vector of state functions. The number of state equations is $N_s = 9N - 1$, and the state variables are P_k, Q_k, $\delta_{\mathrm{M}k}$, $y_{\omega\ell}$, $v_{\gamma\ell}$, $\delta_{\mathrm{L}\ell}$.

8.2.2 Equilibrium Points

Based on the established model, we set $\mathbf{f}(\mathbf{x}) = \mathbf{0}$ to derive the EQPs, as shown in Fig. 8.4. From Fig. 8.4, when $Q_{\mathrm{L1,ref}}$ varies from 0.28 p.u. to 1.79 p.u., there are four EQPs, including an SEP and three UEPs. However, when $Q_{\mathrm{L1,ref}} < 0.28$ p.u. or $Q_{\mathrm{L1,ref}} > 1.79$ p.u., there are only one SEP and one UEP, implying two different generalized saddle-node bifurcations occur when $Q_{\mathrm{L1,ref}} \approx 0.28$ p.u. and $Q_{\mathrm{L1,ref}} \approx 1.79$ p.u., respectively. The aforementioned three UEPs have a significant impact on the system global behavior [97, 136], which will be explained in Sect. 8.4.

8.3 Effect of Grid-Following Reactive Power Injection on Critical Fault Clearing Time

In this section, the mechanism behind the long CFCT corresponding to the medium range of $Q_{\mathrm{L}\ell,\mathrm{ref}}$, as indicated in Sect. 8.1.2, is explained.

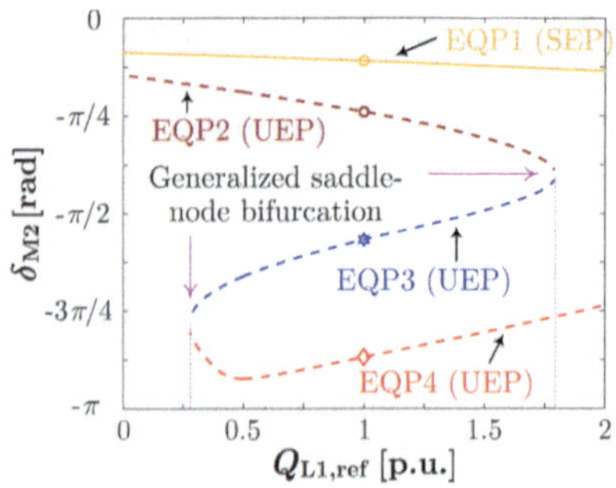

Fig. 8.4 Codimension-1 bifurcation diagram showing the EQPs. Solid curve denotes SEP, and dashed curves denote UEPs. System parameters follow the default set with varying $Q_{\mathrm{L1,ref}}$

8.3.1 Hidden Attractors

In terms of δ_{M2} and δ_{L1}, the islanded microgrid may have four different types of dynamical behavior, corresponding to different attractors, namely, SEP, Type-I hidden periodic orbit, hidden quasi-periodic orbit, and Type-II hidden periodic orbit. Here, the periodic orbits are classified according to the behavior of δ_{M2} and δ_{L1} [53]. For Type-I hidden periodic orbit, only δ_{L1} varies from $-\pi$ to π, while for Type-II hidden periodic orbit, two angles all vary from $-\pi$ to π, and they are phase-locked. Furthermore, the steady-state data of δ_{M2} and δ_{L1} are used to plot the attractor in $(\delta_{\mathrm{L1}}, \delta_{\mathrm{M2}})$-torus. The relationship between the torus and the 3-dim space is presented as [53]

$$\begin{cases} u_x = (R_x + R_z \cos\delta_{\mathrm{M2}})\cos\delta_{\mathrm{L1}} \\ u_y = (R_y + R_z \cos\delta_{\mathrm{M2}})\sin\delta_{\mathrm{L1}} \\ u_z = R_z \sin\delta_{\mathrm{M2}} \end{cases} \tag{8.16}$$

where $R_x = R_y = 5$ and $R_z = 2$. The specific attractors are:

(1) SEP: $\delta_{\mathrm{M2}} \to \delta_{\mathrm{M2,SEP}}$, and $\delta_{\mathrm{L1}} \to \delta_{\mathrm{L1,SEP}}$. Here, all GFMCs and GFLC-1 remain synchronized.
(2) Type-I hidden periodic orbit: Only δ_{L1} varies from $-\pi$ to π periodically, and δ_{M2} oscillates slightly, as shown in Fig. 8.5a, b. From Fig. 8.5b, $y_{\omega 1}$ diverges to $-\infty$, implying a malfunction of the PLL.
(3) Hidden quasi-periodic orbit: δ_{M2} and δ_{L1} all vary from $-\pi$ to π periodically, but they are uncoupled, as shown in Fig. 8.5c, d. Besides, the ratio of their frequencies is an irrational constant. From Fig. 8.5d, $y_{\omega 1}$ also diverges to $-\infty$, implying a malfunction of the PLL.
(4) Type-II hidden periodic orbit: δ_{M2} and δ_{L1} all vary from $-\pi$ to π periodically; however, they are phase-locked, as shown in Fig. 8.5e, f. From Fig. 8.5f, $y_{\omega 1}$ is very small and close to zero, implying that the PLL operates normally. On the other hand, y_2 oscillates periodically. Therefore, such oscillation initially arises from the unstable interaction between different GFMCs, and then, GFLC-1 is synchronized with this oscillation through the PLL.

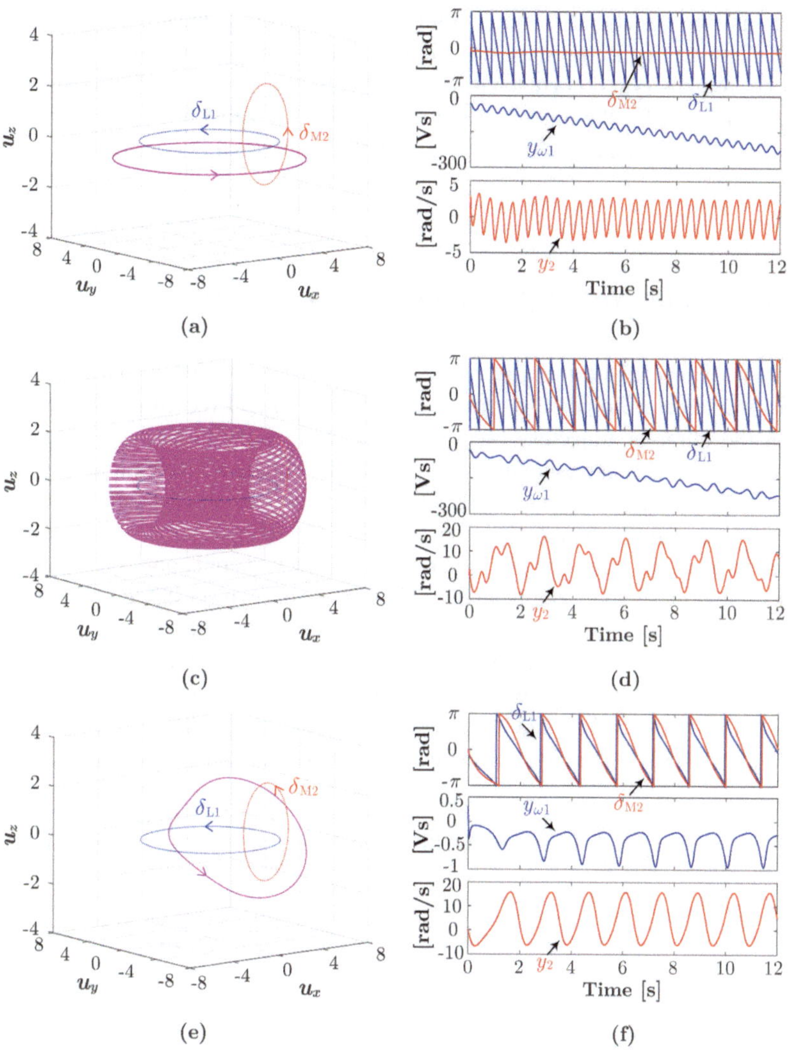

Fig. 8.5 Type-I hidden periodic orbit (**a**) in (δ_{L1}, δ_{M2})-torus and (**b**) in the time domain, where the initial condition $y_2(0) = 0$ and $y_{\omega 1}(0) = -35$; hidden quasi-periodic orbit (**c**) in (δ_{L1}, δ_{M2})-torus and (**d**) in the time domain, where $y_2(0) = -5$ and $y_{\omega 1}(0) = -35$, and Type-II hidden periodic orbit (**e**) in (δ_{L1}, δ_{M2})-torus and (**f**) in the time domain, where $y_2(0) = -5$ and $y_{\omega 1}(0) = 0$. Here, $P_2(0) = (m_{p1}P_1(0) + y_2(0))/m_{p2}$, and $P_1(0)$, $Q_1(0)$, $Q_2(0)$, $\delta_{M2}(0)$, $v_{\gamma 1}(0)$ and $\delta_{L1}(0)$ follow the SEP value. All parameters follow the default set

Usually, the system operates at the SEP. However, under transient disturbances, the trajectory may be perturbed from the SEP and move to one of the aforementioned hidden attractors, which appears as sub-synchronous oscillations (as shown in Fig. 8.5). However, among these three hidden attractors, the Type-II hidden periodic orbit may be beneficial to maintaining the microgrid's synchronization stability upon removal of the transient disturbance. Since the PLL operates normally, the GFLCs remain synchronized with the microgrid for the Type-II hidden periodic orbit. However, for the other two types of hidden attractors, the PLL malfunctions and the GFLCs are desynchronized. This will be analyzed in detail in the next two subsections.

8.3.2 Saddle-Node Bifurcation of Periodic Orbits

In order to investigate the generation mechanism of the Type-II hidden periodic orbit, we select a Poincaré section where $\delta_{\mathrm{M2}} = 0$ and record the value of v_{gd1} when the attractor trajectory crosses this section. Thus, through varying $Q_{\mathrm{L1,ref}}$ and recording the corresponding v_{gd1}, we can obtain a bifurcation diagram on $Q_{\mathrm{L1,ref}}$-v_{gd1} plane, as shown in Fig. 8.6a, from which we observe dense and regular intersection points when $Q_{\mathrm{L1,ref}} < 0.45$ p.u. and $Q_{\mathrm{L1,ref}} > 1.55$ p.u., implying the quasi-periodic behavior. For other $Q_{\mathrm{L1,ref}}$ values, there is only one intersection point, showing the periodic behavior. Therefore, the Type-II hidden periodic orbit and hidden quasi-periodic orbit can transform from each other.

From the previous subsection, δ_{M2} and δ_{L1} are phase-locked in Type-II hidden periodic orbit, but become decoupled in the hidden quasi-periodic orbit. Therefore, the relation between δ_{M2} and δ_{L1} is essential. Denoting $\phi_1 \triangleq \delta_{\mathrm{L1}} - \delta_{\mathrm{M2}}$ and $z_1 \triangleq \dot{\phi}_1 = \omega_{\mathrm{L1}} - \omega_{\mathrm{M2}}$ and applying the long-term average, we have

$$\langle \phi_1 \rangle = \frac{1}{T_{\mathrm{p2}} - T_{\mathrm{p1}}} \int_{T_{\mathrm{p1}}}^{T_{\mathrm{p2}}} (\delta_{\mathrm{L1}} - \delta_{\mathrm{M2}}) \mathrm{d}t, \quad \langle z_1 \rangle = \frac{1}{T_{\mathrm{p2}} - T_{\mathrm{p1}}} \int_{T_{\mathrm{p1}}}^{T_{\mathrm{p2}}} (\omega_{\mathrm{L1}} - \omega_{\mathrm{M2}}) \mathrm{d}t \tag{8.17}$$

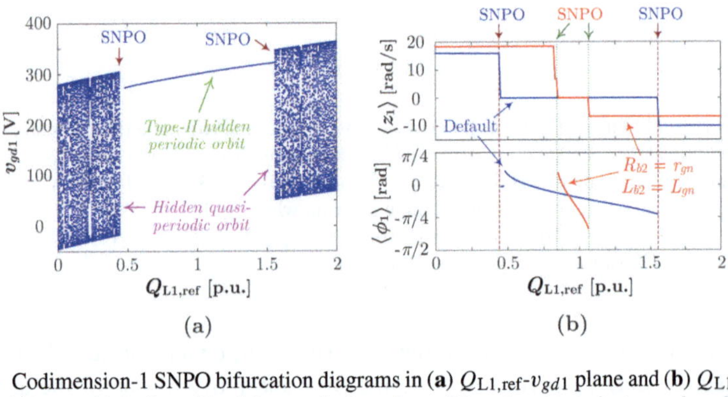

Fig. 8.6 Codimension-1 SNPO bifurcation diagrams in (**a**) $Q_{\mathrm{L1,ref}}$-v_{gd1} plane and (**b**) $Q_{\mathrm{L1,ref}}$-$\langle z_1 \rangle$ plane and $Q_{\mathrm{L1,ref}}$-$\langle \phi_1 \rangle$ plane. For (**a**), v_{gd1} is the value of the attractor trajectory when δ_{M2} each crosses the zero value. For (**a**) and the blue curve in (**b**), system parameters follow the default set with varying $Q_{\mathrm{L1,ref}}$. For the red curve in (**b**), $R_{b2} = r_{gn}$ and $L_{b2} = L_{gn}$, while other parameters follow the default set with varying $Q_{\mathrm{L1,ref}}$

where $T_{\mathrm{p1}} = 30\,\mathrm{s}$ and $T_{\mathrm{p2}} = 330\,\mathrm{s}$. On this basis, we derive the values of $\langle\phi_1\rangle$ and $\langle z_1\rangle$ when varying $Q_{\mathrm{L1,ref}}$, as shown in the blue curve of Fig. 8.6b. When $Q_{\mathrm{L1,ref}} = 0.45 \sim 1.55\,\mathrm{p.u.}$, $\langle z_1\rangle \approx 0$ and $\langle\phi_1\rangle$ is constant, implying that δ_{M2} and δ_{L1} are phase-locked, and a Type-II hidden periodic orbit exists. When $Q_{\mathrm{L1,ref}} < 0.45$ or $> 1.55\,\mathrm{p.u.}$, $|\langle z_1\rangle| \gg 0$, while $\langle\phi_1\rangle$ is not constant and very large. In this case, δ_{L1} and δ_{M2} are decoupled, and the Type-II hidden periodic orbit is replaced by the hidden quasi-periodic orbit.

For specific $Q_{\mathrm{L1,ref}}$, constant $\langle\phi_1\rangle$ corresponds to the stable locked periodic orbit. Near this stable locked periodic orbit, some unstable locked periodic orbits may exist. When varying $Q_{\mathrm{L1,ref}}$, the stable locked periodic orbit may merge with one specific unstable locked periodic orbit and disappear eventually [53, 128]. In other words, an SNPO occurs. After the onset of this bifurcation, there is no constant value of $\langle\phi_1\rangle$, and no stable locked periodic orbit exists. Hence, for the blue curve shown in Fig. 8.6b, the SNPO occurs near $Q_{\mathrm{L1,ref}} = 0.45\,\mathrm{p.u.}$ and $Q_{\mathrm{L1,ref}} = 1.55\,\mathrm{p.u.}$

8.3.3 *Impact of Type-II Hidden Periodic Orbit on Maintaining Synchronization After Removing Transient Disturbances*

Here, we will investigate the impact of the Type-II hidden periodic orbit on the microgrid's synchronization stability under a temporary SCF.

During the SCF, the local bus is connected to the ground through $r_{gn} + j\omega_0 L_{gn}$. As indicated in Sect. 8.1.2, the SCF that occurs in Bus-2 is the worst. In order to study the nonlinear behavior during the SCF, we set $R_{b2} = r_{gn}$ and $L_{b2} = L_{gn}$ in the model established in Sect. 8.2.1. In this case, no EQP exists. However, the hidden attractors may exist. We also derive the values of $\langle\phi_1\rangle$ and $\langle z_1\rangle$ when varying $Q_{\mathrm{L1,ref}}$, as shown in the red curve of Fig. 8.6b. From Fig. 8.6b, $\langle z_1\rangle \approx 0$ and $\langle\phi_1\rangle$ is constant when $Q_{\mathrm{L1,ref}} = 0.85 \sim 1.06\,\mathrm{p.u.}$, implying that a Type-II hidden periodic orbit exists. However, $|\langle z_1\rangle| \gg 0$ and $\langle\phi_1\rangle$ is not constant for other values of $Q_{\mathrm{L1,ref}}$, implying that the Type-II hidden periodic orbit is replaced by the hidden quasi-periodic orbit.

8.3.3.1 Mechanism

For the sake of clarity, relevant basins of attraction in the (δ_{L1}, $y_{\omega 1}$)-plane and (δ_{M2}, y_2)-plane are derived, as shown in Fig. 8.7. These basins of attraction are found through the following procedure: (1) The initial conditions of all state variables follow the corresponding SEP values, except δ_{L1} and $y_{\omega 1}$, or δ_{M2} and y_2. (2) For given initial point (δ_{L1}, $y_{\omega 1}$) or (δ_{M2}, y_2), if the trajectory converges to a specific attractor, the aforementioned point belongs to such attractor's basin of attraction.

In Fig. 8.7, the blue area is the Type-I hidden periodic orbit's basin of attraction, the purple area is the hidden quasi-periodic orbit's basin of attraction, and the green

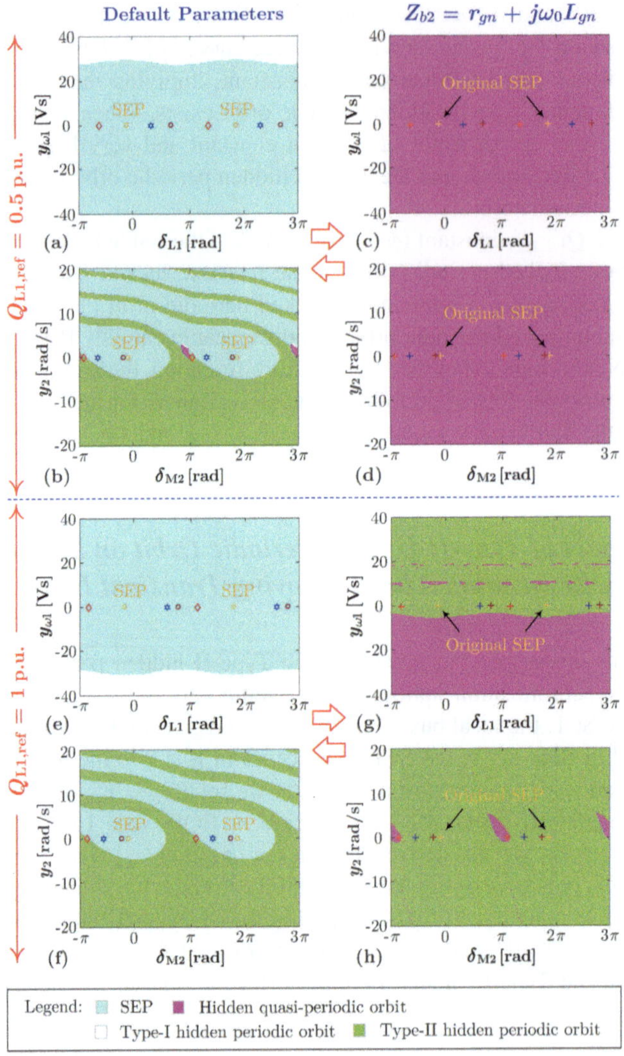

Fig. 8.7 Basins of attraction in (**a**) the (δ_{L1}, $y_{\omega 1}$)-plane for $Q_{\mathrm{L1,ref}} = 0.5$ p.u., and default R_{b2} and L_{b2}; (**b**) the (δ_{M2}, y_2)-plane for $Q_{\mathrm{L1,ref}} = 0.5$ p.u., and default R_{b2} and L_{b2}; (**c**) the (δ_{L1}, $y_{\omega 1}$)-plane for $Q_{\mathrm{L1,ref}} = 0.5$ p.u., $R_{b2} = r_{gn}$, and $L_{b2} = L_{gn}$; (**d**) the (δ_{M2}, y_2)-plane for $Q_{\mathrm{L1,ref}} = 0.5$ p.u., $R_{b2} = r_{gn}$, and $L_{b2} = L_{gn}$; (**e**) the (δ_{L1}, $y_{\omega 1}$)-plane for $Q_{\mathrm{L1,ref}} = 1$ p.u., and default R_{b2} and L_{b2}; (**f**) the (δ_{M2}, y_2)-plane for $Q_{\mathrm{L1,ref}} = 1$ p.u., and default R_{b2} and L_{b2}; (**g**) the (δ_{L1}, $y_{\omega 1}$)-plane for $Q_{\mathrm{L1,ref}} = 1$ p.u., $R_{b2} = r_{gn}$, and $L_{b2} = L_{gn}$; (**h**) the (δ_{M2}, y_2)-plane for $Q_{\mathrm{L1,ref}} = 1$ p.u., $R_{b2} = r_{gn}$, and $L_{b2} = L_{gn}$. Other parameters follow the default set

area is the Type-II hidden periodic orbit's basin of attraction. At first, for the default values of R_{b2} and L_{b2}, from Fig. 8.7b where $Q_{\mathrm{L1,ref}} = 0.5$ p.u. and from Fig. 8.7f where $Q_{\mathrm{L1,ref}} = 1$ p.u., a Type-II hidden periodic orbit exists, which is consistent with the blue curve shown in Fig. 8.6b. Besides, the Type-II hidden periodic orbit occupies almost the half bottom of the (δ_{M2}, y_2)-plane. However, for $R_{b2} = r_{gn}$, and $L_{b2} = L_{gn}$, from Fig. 8.7c, d where $Q_{\mathrm{L1,ref}} = 0.5$ p.u., the whole space is the hidden quasi-periodic orbit's basin of attraction, implying the nonexistence of Type-II hidden periodic orbit. This is consistent with the red curve shown in Fig. 8.6b. Furthermore, from Fig. 8.7g, h where $Q_{\mathrm{L1,ref}} = 1$ p.u., the area near the SEP is the Type-II hidden periodic orbit's basin of attraction. In particular, for Fig. 8.7h, almost the whole (δ_{M2}, y_2)-plane is the Type-II hidden periodic orbit's basin of attraction.

For $Q_{\mathrm{L1,ref}} = 0.5$ p.u., Fig. 8.7a, b represents the basins of attraction before and after a temporary SCF, while Fig. 8.7c, d represents the basins of attraction during a temporary SCF. Initially, the system operates at the SEP. During the SCF, the SEP disappears, and the area near the original SEP becomes the basin of attraction of hidden quasi-periodic orbit (i.e., the purple area), as shown in Fig. 8.7c, d. In this case, $y_{\omega 1}$ diverges very fast, as shown in Fig. 8.5d, and the trajectory will be far from the original SEP very soon. So, if the fault-clearing time is relatively large, $|y_{\omega 1}|$ will become quite large once the SCF is removed. In this case, the trajectory is out of the SEP's basin of attraction, but in the hidden quasi-periodic orbit's basin of attraction, leading to the desynchronization of the post-fault system. Therefore, the CFCT is relatively small due to diverging $y_{\omega 1}$ under a temporary SCF, as shown in Fig. 8.3.

For $Q_{\mathrm{L1,ref}} = 1$ p.u., Fig. 8.7e, f represents the basins of attraction before and after a temporary SCF, while Fig. 8.7g, h represents the basins of attraction during a temporary SCF. Initially, the system operates at the SEP. During the SCF, the SEP disappears, and the area near the original SEP becomes the basin of attraction of Type-II hidden periodic orbit, as shown in Fig. 8.7g, h. Therefore, the trajectory will converge to this Type-II hidden periodic orbit, where δ_{M2}, δ_{L1}, y_2, and $y_{\omega 1}$ all oscillate periodically. In this case, the system exhibits a sub-synchronous oscillation, and the inverters are desynchronized. However, it is noted that $y_{\omega 1}$ is small, as shown in Fig. 8.5f, while y_2 varies slowly in transient due to small ω_{pk}. Thus, the trajectory will wander near the SEP. After the SCF is removed, the basins of attraction become the original ones, as shown in Fig. 8.7e, f. Depending on the value of y_2, the trajectory may be in the SEP's basin of attraction for small y_2 (i.e., the inverters will be resynchronized with each other), or in the post-fault Type-II hidden periodic orbit's basin of attraction for large y_2 (i.e., the inverters remain desynchronized). The value of y_2 once removing the SCF is determined by the fault-clearing time. Large fault-clearing time leads to large y_2. Therefore, due to the slow variation of y_2, the CFCT is relatively large in this scenario, as shown in Fig. 8.3.

In summary, if a Type-II hidden periodic orbit exists during the SCF, the trajectory will wander around the original SEP for a long time and may be still in the SEP's basin of attraction once the SCF is removed. However, if no Type-II hidden periodic orbit exists during the SCF, the trajectory will move far away from the SEP very quickly. Therefore, the CFCT is relatively large (small) if a Type-II hidden periodic orbit exists (does exist) during the SCF.

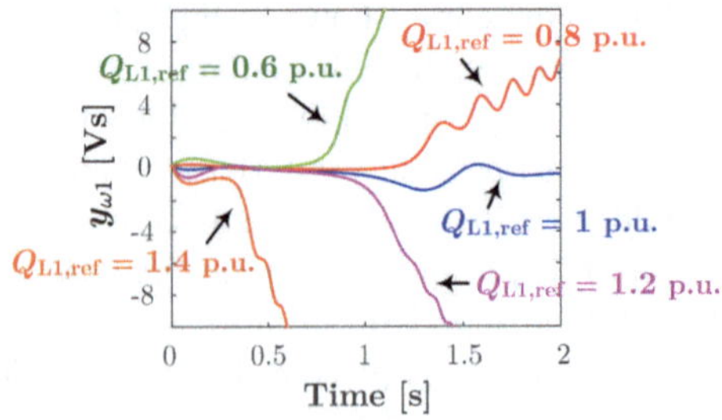

Fig. 8.8 Waveforms of y_ω for different $Q_{L1,ref}$. System parameters adopt the default set with $R_{b2} = r_{gn}$ and $L_{b2} = L_{gn}$. Besides, the initial condition of the trajectories is the value of the original SEP

8.3.3.2 Ghost Transient

From Figs. 8.3 and 8.6b, it is observed that the CFCT may still be long even when the Type-II hidden periodic orbit is absent during the SCF. For example, $\min(\text{CFCT}_\zeta) > 0.6$ for $Q_{L1,ref} = 0.5 \sim 1.4$ p.u. However, during the SCF, the Type-II hidden periodic orbit only exists for $Q_{L1,ref} = 0.85 \sim 1.06$ p.u. The mechanism is explained below.

During the SCF, although the Type-II hidden periodic orbit disappears for $Q_{L1,ref} = 0.5 \sim 0.85$ p.u. and $Q_{L1,ref} = 1.06 \sim 1.4$ p.u., the vector field associated with the original Type-II hidden periodic orbit still exists, which forces the trajectory to behave as before in transient. Such behavior is called *ghost transient*, which is commonly found near the bifurcation onset point, especially in high-order nonlinear systems [137]. This effect is more obvious near the onset point of SNPO, as shown in Fig. 8.8. For Fig. 8.8, the trajectories start from the original SEP, and $R_{b2} = r_{gn}$ and $L_{b2} = L_{gn}$ are selected to emulate the SCF. From Fig. 8.8, $y_{\omega 1}$ oscillates near the zero value for $Q_{L1,ref} = 1$ p.u., implying the existence of a Type-II hidden periodic orbit, while $y_{\omega 1}$ wanders around the zero value for a period of time, but eventually diverges for other values of $Q_{L1,ref}$, implying the nonexistence of a Type-II hidden periodic orbit. Moreover, the divergence rate of $y_{\omega 1}$ becomes higher if $Q_{L1,ref}$ is farther from the SNPO onset point value. Therefore, the CFCT increases when $Q_{L1,ref}$ is near the SNPO onset point value, but shortens when $Q_{L1,ref}$ is far from the SNPO onset point.

8.3.4 *Bifurcation Diagrams in PQ-Plane*

Based on the above analytical results, we can also derive the SNPO bifurcation diagrams on the $P_{L1,ref}$-$Q_{L1,ref}$ plane, as shown in Fig. 8.9. The blue solid curves are derived using the default R_{b2} and L_{b2}, and a Type-II hidden periodic orbit exists in the area between these two blue curves (denoted as Area-1). The red dashed curves are derived using $R_{b2} = r_{gn}$ and $L_{b2} = L_{gn}$, and a Type-II hidden periodic orbit exists in the area between these two red curves (denoted as Area-2). Based on the analyses in previous subsections, it is much better to select $(P_{L1,ref}, Q_{L1,ref})$ in

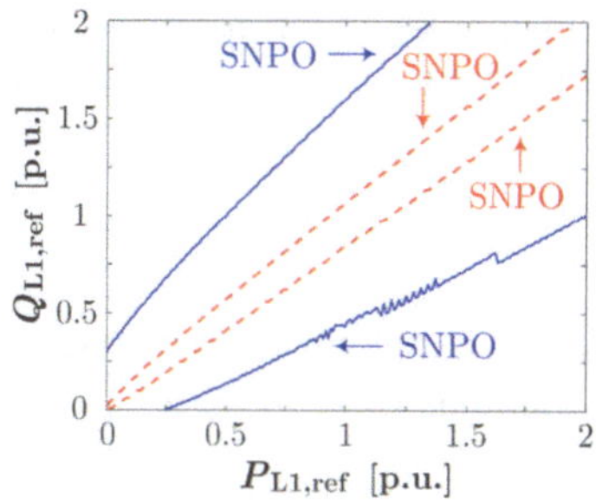

Fig. 8.9 Codimension-2 bifurcation diagram corresponding to SNPO on $P_{\mathrm{L1,ref}}$-$Q_{\mathrm{L1,ref}}$ plane. Other parameters adopt the default set

Area-2. However, as Area-2 is too narrow, an alternative is to select ($P_{\mathrm{L1,ref}}$, $Q_{\mathrm{L1,ref}}$) in Area-1 to guarantee a relatively long CFCT.

8.4 Effects of Grid-Forming Low-Pass Cutoff Frequency on Critical Fault Clearing Time

This section attempts to explain the mechanism behind the much longer CFCT when ω_{pk} is relatively large for a medium range of $Q_{\mathrm{L}\ell,\mathrm{ref}}$ and the shorter CFCT with larger ω_{pk} for low and high ranges of $Q_{\mathrm{L}\ell,\mathrm{ref}}$, as given in Sect. 8.1.2.

8.4.1 Preliminaries of Homoclinic Bifurcation

8.4.1.1 Manifolds

For (8.15), we can derive the Jacobian

$$\mathbf{J} = \begin{bmatrix} \frac{\partial f_1}{\partial x_1} & \cdots & \frac{\partial f_1}{\partial x_{N_s}} \\ \vdots & \ddots & \vdots \\ \frac{\partial f_{N_s}}{\partial x_1} & \cdots & \frac{\partial f_{N_s}}{\partial x_{N_s}} \end{bmatrix}. \tag{8.18}$$

From Fig. 8.4, three distinct UEPs may exist. For each UEP, we can use the corresponding Jacobian to derive the eigenvalues and eigenvectors. For example, λ_u and λ_s are two of the eigenvalues of a specific UEP, where $\lambda_u > 0 > \lambda_s$. Moreover, $\mathbf{u}^u$ and $\mathbf{u}^s$ are the unstable and stable eigenvectors, respectively, where $\mathbf{u}^u$ corresponds to the positive eigenvalue λ_u, while $\mathbf{u}^s$ corresponds to the negative eigenvalue λ_s. Then, we can select the initial condition as

$$\mathbf{x}|_{t=0} = \mathbf{x}|_{\mathrm{UEP}} + \mathbf{u}^u \times \varepsilon, \tag{8.19}$$

which can be used to derive the unstable manifolds W^u. Here $\varepsilon = 10^{-3}$ or $\varepsilon = -10^{-3}$. Moreover, in order to derive the stable manifolds, (8.15) is adjusted as

$$\frac{\mathrm{d}\mathbf{x}}{\mathrm{d}t} = -\mathbf{f}(\mathbf{x}), \tag{8.20}$$

which is the time-reversal version of (8.15). Then, using (8.20) and the initial condition

$$\mathbf{x}|_{t=0} = \mathbf{x}|_{\mathrm{UEP}} + \mathbf{u}^s \times \varepsilon, \tag{8.21}$$

the stable manifolds W^s can be derived. The UEP, eigenvectors, and manifolds are shown in Fig. 8.10.

Using the manifolds, we can study the homoclinic bifurcation behavior, as shown in Fig. 8.11. Figure 8.11a shows the manifolds before the onset of a homoclinic bifurcation. The SEP's basin of attraction is surrounded by the stable manifolds W^s, and one of the unstable manifolds W^u diverges to infinity or converges to a specific hidden attractor. Figure 8.11b shows the manifolds after the onset of a homoclinic bifurcation. Some UPOs may exist and serve as the boundaries to separate the SEP's basin of attraction and the hidden attractors' basin of attraction. In this case, both unstable manifolds W^u converge to the SEP [97].

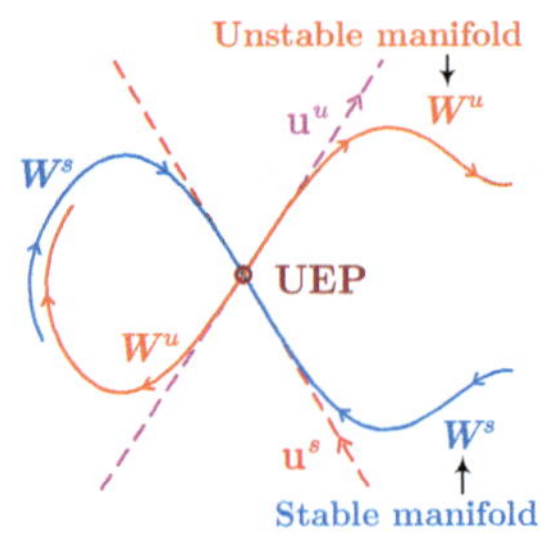

Fig. 8.10 Manifolds and eigenvectors of UEP

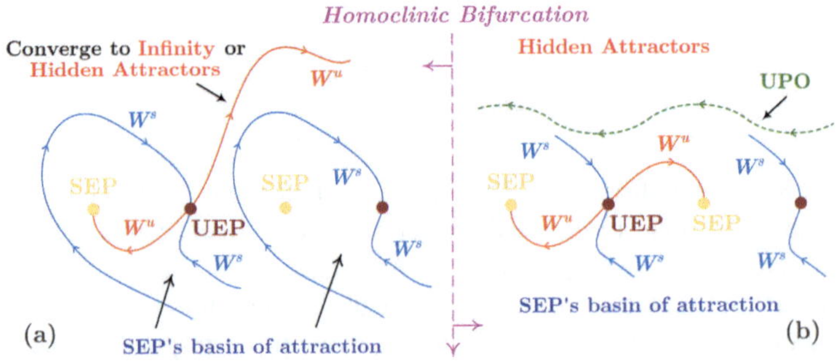

Fig. 8.11 Unstable and stable manifolds: (**a**) before the onset of homoclinic bifurcation and (**b**) after the onset of homoclinic bifurcation

8.4.1.2 Homoclinic Center Manifolds and Shil'nikov Theorem

In high-order systems, the homoclinic bifurcation is characterized by the homoclinic center manifolds and Shil'nikov theorem. For a specific UEP, the eigenvalues with positive (negative) real part that are closest to the imaginary axis are called the unstable (stable) leading eigenvalues. If all of the leading eigenvalues are real numbers, the UEP is a saddle. These leading eigenvalues are denoted as λ_u and λ_s, where $\lambda_u > 0 > \lambda_s$. The saddle quantity is $\sigma = \lambda_u + \lambda_s$. If $\sigma > 0$ at the homoclinic bifurcation, a unique UPO will generate after the onset of homoclinic bifurcation (see Fig. 8.11b). If $\sigma < 0$ at the homoclinic bifurcation, a unique SPO will generate before the onset of homoclinic bifurcation [54] (see Fig. 8.11a).

8.4.2 *Medium Injection Reactive Power*

8.4.2.1 Homoclinic Bifurcation

The unstable manifolds of three UEPs for different ω_{pk} are shown in Fig. 8.12. When $\omega_{pk} = 0.1$ p.u., only one unstable manifold converges to the Type-II hidden periodic orbit, while other unstable manifolds all converge to the SEP, as shown in Fig. 8.12a. When ω_{pk} is increased to 0.3 p.u., all unstable manifolds converge to the SEP, as shown in Fig. 8.12b. Therefore, between Fig. 8.12a, b, a homoclinic bifurcation occurs. We also find that the homoclinic bifurcation occurs near $\omega_{pk} = 0.1733$ p.u., where the leading eigenvalues are $\lambda_u = 2.9284$ and $\lambda_s = -0.7897$. Therefore, the saddle quantity $\sigma > 0$, implying the generation of a UPO. This UPO serves as the boundary to separate the SEP's basin of attraction and the Type-II hidden periodic orbit's basin of attraction. Therefore, due to such UPO, all unstable manifolds converge to the SEP, as shown in Fig. 8.12b.

Using the unstable manifolds of UEPs, we derive the homoclinic bifurcation diagrams on $Q_{\text{L1,ref}}$-ω_{pk} plane, as shown in Fig. 8.13. The red, blue, and green

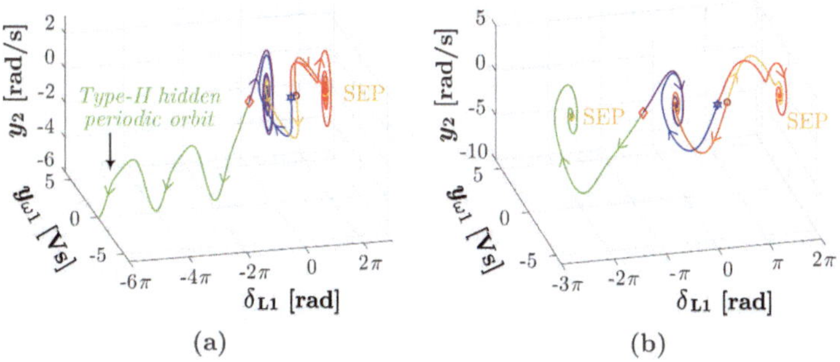

Fig. 8.12 Unstable manifolds of three UEPs for (**a**) $\omega_{pk} = 0.1$ p.u. and (**b**) $\omega_{pk} = 0.3$ p.u. Other parameters adopt the default set (i.e., $Q_{\text{L1,ref}} = 1$ p.u.)

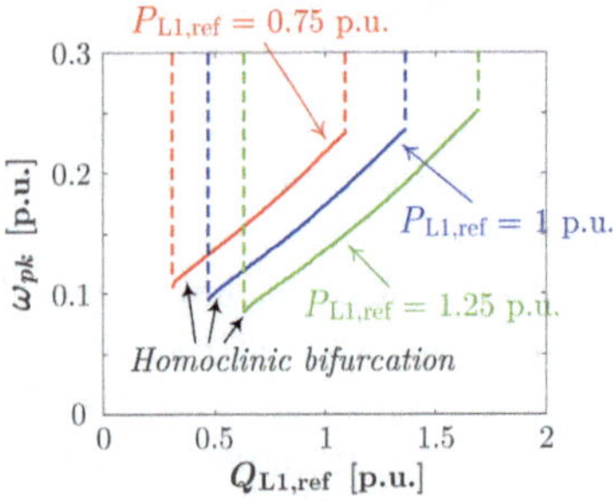

Fig. 8.13 Codimensional-2 homoclinic bifurcation diagrams on the $Q_{\mathrm{L1,ref}}$-ω_{pk} plane. Other parameters adopt the default set with varying $P_{\mathrm{L1,ref}}$

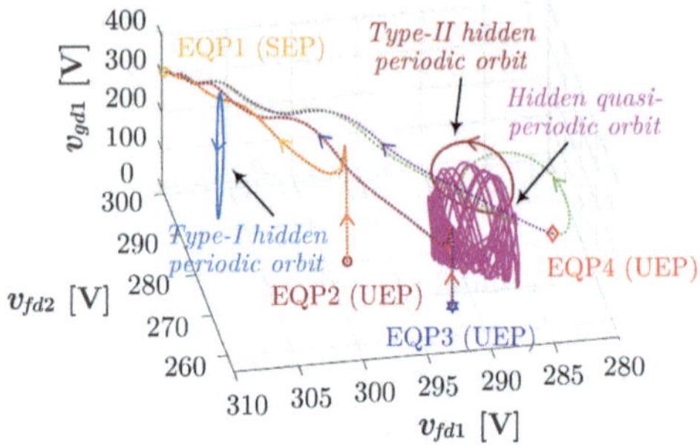

Fig. 8.14 SEP, UEPs, several hidden attractors, and unstable manifolds of UEPs in $(v_{fd1}, v_{fd2}, v_{gd1})$-space. All parameters follow the default set with $\omega_{pk} = 0.1735$ p.u.

curves correspond to $P_{\mathrm{L1,ref}} = 0.75$ p.u., $P_{\mathrm{L1,ref}} = 1$ p.u., and $P_{\mathrm{L1,ref}} = 1.25$ p.u., respectively. On each solid curves, the homoclinic bifurcation occurs. For the area surrounded by the dashed and solid curves with the same color, all unstable manifolds converge to the SEP. For example, we select a group of parameters in the aforementioned area, i.e., $P_{\mathrm{L1,ref}} = 1$ p.u., $Q_{\mathrm{L1,ref}} = 1$ p.u., and $\omega_{pk} = 0.1735$ p.u. In this case, all of the unstable manifolds of UEPs converge to the SEP, as shown in Fig. 8.14. However, Type-I hidden periodic orbit, hidden quasi-periodic orbit, and Type-II hidden periodic orbit still exist, but starting from the SEP or UEPs, the trajectories cannot reach any of the hidden attractors [69].

8.4.2.2 Impact of Homoclinic Bifurcation on Maintaining Synchronization

After the onset of the above homoclinic bifurcation, the SEP's basin of attraction will be enlarged, which may have a significant impact on the synchronization stability of the post-fault system. For the sake of clarity, relevant basins of attraction for $\omega_{pk} = 0.3$ p.u. are derived and shown in Fig. 8.15. Figure 8.15a shows the basins of attraction before and after a temporary SCF, while Fig. 8.15b represents the basins of attraction during the SCF. Initially, the system operates at the SEP. During the SCF, the SEP disappears, and the original SEP's basin of attraction becomes the Type-II hidden attractor's basin of attraction. Therefore, the trajectory will move to the Type-II hidden attractor. As shown in Fig. 8.5f, y_2 oscillates in the range of $-10 \sim 20$ for Type-II hidden attractor. In this case, when the SCF is removed, the

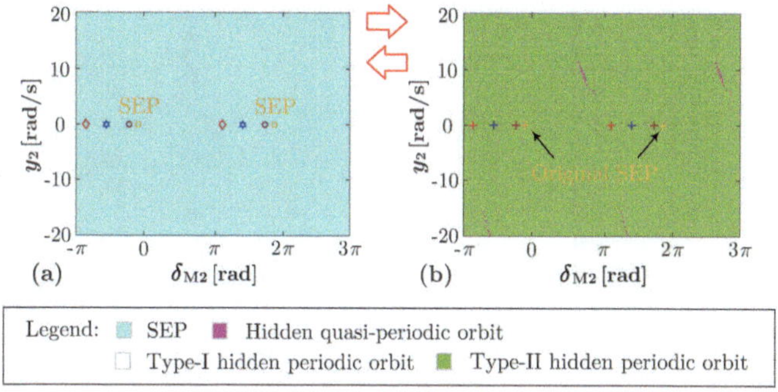

Fig. 8.15 Basins of attraction in (**a**) the (δ_{M2}, y_2)-plane for default R_{b2} and L_{b2}; (**b**) the (δ_{M2}, y_2)-plane for $R_{b2} = r_{gn}$ and $L_{b2} = L_{gn}$. Other parameters follow the default set with $Q_{L1,ref} = 1$ p.u. and $\omega_{pk} = 0.3$ p.u.

original space occupied by the Type-II hidden attractor becomes the SEP's basin of attraction, as shown in Fig. 8.15a. Therefore, the trajectory will return to the SEP, instead of staying at the Type-II hidden attractor. Hence, the CFCT can be further increased, in contrary to the case of $\omega_{pk} = 0.1$ p.u., as shown in Fig. 8.3.

8.4.3 *Small and Large Injection Reactive Power*

8.4.3.1 Another Two Homoclinic Bifurcations

In this subsection, we show that for relatively small or large $Q_{L1,ref}$, homoclinic bifurcations may also occur. As shown in Fig. 8.16a, an unstable manifold converges to the Type-I hidden periodic orbit, while an unstable manifold converges to the hidden quasi-periodic orbit. However, from Fig. 8.16b, only one unstable manifold converges to the Type-I hidden periodic orbit. Hence, a homoclinic bifurcation occurs between Fig. 8.16a, b. Furthermore, from Fig. 8.16c, all unstable manifold converges to the SEP. Therefore, another homoclinic bifurcation occurs between Fig. 8.16b, c. It is also found that the saddle quantity is positive for the aforementioned two homoclinic bifurcations, implying the generation of corresponding UPOs. Using the unstable manifold, we derive the homoclinic bifurcation diagrams, as shown in Fig. 8.17. From Fig. 8.17, the first homoclinic bifurcation occurs when $\omega_{pk} = 0.15 \sim 0.2$ p.u. However, for the second homoclinic bifurcation, the critical ω_{pk} becomes very large when $Q_{L1,ref}$ decreases.

8.4.3.2 Transient Synchronization Stability

As indicated in Sect. 8.4.1, the SEP's basin of attraction is relatively small and surrounded by the stable manifolds of the UEPs before the onset of homoclinic

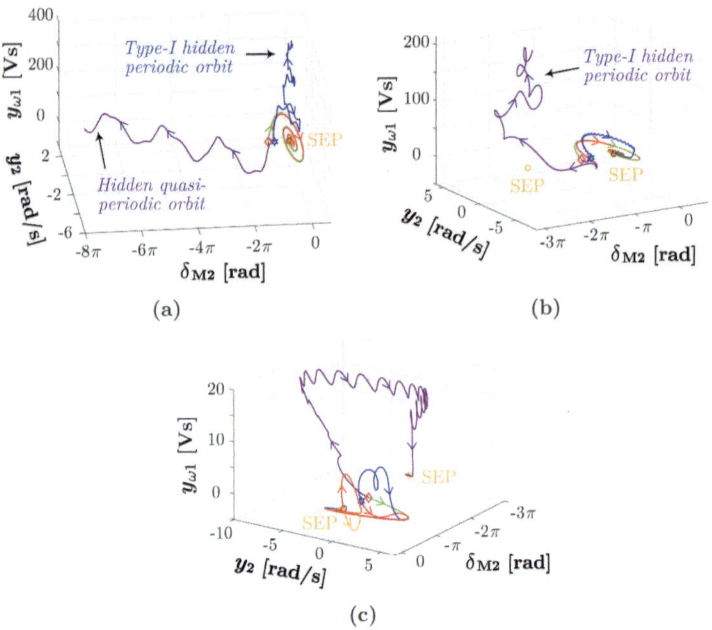

Fig. 8.16 Unstable manifolds of three UEPs for (**a**) $\omega_{pk} = 0.1$ p.u., (**b**) $\omega_{pk} = 0.3$ p.u., and (**b**) $\omega_{pk} = 1.2$ p.u. Other parameters adopt the default set with $Q_{L1,ref} = 0.4$ p.u.

Fig. 8.17 Codimensional-2 homoclinic bifurcation diagrams on the $Q_{L1,ref}$-ω_{pk} plane. Other parameters adopt the default set

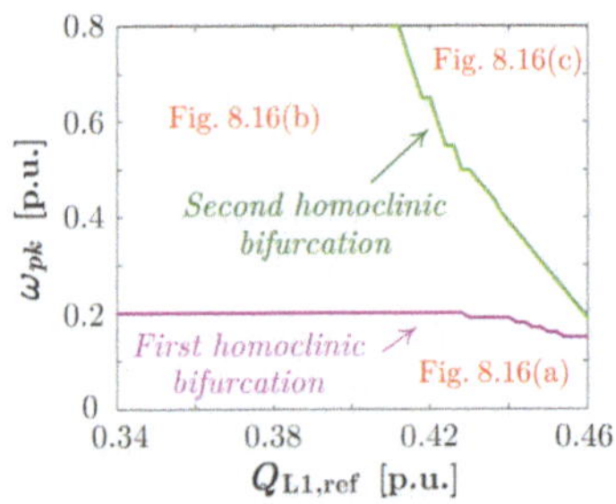

bifurcation. From Fig. 8.17, it is hard to trigger the onset of the second homoclinic bifurcation, due to large critical ω_{pk}. Therefore, the SEP's basin of attraction cannot be significantly enlarged only by triggering the onset of the first homoclinic bifurcation. In other words, further increasing ω_{pk} may not enlarge the SEP's basin of attraction dramatically. In this case, heavy damping (i.e., small ω_{pk}) may prevent the trajectory to move out of the SEP's basin of attraction. As shown in Fig. 8.18, the trajectories start from the SEP, and it is set that $R_{b2} = r_{gn}$ and $L_{b2} = L_{gn}$ initially to emulate the SCF. After 0.4 s, the default values of R_{b2} and L_{b2} are set. The blue and red trajectories correspond to $\omega_{pk} = 0.1$ p.u. and $\omega_{pk} = 0.3$ p.u., respectively. Besides, the dashed curves correspond to $R_{b2} = r_{gn}$ and $L_{b2} = L_{gn}$, while the solid curves correspond to the default R_{b2} and L_{b2}. From Fig. 8.18, the

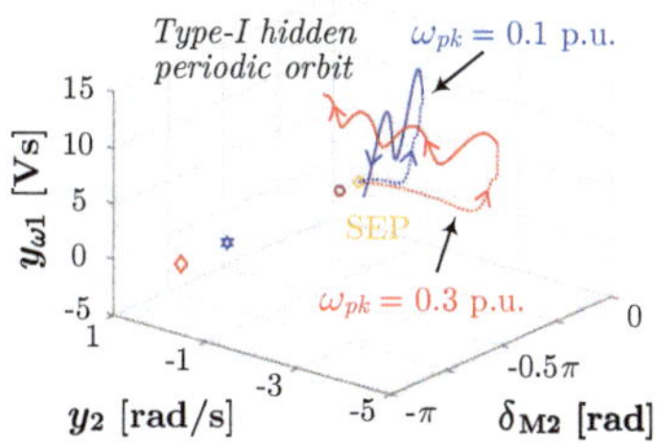

Fig. 8.18 Trajectories for different ω_{pk}. Other parameters adopt the default set with $Q_{L1,ref} = 0.4$ p.u.

variation range of y_2 for the blue trajectory is much smaller than the one for the red trajectory, due to different damping. On this basis, the blue trajectory converges to the SEP eventually, implying that the trajectory is in the SEP's basin of attraction. However, the red trajectory diverges to the Type-I hidden periodic orbit, implying that the trajectory is out of the SEP's basin of attraction.

Therefore, for relatively small or large $Q_{L1,ref}$, increasing ω_{pk} may not significantly enlarge the SEP's basin of attraction, but reduce the damping, which leads to the shortening of the CFCT.

8.5 Design Guidelines

In this section, we develop a set of design guidelines for islanded microgrids containing GFMCs and GFLCs by applying the available research results in the microgrid's bifurcation behavior [138].

From the previous sections, in order to guarantee long CFCTs, it is desirable to select system parameters to (1) make sure the existence of a Type-II hidden periodic orbit in the normal operation and (2) let all unstable manifolds of UEPs converge to the SEP eventually. Therefore, we need to derive the homoclinic bifurcation diagram shown in Fig. 8.13 and put the operating point in the area surrounded by the curves with the same color. As a practical and convenient method to derive such bifurcation diagrams, we can select very small ω_{pk} initially (e.g., 0.05 p.u.). If only one unstable manifold of UEPs converges to the Type-II hidden periodic orbit, and others converge to the SEP, the point ($Q_{L1,ref}$, ω_{pk}) is below the curve that the homoclinic bifurcation occurs. Otherwise, we should adjust $Q_{L1,ref}$ until the aforementioned condition for the unstable manifold is satisfied. On this basis, we can increase ω_{pk} until all unstable manifolds converge to the SEP, and then we find the critical ω_{pk} where the homoclinic bifurcation occurs. With this principle, the homoclinic bifurcation diagram shown in Fig. 8.13 can be derived quickly.

For Fig. 8.13, the operating point should be selected in the area surrounded by the curves with the same color. However, from Figs. 8.6b and 8.9, medium $Q_{L1,ref}$ may guarantee the existence of Type-II hidden periodic orbit during the SCF, which can further increase the CFCTs. Besides, ω_{pk} is associated to the virtual inertia, where large ω_{pk} leads to small virtual inertia [53]. Combining these restrictions, it

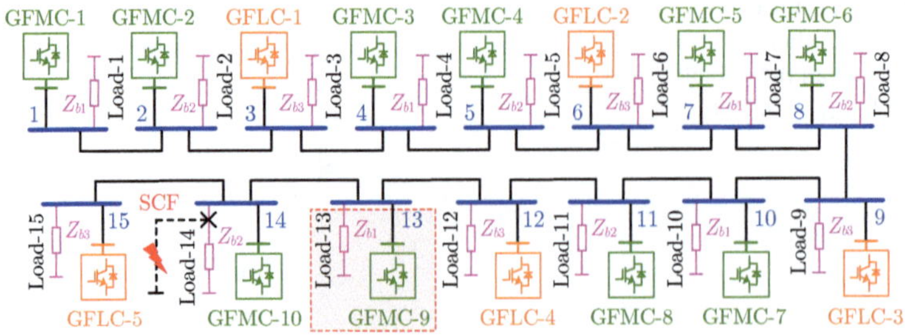

Fig. 8.19 Circuit diagram of the 15-bus islanded microgrid. A temporary SCF may occur in Bus-14

is recommended to select medium $Q_{\mathrm{L1,ref}}$ and small ω_{pk} in the area surrounded by the curves with the same color, as shown in Fig. 8.13.

8.6 Validation

8.6.1 Simulation Results

In this section, exact full-circuit cycle-by-cycle simulations based on MATLAB/Simulink are presented to verify the analytical findings. Here, the 15-bus microgrid shown in Fig. 8.19 is studied. This microgrid contains 10 GFMCs and 5 GFLCs, and a temporary SCF may be applied to Bus-14. The default parameters given in Tables 8.1, 8.2, and 8.3 are adopted with $\omega_{pk} = 0.3$ p.u.

The simulation results are shown in Fig. 8.20, where the microgrid contains 10 GFMCs and 5 GFLCs for Fig. 8.20a, while the microgrid contains 9 GFMCs and 5 GFLCs for Fig. 8.20b. For the latter case, GFMC-9 and Load-13 are shut down, as shown in Fig. 8.19. In Fig. 8.20, a temporary SCF occurs in Bus-14 when $t = 2 \sim 7$ s. From Fig. 8.20a, b, for both microgrids, δ_{M10} and δ_{L5} are phase-locked during the SCF, implying the existence of a Type-II hidden periodic orbit. Then, after clearing the SCF, the oscillations in δ_{M10} and δ_{L5} subside, and $p_{\mathrm{M}k}$, $q_{\mathrm{M}k}$, and $v_{gd\ell}$ return to the normal operation state. All GFMCs and GFLCs remain synchronized with each other. These results confirm that the Type-II hidden periodic orbit is beneficial to maintain the synchronization of GFMCs and GFLCs after removing the SCF. Moreover, the role of the Type-II hidden periodic orbit in enhancing the stability is verified on a microgrid containing various numbers of GFMCs and GFLCs.

8.6.2 Experimental Results

In this section, hardware-in-loop experiments are presented to verify the theoretical findings. The power circuit of the 3-bus microgrid shown in Fig. 8.1

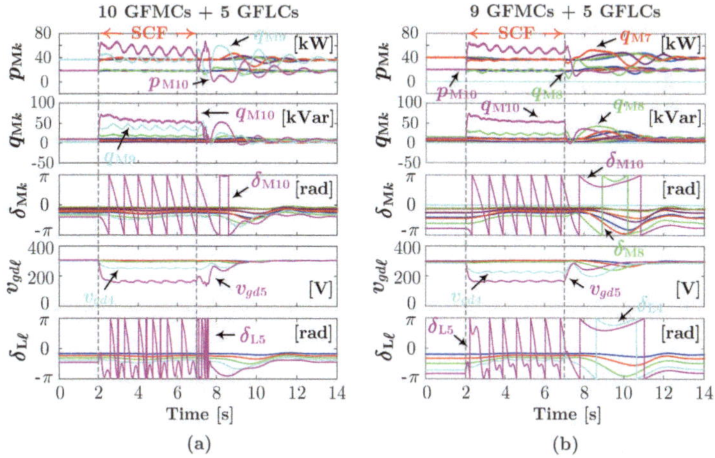

Fig. 8.20 Simulation waveforms of p_{Mk}, q_{Mk}, δ_{Mk}, $v_{gd\ell}$, and $\delta_{L\ell}$, where $\ell = 1, 2, \ldots, 5$, and $k = 1, 2, \ldots, 10$. A temporary SCF occurs in Bus-14 when $t = 2 \sim 7$ s. This fault is cleared at other times. (**a**) There are 10 GFMCs and 5 GFLCs in the microgrid. (**b**) There are 9 GFMCs and 5 GFLCs in the microgrid, where GFMC-9 and Load-13 are shut down. Other parameters adopt the default set with $\omega_{pk} = 0.3$ p.u.

was emulated by OP4512 with eHS toolbox, while the sampling, PWM signal generation and all control algorithms of each inverter were implemented using a DSP TMS320F28377D (i.e., three separated DSPs were adopted). The default parameters are given in Tables 8.1, 8.2, and 8.3, unless otherwise specified.

The experimental results are shown in Fig. 8.21. In these experiments, the system operates normally at the beginning, and then an SCF with duration Δt_{gf} is applied, where Bus-2 is connected to the ground through $r_{gn} + j\omega_0 L_{gn}$. The system may exhibit different behaviors for different values of $Q_{L1,\text{ref}}$, ω_{pk} and Δt_{gf}.

The results given in Fig. 8.21a correspond to $Q_{L1,\text{ref}} = 0$, $\omega_{pk} = 0.1$ p.u., and $\Delta t_{gf} = 0.2$ s. After removing the SCF, δ_{L1} exhibits high-frequency oscillation initially, implying that the PLL output reaches its limit. After a period of time, sustained oscillation emerges in v_{gm1} and i_{gm1}, and only δ_{L1} oscillates from $-\pi$ to π. Here, v_{gm1} is the magnitude of v_{g1}, while i_{gm1} is the magnitude of i_{g1}. These results indicate that the CFCT to maintain the stability of post-fault system is relatively small.

The results shown in Fig. 8.21b correspond to $Q_{L1,\text{ref}} = 1$ p.u., $\omega_{pk} = 0.1$ p.u., and $\Delta t_{gf} = 0.65$ s. Upon removing the SCF, all variables converge to their normal values after a few cycles of oscillations. It is noted that there is no high-frequency ripple in δ_{L1}, implying that the PLL limit is not triggered. Also, the results shown in Fig. 8.21c correspond to $Q_{L1,\text{ref}} = 1$ p.u., $\omega_{pk} = 0.1$ p.u., and $\Delta t_{gf} = 0.75$ s. After the SCF is removed, sub-synchronous oscillation emerges in v_{gm1} and i_{gm1}, whereas δ_{M2} and δ_{L1} all vary from $-\pi$ to π periodically and are phase-locked. The experimental waveforms shown in Fig. 8.21b, c confirm the existence of a Type-II hidden periodic orbit.

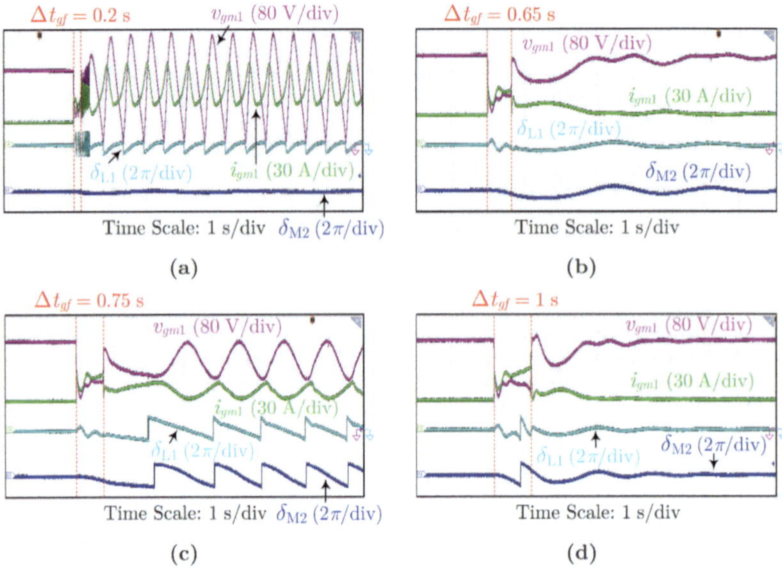

Fig. 8.21 Experimental waveforms of v_{gm1}, i_{gm1}, δ_{M2}, and δ_{L1} before, during, and after the SCF that occurs in Bus-2. Here, v_{gm1} is the magnitude of v_{g1}, while i_{gm1} is the magnitude of i_{g1}. (**a**) $Q_{\mathrm{L1,ref}} = 0$, $\omega_{pk} = 0.1$ p.u., and $\Delta t_{gf} = 0.2$ s; (**b**) $Q_{\mathrm{L1,ref}} = 1$ p.u., $\omega_{pk} = 0.1$ p.u., and $\Delta t_{gf} = 0.65$ s; (**c**) $Q_{\mathrm{L1,ref}} = 1$ p.u., $\omega_{pk} = 0.1$ p.u., and $\Delta t_{gf} = 0.75$ s; (**d**) $Q_{\mathrm{L1,ref}} = 1$ p.u., $\omega_{pk} = 0.3$ p.u., and $\Delta t_{gf} = 1$ s. Other parameters adopt the default set

Finally, the results shown in Fig. 8.21d correspond to $Q_{\mathrm{L1,ref}} = 1$ p.u., $\omega_{pk} = 0.3$ p.u., and $\Delta t_{gf} = 1$ s. During the SCF, δ_{M2} and δ_{L1} vary from $-\pi$ to π and are phase-locked. After clearing the SCF, all variables converge to their normal values.

Comparing the above results, the benefit of Type-II hidden periodic orbit on maintaining the microgrid's synchronization is confirmed. It is also verified that the CFCT can be significantly increased if appropriate reactive power $Q_{\mathrm{L1,ref}}$ is injected by GFLCs, and the cutoff frequency ω_{pk} of GFMCs is slightly increased.

8.7 Recapitulation

In this chapter, the transient synchronization stability of the islanded microgrid containing multiple GFMCs and GFLCs is studied. The microgrid's transient stability is highly relevant to the hidden attractors and the SEP's basin of attraction. Thus, this chapter attempts to develop appropriate design guidelines for transient stability enhancement by analyzing the hidden attractors' characteristics and the related bifurcation behaviors. The main findings of this chapter are as follows:

1. *Characteristics of hidden attractors:* In this chapter, three distinct hidden attractors are identified, including a Type-I hidden periodic orbit, a hidden

quasi-periodic orbit, and a Type-II hidden periodic orbit. Among these three hidden attractors, the Type-II hidden periodic orbit may show more benefits to maintain the microgrid's synchronization stability after removing the transient disturbance. Because the PLL functions properly, and the GFLCs remain synchronized with the microgrid for Type-II hidden periodic orbit, however, the PLL malfunctions, and the GFLCs are desynchronized for the other two hidden attractors. Therefore, compared with the other two hidden attractors, the existence of a Type-II hidden periodic orbit may allow the converters to maintain synchronization with the microgrid, even during prolonged temporary SCFs. This finding highlights the positive impact of hidden attractors on microgrid stability, which is a departure from previous studies that primarily focus on the detrimental effects of hidden attractors.

2. *Bifurcation behaviors:* It is found that the Type-II hidden periodic orbit emerges and subsides through an SNPO. Therefore, triggering the onset of an SNPO to maintain the existence of a Type-II hidden periodic orbit is beneficial to enhance the microgrid's transient stability. Besides, the change of the SEP's basin of attraction is associated with three different homoclinic bifurcations. Using the saddle quantity and unstable manifolds, it is found that for each homoclinic bifurcation, a corresponding UPO emerges after the onset of the bifurcation, which enlarges the SEP's basin of attraction and guarantees the converters' successful resynchronization after clearing the SCF.
3. *Design guidelines:* By applying the aforementioned research results, a set of design guidelines to enhance the microgrid's transient stability is developed. Based on these design guidelines, medium reactive powers are injected by GFLCs to maintain the existence of a Type-II hidden periodic orbit, while slightly large cutoff frequencies of low-pass filters are selected by GFMCs to trigger the onset of homoclinic bifurcation. These control parameters can be quantitatively chosen according to the bifurcation diagrams, which can be derived conveniently through the unstable manifolds.

Chapter 9
Stability of Islanded Microgrids Considering Distributed Secondary Control

For an islanded microgrid containing multiple GFMCs, *distributed secondary control* can be used to restore the microgrid's frequency, compensate any voltage deviation, and achieve reactive power sharing. The distributed secondary control is commonly known as *leader-following consensus control* [139], or *pinning-based control* [37], where at least one specific GFMC's controller (called the driver controller, a term borrowed from network science [140]) is selected to receive the reference control signals from the microgrid controller. Although there can be numerous driver controllers, most studies have focused on one driver controller, leaving the aggregated effects of driver controllers on the system controllability and stability unexplored. The controllability is defined as the ability to reach any desired final state under a suitable choice of control inputs [141]. The network pinning control was first proposed in 2002 to control a large complex network of dynamical systems by sending control signals to a small fraction of network nodes (i.e., driver nodes) [140, 142]. The fundamental question of such control is the network controllability problem, i.e., the number of driver nodes, and the location of driver nodes [140, 141]. However, such a controllability problem for microgrid applications has not been properly considered. Furthermore, due to the asymmetrical power network and the heterogenous GFMCs, the topologies of communication network and specific choices of GFMCs' controllers as driver controllers may have a significant effect on the system stability (i.e., small-signal stability, transient stability, and frequency stability), which remain unexplored in the literature.

Models used in much of the prior work have neglected the dynamics of GFMCs' inner voltage loops [37, 139, 143–146] and the communication delay [37, 144–147]. However, the omission of inner voltage loops may lead to optimistic stability evaluation results [52]. Moreover, the microgrid's hierarchical control system is usually a multirate system [148], where the execution frequency of the controllers is relatively high in comparison with the frequency of the communication between these controllers. Therefore, communication delay may include the delay due to the data transmission and the multirate process. However, only the first delay was

J. Yang, C. K. Tse, *Complex Behavior of Grid-Connected Power Electronics Systems*, https://doi.org/10.1007/978-3-031-69340-3_9

commonly considered [139, 143]. Besides, the complex nonlinear behavior of the microgrid using distributed secondary control was seldom explored. The complex behavior, including the saddle-node bifurcation, Hopf bifurcation, hidden attractors, etc., is the essential cause for the loss of stability [1, 5, 138]. So far, only the stability status of the usual EQP has been studied [37, 139, 143–149], which is associated with Hopf bifurcation. Moreover, using the Hopf bifurcation behavior to evaluate the impact of different basic secondary control structures remains unexplored.

In this chapter, the impact of the distributed secondary control on the microgrid stability (i.e., small-signal stability, and transient stability) is studied.

9.1 Control Structure and Modeling

9.1.1 Basic Structures

The distributed secondary control is targeted to restore the microgrid frequency, compensate for voltage deviation, and achieve reactive power sharing. For the latter two control tasks, various control schemes have been proposed [31, 33, 37, 139, 143–145, 147, 149–152]. In this section, two basic control structures are described to facilitate the comparison of the various control schemes.

9.1.1.1 First Type

The reactive power sharing and the voltage deviation compensation are achieved using separate control loops, as shown in Fig. 9.1a and represented by

$$\frac{\mathrm{d}e_\ell}{\mathrm{d}t} = \gamma_{mc} g_\ell (V_{\mathrm{MG0}} - v_{\mathrm{MG}}) + \zeta_{lc} \sum_{k=1}^{N} a_{\ell k}(e_k - e_\ell) \tag{9.1}$$

$$\frac{\mathrm{d}z_\ell}{\mathrm{d}t} = \gamma_{lc} \sum_{k=1}^{N} a_{\ell k}\left(\frac{n_{qk}}{n_{q1}} q_k - \frac{n_{q\ell}}{n_{q1}} q_\ell\right) \tag{9.2}$$

where e_ℓ and z_ℓ can be the increment of the voltage reference $v^*_{fd\ell}$, or the decrement of the virtual impedance $(r_{p\ell} + j\omega_\ell L_{p\ell})$. Here, $\ell = 1, 2, \cdots, N$, where N is the number of GFMCs in the microgrid. Notably, (9.1) corresponds to the voltage deviation compensation, and (9.2) corresponds to the reactive power sharing. Besides, $n_{q\ell}$ is the droop coefficient of GFMC-ℓ, q_ℓ is the reactive power of GFMC-ℓ, v_{MG} is the voltage magnitude of a selected bus of the microgrid, and V_{MG0} is the target value of v_{MG}. Moreover, γ_{mc}, ζ_{lc}, and γ_{lc} are the proportional coefficients. Furthermore, $g_\ell = 1$ if there is a communication link between the local controller and the microgrid controller, and $g_\ell = 0$ otherwise. Likewise, $a_{\ell k} = 1$ if there is a communication link between the two local controllers (i.e., local controller

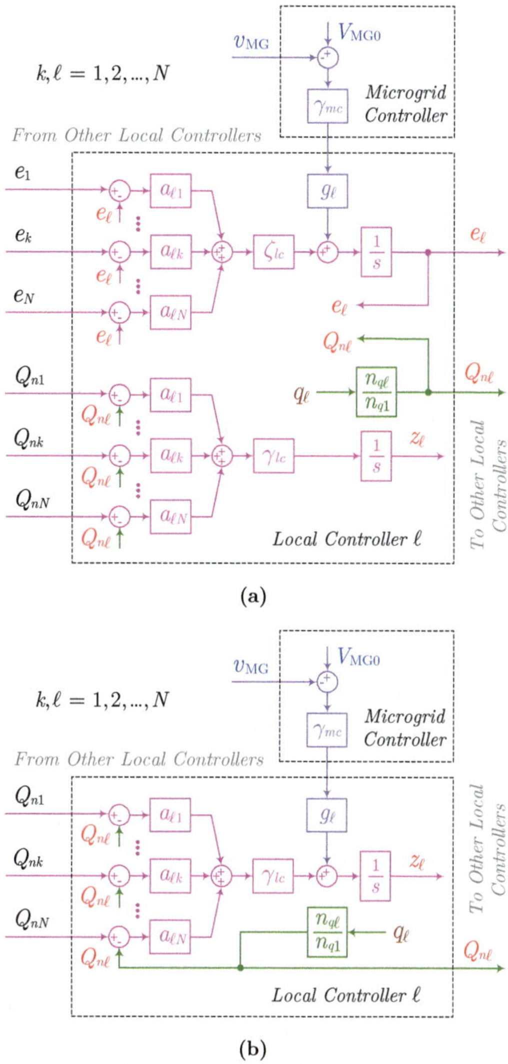

Fig. 9.1 Basic structures of the distributed control to regulate the microgrid voltage and the reactive powers: (**a**) first type and (**b**) second type

ℓ and local controller k); and $a_{\ell k} = 0$ otherwise. It should be noted that $a_{\ell\ell}$ is always 0.

9.1.1.2 Second Type

The sharing of reactive power and the compensation for voltage deviation are achieved using the same control loop, as shown in Fig. 9.1b, i.e.,

Table 9.1 Comparison between basic structures

	Eq.	Data[a]	Output control variable		Control schemes in the literature
			Symbol	Meaning	
First type	(1)	e_ℓ	e_ℓ	$v^*_{fd\ell}$	[37, 139, 143, 147] [145, 149]
				$r_{p\ell} + j\omega_\ell L_{p\ell}$	/
	(2)	$\frac{n_{q\ell}}{n_{q1}} q_\ell$	z_ℓ	$v^*_{fd\ell}$	[37, 143, 147, 150]
				$r_{p\ell} + j\omega_\ell L_{p\ell}$	[145, 151, 152]
Second type	(3)	$\frac{n_{q\ell}}{n_{q1}} q_\ell$	z_ℓ	$v^*_{fd\ell}$	[144]
				$r_{p\ell} + j\omega_\ell L_{p\ell}$	/

[a] The data being exchanged among the local controllers

$$\frac{\mathrm{d}z_\ell}{\mathrm{d}t} = \underbrace{\gamma_{mc} g_\ell (V_{\mathrm{MG0}} - v_{\mathrm{MG}})}_{\text{Pinning-Based}} + \underbrace{\gamma_{lc} \sum_{k=1}^{N} a_{\ell k} (\frac{n_{qk}}{n_{q1}} q_k - \frac{n_{q\ell}}{n_{q1}} q_\ell)}_{\text{Consensus-Based}} \tag{9.3}$$

where z_ℓ can be the increment of $v^*_{fd\ell}$, or the decrement of $(r_{p\ell} + j\omega_\ell L_{p\ell})$. From (9.3), the distributed control is made up of two parts, i.e., pinning-based algorithm [140] and consensus-based algorithm [153]. For the former one, only some of the local controllers will receive the control signals from the microgrid controller (i.e., $\gamma_{mc}(V_{\mathrm{MG0}} - v_{\mathrm{MG}})$) to compensate for the voltage deviation. For the latter one, the information of $(n_{q\ell} q_\ell)/n_{q1}$ is shared among all the local controllers to reach an agreement that $n_{q1} q_1 = n_{q2} q_2 = \cdots = n_{qN} q_N$.

Remark Judging from the control equations and control output variables, the aforementioned two basic structures cover most of the existing control schemes, as shown in Table 9.1. Compared with the first type of control structure, the second type can reduce the communication traffic among the local controllers and simplify the secondary control algorithms in each local controller. Thus, the second type is adopted in the following to establish a model for studying the impact of the secondary control on the system's stability.

9.1.2 Communication Delay

As shown in Fig. 9.1, a specific local controller may need to communicate with other local controllers and/or the microgrid controller. Such a communication process may generate extra time delay. As shown in Fig. 9.2, the control algorithms are executed in the digital controllers at a frequency of f_{ctr}, which is always much higher than the frequency of transmitting data from one controller to another (i.e.,

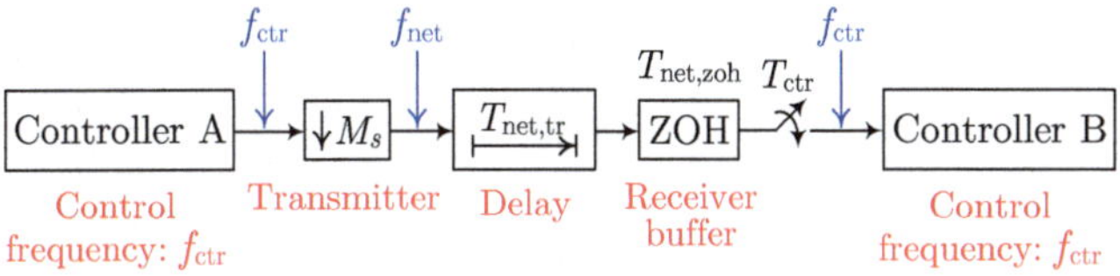

Fig. 9.2 Communication between two controllers

f_{net}). Thus, the transmitter involves a decimation and down-sampling process [154]. Due to the limited bit rate, a time delay (i.e., $T_{net,tr}$) exists in transmitting the data. Moreover, when the data are received by another controller, they are stored in the receiver buffer until the next data sequence arrives. This process resembles a zero-order hold (ZOH) [154], where the corresponding delay is

$$T_{net,zoh} = \frac{T_{net}}{2} = \frac{1}{2} \times \frac{1}{f_{net}}. \tag{9.4}$$

Due to the discrepancy between f_{net} and f_{ctr}, the receiving-side controller will access the receiver buffer at f_{ctr}, representing a resampling process [154]. The exact value of $T_{net,tr}$ is determined by the adopted communication protocol. It may be very large for the Ethernet [143]. However, in our prototype, the bit rate of the communication is relatively high, e.g., 500 kbit/s for the CAN communication. Therefore, $T_{net,zoh} \gg T_{net,tr}$, and we denote the overall time delay as

$$T_{dn} = T_{net,tr} + T_{net,zoh} \approx T_{net,zoh}. \tag{9.5}$$

9.1.3 Complete Control Configuration

The complete configuration of the control strategy adopted by the multi-GFMC microgrid is shown in Fig. 9.3. The control structure is hierarchical, having a primary control level and a secondary control level.

The primary control includes a droop control, a voltage-inner-loop control (using virtual impedances), and a current-inner-loop control. However, as revealed in a previous study [52], the current loop has insignificant impact on the nonlinear behavior associated with sub-synchronous oscillations. Therefore, the current-loop dynamics can be neglected in Fig. 9.3. In this chapter, we focus on conventional voltage control strategy [52], while other advanced voltage control schemes [155, 156] will be left for future study. To ensure accurate stability evaluation results, the voltage-loop dynamics are preserved. For the voltage control, $h_{p\ell}$ and $h_{i\ell}$ are the proportional and integral coefficients, respectively; $C_{f\ell}$ is the ac-side filter capacitance of GFMC-ℓ (see Fig. 9.4); $v_{fd\ell}$ and $v_{fq\ell}$ are the d-axis and q-axis output voltage of GFMC-ℓ, respectively; and $i_{ld\ell}$ and $i_{lq\ell}$ are the d-axis and q-axis output current of GFMC-ℓ,

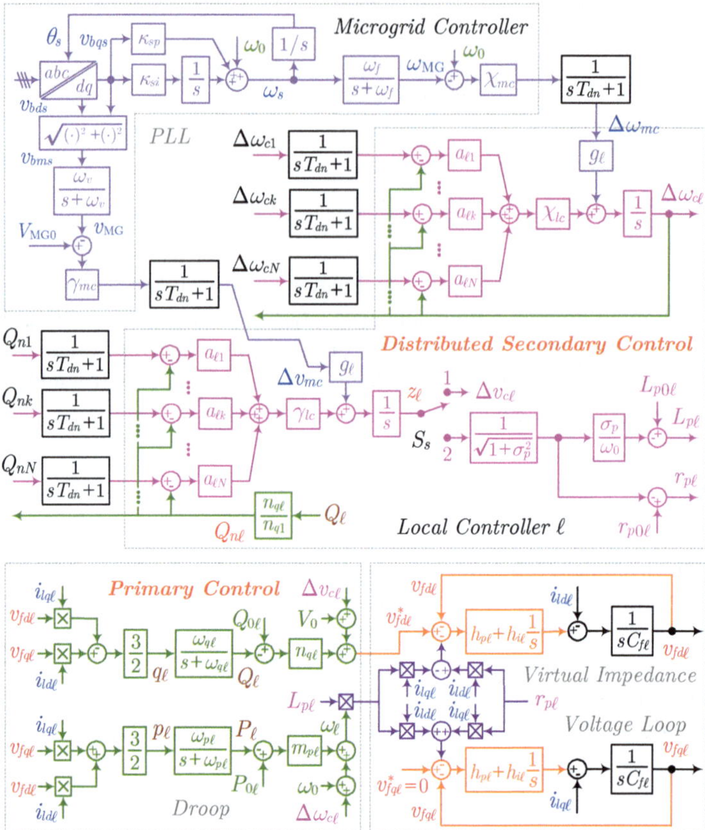

Fig. 9.3 Control block diagram of the distributed secondary control and the primary control

respectively. For the power control, p_ℓ and q_ℓ are the active and reactive powers, respectively; $m_{p\ell}$ and $n_{q\ell}$ are the droop coefficient; and the cutoff frequencies of the low-pass filters are $\omega_{p\ell}$ and $\omega_{q\ell}$. Moreover, $\Delta v_{c\ell}$ and $\Delta\omega_{c\ell}$ are the voltage regulation term and the frequency regulation term, respectively, generated from the secondary control. In addition, $L_{p\ell}$ and $r_{p\ell}$ may be updated through the secondary control, depending on the setting of z_ℓ. The primary control is executed in each local controller, where one local controller corresponds to one GFMC.

Based on (9.3), the secondary control consists of two parts, namely, the generation of pinning-based control signals (which is executed in the microgrid controller) and the combination of pinning-based control signals and consensus-based control signals (which is executed in each local controller). In the microgrid controller, the voltage of a specific bus (e.g., Bus-1) is selected to extract the frequency ω_{MG} and the voltage v_{MG} using a PLL and low-pass filters. For the PLL, κ_{sp} and κ_{si} are the proportional and integral coefficients, respectively. For the low-pass filters, ω_f and ω_v are the cutoff frequencies. For the pinning-based control, χ_{mc} and γ_{mc} are the proportional coefficients, and $\Delta\omega_{mc}$ and Δv_{mc} are the pinning-based control signals transmitted from the microgrid controller to some specific local controllers. In each local controller, $Q_{n\ell} = (n_{q\ell}Q_\ell)/n_{q1}$ is used as the consensus-based control signals, and χ_{lc} and γ_{lc} are the consensus-based proportional coefficients. The setting of z_ℓ is determined by switch S_s, as detailed below.

If $S_s = 1$ (i.e., voltage-increment-based secondary control),

$$\Delta v_{c\ell} = z_\ell, \quad L_{p\ell} = L_{p0\ell}, \quad r_{p\ell} = r_{p0\ell}. \tag{9.6}$$

If $S_s = 2$ (i.e., virtual-impedance-based secondary control),

$$\begin{aligned} &\Delta v_{c\ell} = 0, \\ &L_{p\ell} = L_{p0\ell} - \frac{\sigma_p}{\omega_0\sqrt{1+\sigma_p^2}} z_\ell, \quad r_{p\ell} = r_{p0\ell} - \frac{1}{\sqrt{1+\sigma_p^2}} z_\ell \end{aligned} \tag{9.7}$$

where σ_p is the fixed ratio of reactance to resistance. It should be noted that z_ℓ is the decrement of the virtual inductance if $\sigma_p \to +\infty$, while z_ℓ is the decrement of the virtual resistance if $\sigma_p \to 0$. In this chapter, we use a large virtual inductance to ensure an inductive microgrid so that an active power-frequency droop control and reactive power-voltage droop control can be adopted [113, 157]. Thus, $\omega_0 L_{p0\ell} \gg r_{p0\ell}$, and $\sigma_p > 1$.

9.1.4 Modeling

In an islanded microgrid, there are N local buses, and each bus is connected to a GFMC through a connecting line with the impedance $(r_{l\ell} + j\omega_1 L_{l\ell})$. Moreover, each bus has a constant impedance load $Z_{b\ell} = R_{b\ell} + j\omega_1 L_{b\ell}$. Besides, each bus may connect to other buses to form a power network, where the impedance of the connecting line between two different buses is $(r_{c\rho} + j\omega_1 L_{c\rho})$. Here, $\rho = 1, 2, \cdots, N_c$, where N_c is the number of such connecting lines.

9.1.4.1 Primary Control Level

From Fig. 9.3, the power loop can be modeled as

$$\frac{\mathrm{d}}{\mathrm{d}t}\begin{bmatrix} P_\ell \\ Q_\ell \end{bmatrix} = \begin{bmatrix} \omega_{p\ell} & 0 \\ 0 & \omega_{q\ell} \end{bmatrix}\left(\begin{bmatrix} p_\ell \\ q_\ell \end{bmatrix} - \begin{bmatrix} P_\ell \\ Q_\ell \end{bmatrix}\right) \tag{9.8}$$

$$\omega_\ell = \omega_0 + \Delta\omega_{c\ell} + m_{p\ell}(P_{0\ell} - P_\ell) \tag{9.9}$$

$$v_{fd\ell}^* = V_0 + \Delta v_{c\ell} + n_{q\ell}(Q_{0\ell} - Q_\ell), \quad v_{fq\ell}^* = 0 \tag{9.10}$$

where

$$p_\ell = \frac{3}{2}(v_{fd\ell}i_{ld\ell} + v_{fq\ell}i_{lq\ell}), \quad q_\ell = \frac{3}{2}(-v_{fd\ell}i_{lq\ell} + v_{fq\ell}i_{ld\ell}).$$

The voltage loop can be modeled as

$$\frac{\mathrm{d}}{\mathrm{d}t}\begin{bmatrix} x_{vd\ell} \\ x_{vq\ell} \end{bmatrix} = \begin{bmatrix} \Delta v_{fd\ell} \\ \Delta v_{fq\ell} \end{bmatrix} \tag{9.11}$$

$$\frac{\mathrm{d}}{\mathrm{d}t}\begin{bmatrix} v_{fd\ell} \\ v_{fq\ell} \end{bmatrix} = \frac{1}{C_{f\ell}}\left(h_{p\ell}\begin{bmatrix} \Delta v_{fd\ell} \\ \Delta v_{fq\ell} \end{bmatrix} + h_{i\ell}\begin{bmatrix} x_{vd\ell} \\ x_{vq\ell} \end{bmatrix} - \begin{bmatrix} i_{ld\ell} \\ i_{lq\ell} \end{bmatrix}\right) \tag{9.12}$$

where $x_{vd\ell}$ and $x_{vq\ell}$ are the state variables of the voltage controller, and

$$\begin{bmatrix} \Delta v_{fd\ell} \\ \Delta v_{fq\ell} \end{bmatrix} = \begin{bmatrix} v^*_{fd\ell} \\ v^*_{fq\ell} \end{bmatrix} - \begin{bmatrix} v_{fd\ell} \\ v_{fq\ell} \end{bmatrix} - \begin{bmatrix} r_{p\ell} & -\omega_\ell L_{p\ell} \\ \omega_\ell L_{p\ell} & r_{p\ell} \end{bmatrix}\begin{bmatrix} i_{ld\ell} \\ i_{lq\ell} \end{bmatrix}.$$

For GFMC-ℓ, each set of equations shown above is represented in its dq-frame with angular frequency ω_ℓ, which is denoted as $d_\ell q_\ell$-frame. The coordinate frame of GFMC-1 is selected as the common coordinate frame, which is denoted as DQ-frame [52]. The angle between $d_\ell q_\ell$-frame and DQ-frame is δ_ℓ. Therefore,

$$\frac{\mathrm{d}\delta_\ell}{\mathrm{d}t} = \omega_\ell - \omega_1, \quad \delta_1 = 0. \tag{9.13}$$

The transformation between the $d_\ell q_\ell$-frame and DQ-frame is

$$\begin{bmatrix} v_{fD\ell} \\ v_{fQ\ell} \end{bmatrix} = \mathbf{T}(\delta_\ell)\begin{bmatrix} v_{fd\ell} \\ v_{fq\ell} \end{bmatrix}, \quad \begin{bmatrix} i_{ld\ell} \\ i_{lq\ell} \end{bmatrix} = \mathbf{T}^{-1}(\delta_\ell)\begin{bmatrix} i_{lD\ell} \\ i_{lQ\ell} \end{bmatrix} \tag{9.14}$$

where

$$\mathbf{T}(\vartheta) = \begin{bmatrix} \cos\vartheta & -\sin\vartheta \\ \sin\vartheta & \cos\vartheta \end{bmatrix}, \quad \mathbf{T}^{-1}(\vartheta) = \begin{bmatrix} \cos\vartheta & \sin\vartheta \\ -\sin\vartheta & \cos\vartheta \end{bmatrix}.$$

9.1.4.2 Secondary Control Level

The coordinate frame of the microgrid controller is denoted as $d_s q_s$-frame, and the angle between this frame and DQ-frame is δ_s. From Fig. 9.3,

$$\frac{\mathrm{d}y_s}{\mathrm{d}t} = v_{bqs} \tag{9.15}$$

$$\frac{\mathrm{d}\delta_s}{\mathrm{d}t} = \omega_s - \omega_1 \tag{9.16}$$

$$v_{bms} = \sqrt{v_{bds}^2 + v_{bqs}^2} \tag{9.17}$$

$$\omega_s = \kappa_{sp} v_{bqs} + \kappa_{si} y_s + \omega_0 \tag{9.18}$$

where y_s is the state variable of PLL. If the voltage of Bus-1 is sampled by the microgrid controller:

$$\begin{bmatrix} v_{bds} \\ v_{bqs} \end{bmatrix} = \mathbf{T}^{-1}(\delta_s) \begin{bmatrix} v_{bD1} \\ v_{bQ1} \end{bmatrix}, \tag{9.19}$$

where v_{bD1} and v_{bQ1} is the DQ-frame variables of Bus-1 voltage. Then, we get

$$\frac{\mathrm{d}}{\mathrm{d}t} \begin{bmatrix} \omega_{\mathrm{MG}} \\ v_{\mathrm{MG}} \end{bmatrix} = \begin{bmatrix} \omega_f & 0 \\ 0 & \omega_v \end{bmatrix} \left(\begin{bmatrix} \omega_s \\ v_{bms} \end{bmatrix} - \begin{bmatrix} \omega_{\mathrm{MG}} \\ v_{\mathrm{MG}} \end{bmatrix} \right) \tag{9.20}$$

$$\frac{\mathrm{d}}{\mathrm{d}t} \begin{bmatrix} \Delta\omega_{mc} \\ \Delta v_{mc} \end{bmatrix} = \frac{1}{T_{dn}} \left(\begin{bmatrix} \chi_{mc} & 0 \\ 0 & \gamma_{mc} \end{bmatrix} \left(\begin{bmatrix} \omega_0 \\ V_{\mathrm{MG0}} \end{bmatrix} - \begin{bmatrix} \omega_{\mathrm{MG}} \\ v_{\mathrm{MG}} \end{bmatrix} \right) - \begin{bmatrix} \Delta\omega_{mc} \\ \Delta v_{mc} \end{bmatrix} \right). \tag{9.21}$$

In the local controllers, we have

$$\frac{\mathrm{d}}{\mathrm{d}t} \begin{bmatrix} \Delta\omega_{c\ell} \\ z_\ell \end{bmatrix} = g_\ell \begin{bmatrix} \Delta\omega_{mc} \\ \Delta v_{mc} \end{bmatrix} + \begin{bmatrix} \chi_{lc} & 0 \\ 0 & \gamma_{lc} \end{bmatrix} \sum_{k=1}^{N} a_{\ell k} \left(\begin{bmatrix} \Delta\omega_{ck}^d \\ Q_{nk}^d \end{bmatrix} - \begin{bmatrix} \Delta\omega_{c\ell} \\ Q_{n\ell} \end{bmatrix} \right) \tag{9.22}$$

$$\frac{\mathrm{d}}{\mathrm{d}t} \begin{bmatrix} \Delta\omega_{c\ell}^d \\ Q_{n\ell}^d \end{bmatrix} = \frac{1}{T_{dn}} \left(\begin{bmatrix} \Delta\omega_{c\ell} \\ Q_{n\ell} \end{bmatrix} - \begin{bmatrix} \Delta\omega_{c\ell}^d \\ Q_{n\ell}^d \end{bmatrix} \right) \tag{9.23}$$

where $Q_{n\ell} = n_{q\ell} Q_\ell / n_{q1}$. Moreover, g_ℓ is the element of the pinning vector (i.e., **G**), and $a_{\ell k}$ is the element of the adjacency matrix of the communication network (i.e., **A**) [37, 140, 158].

9.1.4.3 Power Network and Loads

Using the reduced-order model [52], we have

$$\begin{bmatrix} \mathbf{i}_{\mathrm{lDQ1}} \\ \mathbf{i}_{\mathrm{lDQ2}} \\ \vdots \\ \mathbf{i}_{\mathrm{lDQ}N} \end{bmatrix} = (\mathbf{I} + \mathbf{D_N B_L})^{-1} \mathbf{D_N} \begin{bmatrix} \mathbf{v}_{\mathrm{fDQ1}} \\ \mathbf{v}_{\mathrm{fDQ2}} \\ \vdots \\ \mathbf{v}_{\mathrm{fDQ}N} \end{bmatrix}, \tag{9.24}$$

where

$$\mathbf{i_{lDQ\ell}} \triangleq \left[i_{lD\ell}\ i_{lQ\ell}\right]^T, \quad \mathbf{v_{fDQ\ell}} \triangleq \left[v_{fD\ell}\ v_{fQ\ell}\right]^T,$$

$$\mathbf{D_N} = \mathbf{D_B} + \mathbf{D_C}, \quad \mathbf{D_B} = \text{diag}\{\mathbf{D_{b\ell}}\},$$

$$\mathbf{D_{b\ell}} = \frac{1}{R_{b\ell}^2 + (\omega_1 L_{b\ell})^2} \begin{bmatrix} R_{b\ell} & \omega_1 L_{b\ell} \\ -\omega_1 L_{b\ell} & R_{b\ell} \end{bmatrix},$$

$$\mathbf{B_L} = \text{diag}\{\mathbf{B_{l\ell}}\}, \quad \mathbf{B_{l\ell}} = \begin{bmatrix} r_{l\ell} & -\omega_1 L_{l\ell} \\ \omega_1 L_{l\ell} & r_{l\ell} \end{bmatrix}.$$

Here, $\mathbf{D_C}$ is a weighted Laplacian matrix of the power network, and the edge of the graph is weighted by $\mathbf{D_{c\rho}}$ [158]:

$$\mathbf{D_{c\rho}} = \frac{1}{r_{c\rho}^2 + (\omega_1 L_{c\rho})^2} \begin{bmatrix} r_{c\rho} & \omega_1 L_{c\rho} \\ -\omega_1 L_{c\rho} & r_{c\rho} \end{bmatrix}.$$

Besides, the Bus-1 voltage is derived via

$$\begin{bmatrix} v_{bD1} \\ v_{bQ1} \end{bmatrix} = \begin{bmatrix} v_{fD1} \\ v_{fQ1} \end{bmatrix} - \begin{bmatrix} r_{l1} & -\omega_1 L_{l1} \\ \omega_1 L_{l1} & r_{l1} \end{bmatrix} \begin{bmatrix} i_{lD1} \\ i_{lQ1} \end{bmatrix}. \tag{9.25}$$

Combining (9.6) and (9.8)–(9.25), we get the model for $S_s = 1$. Combining (9.7)–(9.25), we get the model for $S_s = 2$. The number of state equations is $N_s = 11N+5$, and the state variables are P_ℓ, Q_ℓ, $x_{vd\ell}$, $x_{vq\ell}$, $v_{fd\ell}$, $v_{fq\ell}$, δ_ℓ ($\ell \neq 1$), y_s, δ_s, ω_{MG}, v_{MG}, $\Delta\omega_{mc}$, Δv_{mc}, $\Delta\omega_{cl}^d$, $Q_{n\ell}^d$, $\Delta\omega_{c\ell}$, and z_ℓ. Then, the model can be represented as

$$\frac{d\mathbf{x}}{dt} = \mathbf{f}(\mathbf{x}) = \left[f_1(x_1, \cdots, x_{N_s}) \ \cdots \ f_{N_s}(x_1, \cdots, x_{N_s})\right]^T \tag{9.26}$$

where $\mathbf{x}$ is the vector of state variables, and $\mathbf{f}(\mathbf{x})$ is the vector of state functions.

9.2 Impact of Distributed Secondary Control on Dynamical Behavior

The three-bus islanded microgrid shown in Fig. 9.4 is taken as an example to study the impact of the distributed secondary control on the complex nonlinear behavior of the microgrid. From Fig. 9.4, $N = 3$ and $N_c = 2$. Besides, local controller 2 communicates with local controller 1 and local controller 3, but there is no communication between local controller 1 and local controller 3. Moreover,

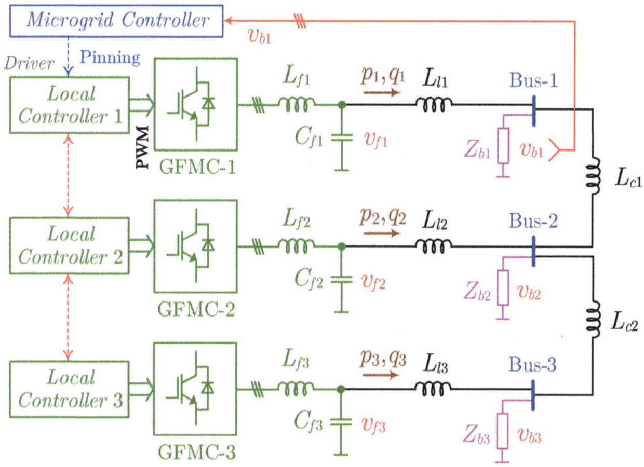

Fig. 9.4 Three-bus islanded microgrid. The ESRs of $L_{f\ell}$, $L_{l\ell}$ and $L_{c\rho}$ are $r_{f\ell}$, $r_{l\ell}$ and $r_{c\rho}$, where $\ell = 1, 2, 3$ and $\rho = 1, 2$

microgrid controller sends the pinning-based control signals to local controller 1. Therefore,

$$\mathbf{G} = \begin{bmatrix} 1 \\ 0 \\ 0 \end{bmatrix}, \quad \mathbf{A} = \begin{bmatrix} 0 & 1 & 0 \\ 1 & 0 & 1 \\ 0 & 1 & 0 \end{bmatrix},$$

$$\mathbf{D_C} = \begin{bmatrix} \mathbf{D_{c1}} & -\mathbf{D_{c1}} & 0 \\ -\mathbf{D_{c1}} & \mathbf{D_{c1}} + \mathbf{D_{c2}} & -\mathbf{D_{c2}} \\ 0 & -\mathbf{D_{c2}} & \mathbf{D_{c2}} \end{bmatrix}.$$

The default circuit and control parameters are shown in Tables 9.2, 9.3, and 9.4. In order to provide sufficient virtual inertia and guarantee the frequency stability, the filters are designed with relatively low cutoff frequencies [18, 159].

9.2.1 Saddle-Node Bifurcation

We reveal in this subsection that *the regulating range of the microgrid voltage* (i.e., the maximum of V_{MG0}) *may be restricted by saddle-node bifurcation.* After the onset of a saddle-node bifurcation, the usual operating point does not exist. Thus, the system becomes unstable (i.e., small-signal instability) [5]. Based on (9.26), the EQP can be derived by setting $\mathbf{f}(\mathbf{x}) = \mathbf{0}$, as shown in Fig. 9.5. For $S_s = 1$, the value of δ_3 remains unchanged when varying V_{MG0}, implying that increasing the microgrid voltage will not trigger a saddle-node bifurcation, and the EQP exists

Table 9.2 Default circuit parameters

Parameter	ℓ or ρ	Value	Parameter	ℓ or ρ	Value
$L_{f\ell}$	$1, 2, \cdots, 9$	1.8 mH	$r_{f\ell}$	$1, 2, \cdots, 9$	0.1 Ω
$C_{f\ell}$	$1, 2, \cdots, 9$	300 μF			
$L_{l\ell}$	$1, 4, 7$	3 mH	$r_{l\ell}$	$1, 4, 7$	0.8 Ω
	$2, 5, 8$	2 mH		$2, 5, 8$	0.5 Ω
	$3, 6, 9$	3 mH		$3, 6, 9$	0.5 Ω
$L_{b\ell}$	$1, 4, 7$	3 mH	$R_{b\ell}$	$1, 4, 7$	6 Ω
	$2, 5, 8$	2 mH		$2, 5, 8$	4.2 Ω
	$3, 6, 9$	1.5 mH		$3, 6, 9$	3 Ω
L_{cp}	$1, 2, 4,$ $5, 7, 8$	2 mH	r_{cp}	$1, 2, 4,$ $5, 7, 8$	0.5 Ω
	3	4 mH		3	1 Ω
	6	3 mH		6	0.75 Ω

Note: These default parameters belong to the three-bus islanded microgrid shown in Fig. 9.4 where $\ell = 1, 2, 3$ and $\rho = 1, 2$ and the nine-bus islanded microgrid shown in Fig. 9.12 where $\ell = 1, 2, \cdots, 9$ and $\rho = 1, 2, \cdots, 8$

Table 9.3 Default primary control parameters

Parameter	ℓ	Value	Parameter	ℓ	Value
$\omega_{p\ell}$	$1, 2, \cdots, 9$	$2\pi \times 0.1$	$\omega_{q\ell}$	$1, 2, \cdots, 9$	$2\pi \times 1$
$P_{0\ell}$	$1, 4, 7$	50 kW	ω_0	/	100π
	$2, 5, 8$	25 kW	V_0	/	311 V
	$3, 6, 9$	100 kW	$Q_{0\ell}$	$1, 2, \cdots, 9$	0
P_m	/	50 kW	Q_m	/	15 kVar
$m_{p\ell}$	$1, 4, 7$	$0.02M_p$	$n_{q\ell}$	$1, 4, 7$	$0.025N_q$
	$2, 5, 8$	$0.04M_p$		$2, 5, 8$	$0.05N_q$
	$3, 6, 9$	$0.01M_p$		$3, 6, 9$	$0.0125N_q$
$h_{p\ell}$	$1, 2, \cdots, 9$	0.21	$h_{i\ell}$	$1, 2, \cdots, 9$	13.43
f_{ctr}	/	4 kHz	f_{net}	/	4 Hz
$L_{p0\ell}$	$1, 4, 7$	3 mH	$r_{p0\ell}$	$1, 2, \cdots, 9$	0
	$2, 5, 8$	2 mH	σ_p	/	3
	$3, 6, 9$	5 mH			

Note: These default parameters belong to the three-bus islanded microgrid shown in Fig. 9.4 where $\ell = 1, 2, 3$ and $\rho = 1, 2$ and the nine-bus islanded microgrid shown in Fig. 9.12 where $\ell = 1, 2, \cdots, 9$ and $\rho = 1, 2, \cdots, 8$. Besides, $M_p = \omega_0/P_m$, and $N_q = V_0/Q_m$

all the time. In this case, the system will not be destabilized via a saddle-node bifurcation. However, for $S_s = 2$, the saddle-node bifurcation may be triggered at a specific value of V_{MG0}, which can be denoted as $V_{\text{MG0,SNB}}$. Furthermore, $V_{\text{MG0,SNB}}$ becomes large if σ_p is decreased, implying that the microgrid voltage can be further increased without losing the usual EQP if more virtual resistance (rather than virtual inductance) is adjusted.

Table 9.4 Default secondary control parameters

Parameter	S_s	Value	Parameter	S_s	Value
κ_{sp}	/	0.413	κ_{si}	/	7.786
ω_f	/	$2\pi \times 0.1$	ω_v	/	$2\pi \times 1$
V_{MG0}	/	280 V			
χ_{mc}	1	0.075	γ_{mc}	1	0.2
	2	0.075		2	2.5×10^{-3}
χ_{lc}	1	0.75	γ_{lc}	1	10^{-3}
	2	0.75		2	3×10^{-6}

Note: The default values of χ_{mc}, γ_{mc}, χ_{lc}, and γ_{lc} shown here belong to the three-bus islanded microgrid shown in Fig. 9.4

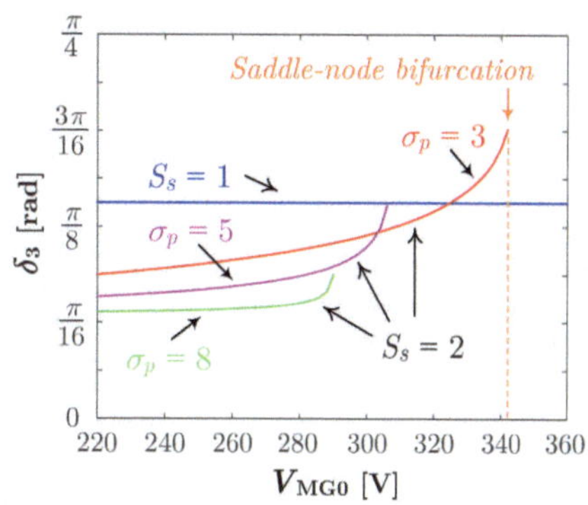

Fig. 9.5 Bifurcation diagram showing the usual EQP for different S_s and varying σ_p. Other parameters are given in Tables 9.2, 9.3, and 9.4

9.2.2 *Subcritical Hopf Bifurcation*

In this subsection, we reveal that *even when the usual EQP exists, it may become unstable due to the onset of a subcritical Hopf bifurcation.* After the onset of a Hopf bifurcation, the usual operating point still exists, but it becomes a UEP (i.e., small-signal instability), which can be verified by inspecting the eigenvalues [5]. We can calculate the eigenvalues $\lambda_{f\xi}$ of the Jacobian of (9.26), where $\xi = 1, 2, \cdots, N_s$. If all $\lambda_{f\xi}$ satisfy $\mathrm{Re}(\lambda_{f\xi}) < 0$, the system is stable. Otherwise, the system is unstable [5]. Besides, by examining the first Lyapunov coefficient, this Hopf bifurcation occurs in subcritical mode [30]. Therefore, after the onset of the aforementioned bifurcation, the SEP absorbs the UPO to become a UEP, and the trajectory may diverge from the SEP to a hidden attractor, inducing a sustained oscillation which is harmful in practice.

The loci of $\lambda_{f\xi}$ for varying specific secondary control parameters are shown in Fig. 9.6, where only the dominant eigenvalue is exhibited. For Fig. 9.6a, a pair of complex eigenvalues crosses the imaginary axis as χ_{mc} increases, where a Hopf bifurcation occurs. From Fig. 9.6, we find that relatively large χ_{mc}, or medium γ_{mc}, or medium γ_{lc} may destabilize the system. However, varying χ_{lc} has negligible effect on the system stability. Using the same procedure, we can also investigate the impact of other secondary control parameters, as summarized in Table 9.5. From the

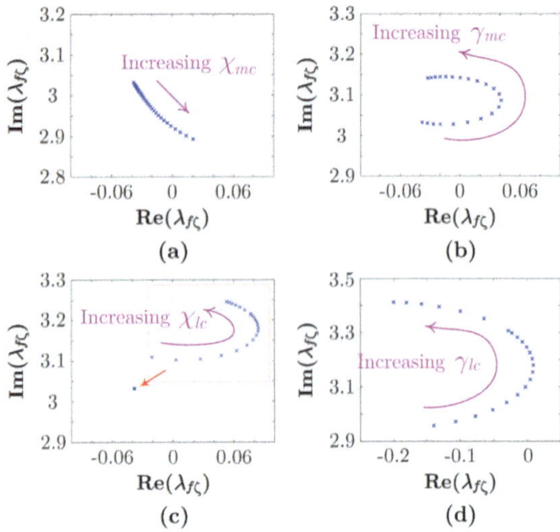

Fig. 9.6 Loci of eigenvalues $\lambda_{f\xi}$ of (9.26) for $S_s = 1$, where **(a)** χ_{mc} varies from 1 p.u. to 150 p.u., **(b)** γ_{mc} varies from 1 p.u. to 100 p.u., **(c)** χ_{lc} varies from 0.1 p.u. to 10 p.u., and **(d)** γ_{lc} varies from 0.1 p.u. to 10 p.u. The base values are the corresponding default values shown in Table 9.4. Other parameters are given in Tables 9.2, 9.3, and 9.4

Table 9.5 Impact of secondary control parameters on Hopf bifurcation

S_s	χ_{mc}	γ_{mc}	χ_{lc}	γ_{lc}	$L_{p0\ell}$	$r_{p0\ell}$	σ_p	ω_f	ω_v
1	↓	↓↑	–	↓↑	↑	↑↓	/	–	–
2	↓	↓↑	–	↓	↑	↓	↑	–	–

Note: "↑" ("↓") denotes that the dominant eigenvalues move away from the imaginary axis by increasing (decreasing) the corresponding parameter. "–" denotes weak or negligible effect on dominant eigenvalues by increasing the corresponding parameter. Besides, for the primary control parameters, increasing $\omega_{p\ell}$ or $\omega_{q\ell}$ may move the dominant eigenvalues away from the imaginary axis [52]

viewpoint of Hopf bifurcation, the impact of secondary control parameters on the stability for $S_s = 1$ is similar to that for $S_s = 2$.

9.2.3 *Hidden Attractor*

In this subsection, we reveal that *sub-synchronous oscillations may be associated with the hidden attractors, where the trajectory may move from the usual SEP to a hidden attractor under a specific transient disturbance (i.e., transient stability).* The hidden attractors coexist with the usual SEP. Therefore, the microgrid transient stability is strong if the SEP's basin of attraction is large, and the hidden attractors are far from the usual SEP [52, 53].

For $S_s = 1$, based on (9.26), we find that the system has four SEPs, a stable hidden periodic orbit (where only δ_2 oscillates from $-\pi$ to π), and no UEP. Among the four SEPs, all $\Delta v_{c\ell}$ are positive for a specific SEP, but at least one $\Delta v_{c\ell}$ is negative for other SEPs (i.e., $\Delta v_{c2} = -607$ V and/or $\Delta v_{c3} = -753$ V). In practice,

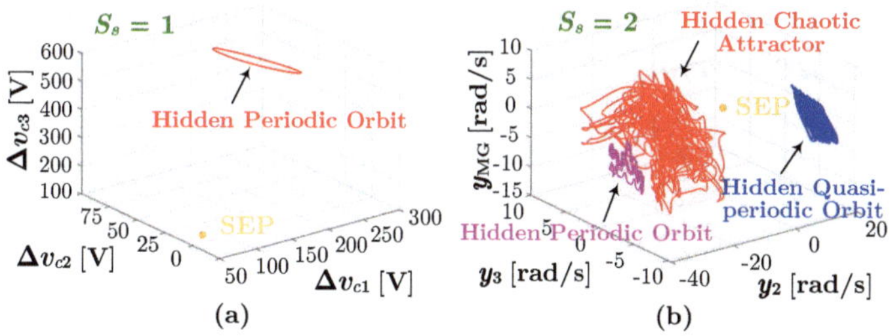

Fig. 9.7 SEP and hidden attractors (**a**) in (Δv_{c1},Δv_{c2},Δv_{c3})-space for $S_s = 1$ and (**b**) in (y_2,y_3,y_{MG})-space for $S_s = 2$. Here, $y_\ell = \omega_\ell - \omega_1$, and $y_{MG} = \omega_{MG} - \omega_1$. All parameters adopt their default values

the system usually operates at the former SEP. The stability analyses given in Sects. 9.2.1 and 9.2.2 are targeted at this SEP. This SEP is far from other SEPs due to the negative $\Delta v_{c\ell}$. As shown in Fig. 9.7a, this SEP is also far from the hidden periodic orbit. It is also found that this SEP's basin of attraction is very large. Therefore, the system is quite robust against transient disturbances even for small cutoff frequencies of filters.

For $S_s = 2$, we find that the system has only one SEP, three UEPs, a stable hidden periodic orbit (where only δ_2 oscillates from $-\pi$ to π), a stable hidden quasi-periodic orbit (where δ_2 and δ_3 all oscillate from $-\pi$ to π, and the ratio between the oscillation frequencies of these angles is constant but irrational [53]), and a stable hidden chaotic attractor (where δ_2 and δ_3 oscillate from $-\pi$ to π, but their oscillations are totally irregular [53]). The SEP and three hidden attractors are shown in Fig. 9.7b. For default parameters, the SEP's basin of attraction is relatively small, implying that the microgrid tends to be desynchronized and exhibits sub-synchronous oscillation under some temporary faults. Although significantly increasing $\omega_{p\ell}$ may eliminate the hidden attractors and enlarge the SEP's basin of attraction, it may reduce the virtual inertia and deteriorate the frequency stability.

Remark Compared with the control strategy $S_s = 2$, the control strategy $S_s = 1$ shows improved robustness to maintain the existence of usual EQP (see Sect. 9.2.1) and higher flexibility to enhance the transient stability and frequency stability (see Sect. 9.2.3).

9.3 Roles of Communication Network and Driver Controllers

Our analysis here is based on the control strategy $S_s = 1$. The analytical results can be extended to the case of $S_s = 2$.

9.3.1 Effect of Communication Network on Consensus-Based Control Performance

We denote

$$\mathbf{z} = \begin{bmatrix} z_1 \ z_2 \cdots z_N \end{bmatrix}^T,$$

$$\mathbf{q} = \frac{1}{n_{q1}} \begin{bmatrix} n_{q1}q_1 \ n_{q2}q_2 \cdots n_{qN}q_N \end{bmatrix}^T.$$

From (9.3), we get

$$\begin{aligned} \frac{\mathrm{d}\mathbf{z}}{\mathrm{d}t} &= \gamma_{mc}\mathbf{G}(V_{\mathrm{MG0}} - v_{\mathrm{MG}}) + \gamma_{lc}(\mathbf{A} - \mathbf{D})\mathbf{q} \\ &= \gamma_{mc}\mathbf{G}(V_{\mathrm{MG0}} - v_{\mathrm{MG}}) - \gamma_{lc}\mathbf{L}\mathbf{q} \end{aligned} \tag{9.27}$$

where $\mathbf{D}$ and $\mathbf{L}$ are the degree matrix and the Laplacian matrix of the communication network, respectively [140, 153, 158]. For example, for Figs. 9.4 and 9.8a, we have

$$\mathbf{D} = \begin{bmatrix} 1 & 0 & 0 \\ 0 & 2 & 0 \\ 0 & 0 & 1 \end{bmatrix}, \quad \mathbf{L} = \begin{bmatrix} 1 & -1 & 0 \\ -1 & 2 & -1 \\ 0 & -1 & 1 \end{bmatrix}.$$

From (9.27),

$$\mathbf{z}(t) = \gamma_{mc}\mathbf{T}_{\mathbf{L}} \int_0^t e^{-\tau\mathcal{J}}\mathbf{T}_{\mathbf{L}}^{-1}\mathbf{G}(V_{\mathrm{MG0}} - v_{\mathrm{MG}}(\tau))\mathrm{d}\tau + \gamma_{lc}\mathbf{T}_{\mathbf{L}}e^{-\tau\mathcal{J}}\mathbf{T}_{\mathbf{L}}^{-1}\mathbf{q}(0) \tag{9.28}$$

where $\mathbf{L} = \mathbf{T}_{\mathbf{L}}\mathcal{J}\mathbf{T}_{\mathbf{L}}^{-1}$, and $\mathcal{J}$ is the real Jordan canonical matrix. In the normal operation state, the communication network of the microgrid is a connected undirected graph. Therefore, $\mathbf{L}$ is symmetrical and semi-positive definite. In other words, the eigenvalues of $\mathbf{L}$ is real and nonnegative, i.e.,

$$0 = \lambda_1 < \lambda_2 < \lambda_3 < \cdots < \lambda_N, \tag{9.29}$$

where λ_2 is the algebraic connectivity of the network [158]. Hence,

$$\mathcal{J} = \mathrm{diag}\{\lambda_\ell\}. \tag{9.30}$$

Therefore, the convergence speed of $\mathbf{z}$ and $\mathbf{q}$ is dominated by λ_2. If λ_2 is large, $\mathbf{z}$ and $\mathbf{q}$ will take less time reaching their steady-state values. As an example, $\lambda_2 = 1$ for Fig. 9.8a, while $\lambda_2 = 3$ for Fig. 9.8b. Therefore, in terms of the convergence speed, the latter structure is better than the former structure. Moreover, for Fig. 9.8a, if one of the two red communication links fails, the system may not operate normally;

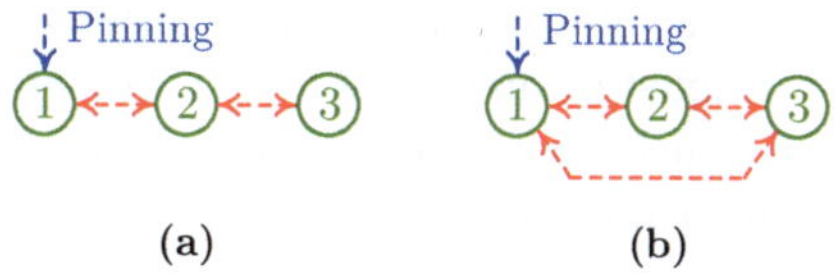

Fig. 9.8 Communication structure (**a**) with no direct communication link between local controller 1 and local controller 3 and (**b**) with a direct communication link between local controller 1 and local controller 3. Here, each circle represents a local controller

however, for Fig. 9.8b, if one of the three red communication links fails, the system can still maintain its stable operation. Hence, in order to maintain the $N - 1$ secure operation [28], it is recommended that each local controller should have at least two distinct communication links.

9.3.2 *Effect of Driver Controllers on Pinning-Based Control Performance*

For the pinning-based control, there are two problems, i.e., the controllability problem and the stability problem.

The controllability problem is about the minimum number n_D of local controllers that receive the control signals from the microgrid controller (i.e., driver controllers). According to the study results in network science [140, 160],

$$n_D = \max\{1, N - \text{rank}(\mathbf{L})\} = \max\{1, N - (N - 1)\} = 1 \tag{9.31}$$

implying that only one driver controller is enough to guarantee the controllability of the microgrid system. However, considering the $N - 1$ secure operation [28], at least two driver controllers may be selected in practice. Otherwise, the pinning-based control may lose its effectiveness if the sole communication link for pinning is broken.

The stability problem is related to the effect of the choice of specific driver controllers on the stability of the usual operating point (i.e., small-signal stability), the transient response (e.g., overshoot and transient oscillations), and the SEP's basin of attraction. The latter two conditions are associated with transient stability.

9.3.2.1 Stability of Usual Operating Point

As analyzed in Sect. 9.2.2, the stability of the EQP can be evaluated by the eigenvalues $\lambda_{f\xi}$ of (9.26). Here, we focus on the maximal value of the real part of $\lambda_{f\xi}$, i.e., $\max(\text{Re}(\lambda_{f\xi}))$. The EQP is stable if $\max(\text{Re}(\lambda_{f\xi})) < 0$, but unstable otherwise. For a specific group of parameters (e.g., specific values of χ_{mc} and γ_{mc}, as shown in Fig. 9.9), the value of $\max(\mathrm{Re}(\lambda_{f\xi}))$ for different driver controllers can be calculated, which is denoted as $\max(\mathrm{Re}(\lambda_{f\xi}))|_{\mathbf{G}}$. For Figs. 9.4 and 9.8a, there are 7 different selections of driver controllers, corresponding to 7 different **G**. For the sake of comparison, the p.u. value of $\max(\mathrm{Re}(\lambda_{f\xi}))$ is calculated as

$$\max(\mathrm{Re}(\lambda_{f\xi}))\ [\text{p.u.}] = \frac{\max(\mathrm{Re}(\lambda_{f\xi}))}{\max\{\mathrm{abs}[\max(\mathrm{Re}(\lambda_{f\xi}))|_{\mathbf{G}}]\}}. \tag{9.32}$$

where $\mathrm{abs}[x]$ represents the absolute value of x.

As shown in Fig. 9.9, different selections of driver controllers may lead to different values of $\max(\mathrm{Re}(\lambda_{f\xi}))$. In particular, when χ_{mc} and γ_{mc} are large, $\max(\mathrm{Re}(\lambda_{f\xi})) > 0$ for specific driver controllers, implying that the system is unstable.

9.3.2.2 Transient Response

For (9.26), the Jacobian is usually a nondiagonal matrix. Therefore, the state variables may interact with each other, leading to fierce transient oscillation due to the conflicts between different control tasks. It is noted that such transient oscillations may finally subside, and the trajectory may converge to the usual SEP. Here, we use the condition number of the Jacobian to measure such interaction. If the condition number is larger, the interaction between different state variables is stronger, and the transient oscillation is fiercer. The Jacobian of (9.26) is denoted as $\mathbf{J_f}$ whose element is $\partial f_k/\partial x_\hbar$, where $k, \hbar = 1, 2, \cdots, N_s$ [5]. The condition number is

$$\gamma_f = \frac{\overline{\sigma}(\mathbf{J_f})}{\underline{\sigma}(\mathbf{J_f})} = \sqrt{\frac{\max[\mathrm{eig}(\mathbf{J_f}^H\mathbf{J_f})]}{\min[\mathrm{eig}(\mathbf{J_f}^H\mathbf{J_f})]}} \tag{9.33}$$

where $\overline{\sigma}(\mathbf{J_f})$ is the largest singular value of $\mathbf{J_f}$, $\underline{\sigma}(\mathbf{J_f})$ is the smallest singular value of $\mathbf{J_f}$, and $\mathrm{eig}(\mathbf{J_f}^H\mathbf{J_f})$ represents the eigenvalues of $\mathbf{J_f}^H\mathbf{J_f}$ [120]. For a specific group of parameters, the value of γ_f for different driver controllers can be calculated, which is denoted as $\gamma_f|_{\mathbf{G}}$. For the sake of comparison, the p.u. value of γ_f is calculated as

$$\gamma_f\ [\text{p.u.}] = \frac{\gamma_f}{\max\{\gamma_f|_{\mathbf{G}}\}} \tag{9.34}$$

As shown in Fig. 9.9, different selections of driver controllers may lead to different values of γ_f. For example, using the default parameters, $\gamma_f = 0.977$ p.u. when the driver controller is local controller 1, while $\gamma_f = 0.688$ p.u. when the driver controller is local controller 2. Thus, the transient oscillation in the former case may be more severe than that in the latter case. Similarly, the topology of communication network may also affect $\max(\mathrm{Re}(\lambda_{f\xi}))$ and γ_f.

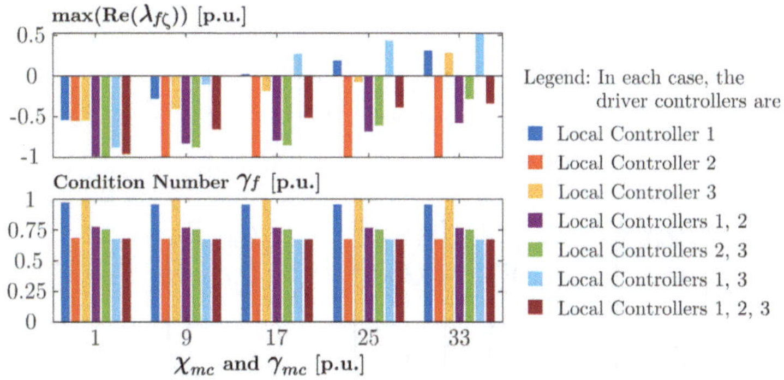

Fig. 9.9 Values of max(Re($\lambda_{f\xi}$)) and γ_f for different driver controllers and varying χ_{mc} and γ_{mc}. Other parameters are given in Tables 9.2, 9.3, and 9.4

9.3.2.3 SEP's Basin of Attraction

The selection of driver controllers may also change the SEP's basin of attraction. However, for the control strategy $S_s = 1$, the effect is negligible because the SEP's basin of attraction is already very large, as indicated in Sect. 9.2.3.

9.4 Selection of Secondary Control Parameters

To facilitate the selection of the secondary control parameters (i.e., χ_{mc}, γ_{mc}, χ_{lc} and γ_{lc}), we establish four respective small-signal models focusing only on the secondary-control dynamics. Based on these simple and focused models, we can derive the transfer functions of the loop gains and select the suitable parameter values by using frequency-domain analysis. These models permit fast selection of the parameter values. While the parameters can be further refined at the final stage, the analytical tools presented in last section can be used to systematically evaluate the system stability.

9.4.1 Consensus-Based Control Parameters

For the consensus-based reactive power regulation, from (9.10), (9.22), and (9.23), we get

$$\begin{bmatrix} \hat{Q}^d_{n1} \\ \hat{Q}^d_{n2} \\ \vdots \\ \hat{Q}^d_{nN} \end{bmatrix} = \frac{1}{T_{dn}s+1}\mathbf{I}\begin{bmatrix} \hat{Q}_{n1} \\ \hat{Q}_{n2} \\ \vdots \\ \hat{Q}_{nN} \end{bmatrix} \tag{9.35}$$

$$\begin{bmatrix} \hat{Q}_{n1} \\ \hat{Q}_{n2} \\ \vdots \\ \hat{Q}_{nN} \end{bmatrix} = \frac{\gamma_{lc}}{n_{q1}}\frac{1}{s}\left(\mathbf{A}\begin{bmatrix} \hat{Q}^d_{n1} \\ \hat{Q}^d_{n2} \\ \vdots \\ \hat{Q}^d_{nN} \end{bmatrix} - \mathbf{D}\begin{bmatrix} \hat{Q}_{n1} \\ \hat{Q}_{n2} \\ \vdots \\ \hat{Q}_{nN} \end{bmatrix}\right) \tag{9.36}$$

where $\hat{Q}_{n\ell}$ and $\hat{Q}^d_{n\ell}$ are the small-signal variables of $Q_{n\ell}$ and $Q^d_{n\ell}$, respectively. Based on (9.35) and (9.36), the block diagram of the small-signal model can be developed, as shown in Fig. 9.10a. Thus, the loop gain is

$$\mathbf{LG_q}(s) = \frac{\gamma_{lc}}{n_{q1}}\frac{1}{T_{dn}s+1}\left(s\mathbf{I} + \frac{\gamma_{lc}}{n_{q1}}\mathbf{D}\right)^{-1}\mathbf{A}. \tag{9.37}$$

The induced 2-norm of $\mathbf{LG_q}$ [120] can be found as

$$||\mathbf{LG_q}(j\omega)||_{i2} = \overline{\sigma}(\mathbf{LG_q}(j\omega)). \tag{9.38}$$

Likewise, we can derive the small-signal model for the consensus-based frequency regulation, as shown in Fig. 9.10b. The loop gain and the corresponding induced 2-norm are

$$\mathbf{LG_\omega}(s) = \frac{\chi_{lc}}{n_{q1}}\frac{1}{T_{dn}s+1}\left(s\mathbf{I} + \chi_{lc}\mathbf{D}\right)^{-1}\mathbf{A} \tag{9.39}$$

$$||\mathbf{LG}_\omega(j\omega)||_{i2} = \overline{\sigma}(\mathbf{LG}_\omega(j\omega)). \tag{9.40}$$

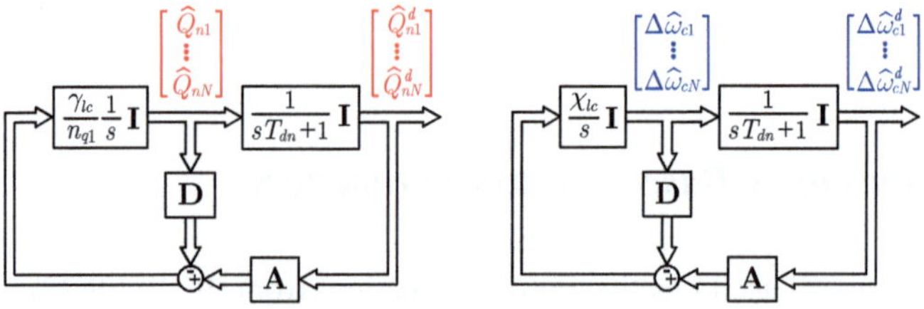

(a) (b)

Fig. 9.10 Small-signal models corresponding to (**a**) consensus-based reactive power regulation and (**b**) consensus-based frequency regulation

9.4.2 Pinning-Based Control Parameters

For the pinning-based voltage regulation, from (9.10), we get

$$\hat{v}_{bms} \approx \frac{1}{N}\sum_{\ell=1}^{N} \hat{v}_{fd\ell} \approx \frac{1}{N}\sum_{\ell=1}^{N} \hat{v}^{*}_{fd\ell} \approx \frac{1}{N}\sum_{\ell=1}^{N} \Delta\hat{v}_{c\ell} \tag{9.41}$$

where $\hat{v}_{bms}$, $\hat{v}_{fd\ell}$, $\hat{v}^{*}_{fd\ell}$, and $\Delta\hat{v}_{c\ell}$ are the small-signal variables of v_{bms}, $v_{fd\ell}$, $v^{*}_{fd\ell}$ and $\Delta v_{c\ell}$, respectively. From (9.6) and (9.20)–(9.22), we have

$$\hat{v}_{\mathrm{MG}} = \frac{\omega_v}{s+\omega_v}, \quad \Delta\hat{v}_{mc} = \frac{1}{T_{dn}s+1}(-\gamma_{mc}\hat{v}_{\mathrm{MG}})\hat{v}_{bms} \tag{9.42}$$

$$\frac{1}{N}\sum_{\ell=1}^{N} \Delta\hat{v}_{c\ell} = \frac{1}{s}\left(\frac{1}{N}\sum_{\ell=1}^{N} g_\ell\right)\Delta\hat{v}_{mc}. \tag{9.43}$$

Based on (9.41)–(9.43), the block diagram of the small-signal model is shown in Fig. 9.11a. Therefore, the loop gain is

$$\mathcal{L}_v(s) = \frac{\omega_v}{s+\omega_v} \times \frac{1}{T_{dn}s+1} \times \frac{1}{s}\left(\frac{1}{N}\sum_{\ell=1}^{N} g_\ell\right)\gamma_{mc}. \tag{9.44}$$

Using a similar procedure, we can derive the small-signal model for the pinning-based frequency regulation, as shown in Fig. 9.11b. The loop gain is

$$\mathcal{L}_\omega(s) = \frac{\omega_f}{s+\omega_f} \times \frac{1}{T_{dn}s+1} \times \frac{1}{s}\left(\frac{1}{N}\sum_{\ell=1}^{N} g_\ell\right)\chi_{mc}. \tag{9.45}$$

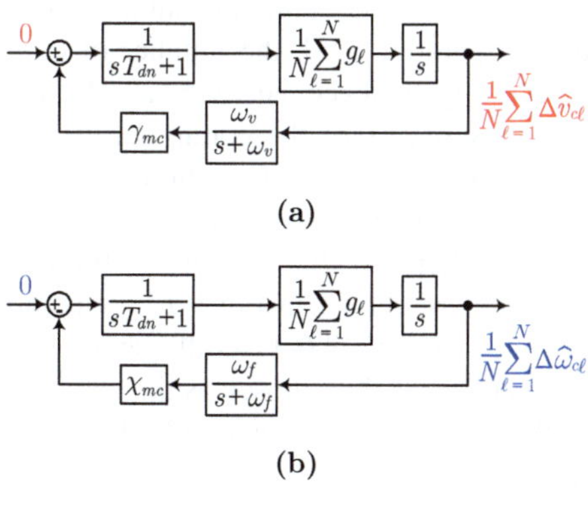

Fig. 9.11 Small-signal models corresponding to (**a**) pinning-based voltage regulation and (**b**) pinning-based frequency regulation

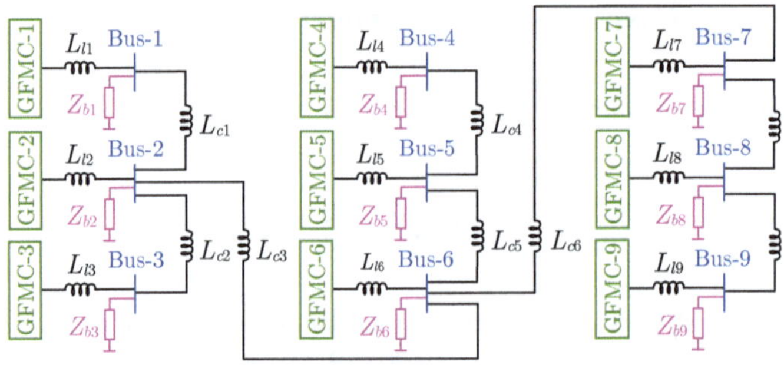

Fig. 9.12 Nine-bus islanded microgrid. The ESRs of $L_{f\ell}$, $L_{l\ell}$, and L_{cp} are $r_{f\ell}$, $r_{l\ell}$, and r_{cp}, respectively, where $\ell = 1, 2, \cdots, 9$ and $\rho = 1, 2, \cdots, 8$

9.5 Application to Design of Distributed Secondary Control

In this section, a nine-bus islanded microgrid is taken as an example to illustrate the design-oriented stability analysis [138], i.e., how to apply the above research results to the design of the distributed secondary control system. The circuit diagram of the nine-bus islanded microgrid is shown in Fig. 9.12, the control strategy is shown in Fig. 9.3, and the default parameters are shown in Tables 9.2, 9.3, and 9.4, except χ_{mc}, γ_{mc}, χ_{lc}, and γ_{lc}.

9.5.1 Design Procedure

The communication network, driver controllers, and the values of χ_{mc}, γ_{mc}, χ_{lc}, and γ_{lc} are to be selected. The design steps are as follows.

Step 1 For a microgrid under study, the weighted Laplacian matrix of the power network $\mathbf{D_C}$ is derived.

Step 2 A suitable communication network is chosen by considering the $N - 1$ secure operation [28], the number of communication links, the maximal degree, and the algebraic connectivity λ_2. Here, λ_2 is derived from the corresponding adjacency matrix of the communication network $\mathbf{A}$ [158]. Moreover, the degree is the number of communication links for a specific local controller. In practice, choosing a small maximal degree may simplify the communication hardware of the controller. For the sake of illustration, four different types of communication networks are shown in Fig. 9.13. A comparison between these four networks is presented in Table 9.6. Overall, the network shown in Fig. 9.13c is more desirable. This network is thus chosen as the communication network of the microgrid. Then, for this communication network, the adjacency matrix $\mathbf{A}$ and the degree matrix $\mathbf{D}$ can be derived.

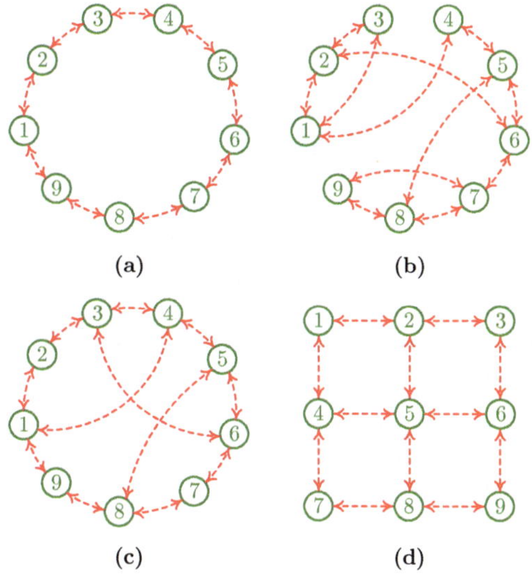

Fig. 9.13 Four types of communication networks for the nine-bus islanded microgrid. Here, each circle represents a local controller. (**a**) Basic ring-shaped network, (**b**) breaking ring-shaped network, (**c**) improved ring-shaped network, and (**d**) lattice-shaped network [158]

Table 9.6 Comparison of communication networks

Fig.	9.13a	9.13b	9.13c	9.13d
$N-1$ secure operation	Yes	Yes	Yes	Yes
Communication links	9	12	12	12
Maximal degree	2	3	3	4
Algebraic connectivity λ_2	0.4679	0.4825	0.9139	1

Step 3 The number of driver controllers is determined by considering the $N-1$ secure operation [28]. Here, the number of driver controllers is selected as 2.

Step 4 The secondary control parameters (i.e., χ_{mc}, γ_{mc}, χ_{lc}, and γ_{lc}) are selected according to Sect. 9.4. To reduce the coupling effect between the consensus-based control and the pinning-based control, the gain crossover frequencies of $||\mathbf{LG}_{\boldsymbol{\omega}}(j\omega)||_{i2}$ and $||\mathbf{LG}_{\mathbf{q}}(j\omega)||_{i2}$ should be much higher than those of $\mathcal{L}_{\omega}(j\omega)$ and $\mathcal{L}_{v}(j\omega)$. Using frequency-domain analysis, the gain crossover frequencies of $||\mathbf{LG}_{\boldsymbol{\omega}}(j\omega)||_{i2}$, $||\mathbf{LG}_{\mathbf{q}}(j\omega)||_{i2}$, $\mathcal{L}_{\omega}(j\omega)$, and $\mathcal{L}_{v}(j\omega)$ are 0.5384 Hz, 0.4467 Hz, 0.0106 Hz, and 0.0283 Hz, respectively. The selected secondary control parameters are $\chi_{mc}=0.3$, $\gamma_{mc}=0.8$, $\chi_{lc}=0.75$, and $\gamma_{lc}=4.2\times10^{-4}$.

Step 5 The values of $\max(\mathrm{Re}(\lambda_{f\xi}))$ and γ_f for all possible selections of driver controllers are calculated. Then, the specific driver controllers are determined, where the corresponding $\max(\mathrm{Re}(\lambda_{f\xi}))$ is negative and its absolute value is relatively large, while the corresponding γ_f is relatively small. According to Fig. 9.14, the local controllers 4 and 8 are selected as the driver controllers, as $\max(\mathrm{Re}(\lambda_{f\xi}))=-1$ p.u. and $\gamma_f=0.6775$ p.u. For comparison, $\max(\mathrm{Re}(\lambda_{f\xi}))=-0.6495$ p.u. and $\gamma_f=0.9738$ p.u. if the driver controllers are local controllers 8 and 9. The selected communication network and driver controllers are shown in Fig. 9.15.

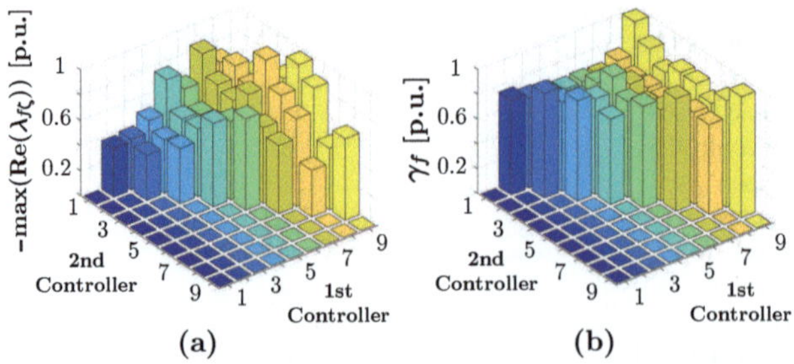

Fig. 9.14 Values of (**a**) $-\max(\text{Re}(\lambda_{f\xi}))$ and (**b**) γ_f for different driver controllers. All parameters adopt their default values with $\chi_{mc} = 0.3$, $\gamma_{mc} = 0.8$, $\chi_{lc} = 0.75$, and $\gamma_{lc} = 4.2 \times 10^{-4}$

Fig. 9.15 Selected communication network and driver controllers

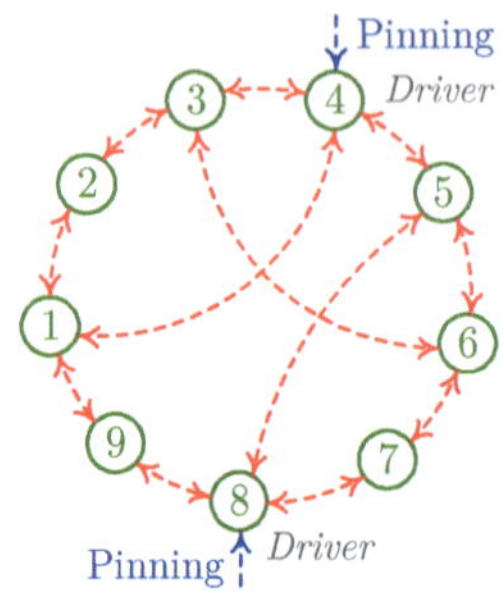

Step 6 The SEP's basin of attraction is examined. If it is not sufficiently large, we need to adjust some primary control parameters to eliminate some hidden attractors and enlarge the SEP's basin of attraction [53]. In this case, we need to go back to *Step 5* to recalculate $\max(\text{Re}(\lambda_{f\xi}))$ and γ_f and choose the suitable driver controllers again.

Step 7 The frequency stability and stabilization time of the microgrid are assessed. If the RoCoF and frequency deviation are too large, the cutoff frequencies of the filters should be lowered [18, 159]. However, if the stabilization time is too long, the cutoff frequencies of the filters should be increased [37, 156]. Then, we need to go back to *Step 4*.

Step 8 The final step is to examine whether the system is stable under the load change. If $\max(\text{Re}(\lambda_{f\xi})) < 0$ when all local loads homogeneously become light, the design procedure is finished. Otherwise, we need to return to *Step 4* to reduce χ_{mc} and γ_{mc}, according to Sect. 9.2.2.

9.5.2 Simulation Verifications

9.5.2.1 Basic Control Performance

Using the default parameters, the simulation results are shown in Fig. 9.16. Here, $f_\ell = \omega_\ell/(2\pi)$, and $v_{bm\ell}$ is the magnitude of the Bus-ℓ voltage. In Fig. 9.16, the secondary control does not operate before $t = 10$ s, i.e., $\Delta\omega_{c\ell}$ and $\Delta v_{c\ell}$ are zero during this period of time. In this case, $m_{p1}p_1 = m_{p2}p_2 = \cdots = m_{p9}p_9$, but $q_8 > q_2 > q_5 \gg q_7 > q_1 > q_4 \gg q_3 > q_6 \approx q_9$, implying the reactive power sharing among all GFMCs is not automatically achieved. Besides, f_ℓ is higher than 50 Hz, and $v_{bm\ell}$ is much smaller than 280 V. After $t = 10$ s, the secondary control starts to operate. However, due to different communication networks and different driver controllers, the performance of the microgrid after $t = 10$ s may be different.

Using the design result shown in Fig. 9.15, the simulation waveforms are shown in Fig. 9.16a. In this case, q_ℓ converges to the new value quickly. In the steady state, the value of q_3, q_6, and q_9 is 20.4 kVar, the value of q_1, q_4, and q_7 is 10.3 kVar, and the value of q_2, q_5, and q_8 is 5.25 kVar, implying that the reactive power sharing is achieved. Besides, f_ℓ converges to the 50 Hz value, and $v_{bm\ell}$ reaches a value close to 280 V.

In contrast to Fig. 9.15, the communication network shown in Fig. 9.13a is adopted, and the results are shown in Fig. 9.16b. In this case, p_ℓ, q_ℓ, f_ℓ, and $v_{bm\ell}$ exhibits drastic oscillations before reaching the normal state. In particular, the reactive power sharing is achieved within 40 s in Fig. 9.16a, but within 80 s in Fig. 9.16b. These results confirm the effect of the communication network structure on the consensus-based control performance, as identified in Sect. 9.3.1.

In contrast to Fig. 9.15, the driver controllers are local controllers 8 and 9. In this case, p_ℓ, q_ℓ, f_ℓ, and $v_{bm\ell}$ exhibits severe oscillations after $t = 10$ s. At $t = 100$ s, p_ℓ and f_ℓ reach the normal state. However, q_ℓ and $v_{bm\ell}$ may take more time converging to the corresponding values in the normal operation sate. The simulation results shown in Fig. 9.16a and 9.16c expose the significant effect of the choice of specific driver controllers on the stable operation of microgrids, as reflected by the value of γ_f (see Sect. 9.3.2.2 and *Step 4* of Sect. 9.5.1).

When the cutoff frequencies of filters are set as 5 Hz, the simulation results are shown in Fig. 9.17. After the secondary control operates, f_ℓ converges to 50 Hz within 18 s, and the RoCoF value is 0.67 Hz/s. However, in Fig. 9.16a, f_ℓ converges to 50 Hz within 40 s, and the RoCoF value is 0.05 Hz/s. Therefore, increasing the cutoff frequencies of filters can reduce the stabilization time, but may deteriorate the frequency stability.

9.5.2.2 Load Change Performance

The performance of the microgrid under the load change is shown in Fig. 9.18. Under the load changes, the active and reactive power sharing is achieved quickly, while f_ℓ and $v_{bm\ell}$ converge to the corresponding rated values eventually, implying that the system remains stable under load changes.

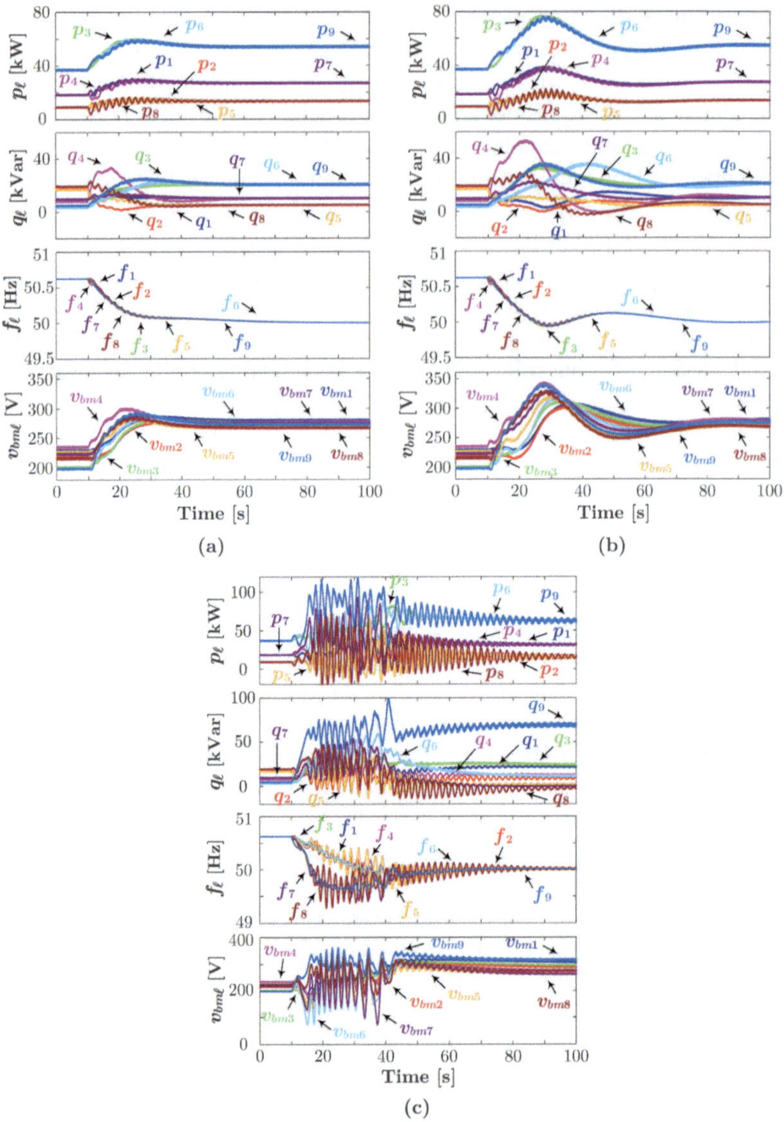

Fig. 9.16 Simulation waveforms of p_ℓ, q_ℓ, f_ℓ, and $v_{bm\ell}$. Secondary control operates after t = 10 s. All parameters adopt their default values with $\chi_{mc} = 0.3$, $\gamma_{mc} = 0.8$, $\chi_{lc} = 0.75$, and $\gamma_{lc} = 4.2 \times 10^{-4}$. (**a**) Results given in Fig. 9.15 are used; (**b**) communication network given in Fig. 9.13a is adopted; (**c**) driver controllers are local controllers 8 and 9

9.5.2.3 Communication Link Failure

The microgrid stability under the communication link failure is examined, as shown in Fig. 9.19, where the communication link between local controllers 1 and 2 fails

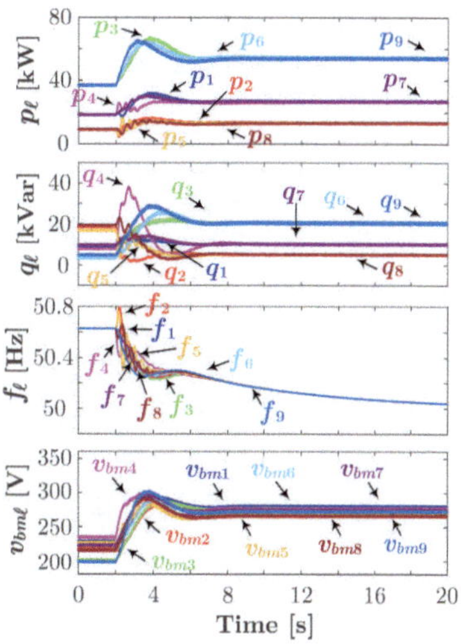

Fig. 9.17 Simulation waveforms of p_ℓ, q_ℓ, f_ℓ, and $v_{bm\ell}$. Secondary control operates after $t = 2$ s. All parameters adopt their default values with $\omega_f = \omega_v = \omega_{p\ell} = \omega_{q\ell} = 2\pi \times 5$ rad/s, $\chi_{mc} = 3$, $\gamma_{mc} = 8$, $\chi_{lc} = 7.5$, and $\gamma_{lc} = 4.2 \times 10^{-3}$. Selected communication network and driver controllers are shown in Fig. 9.15

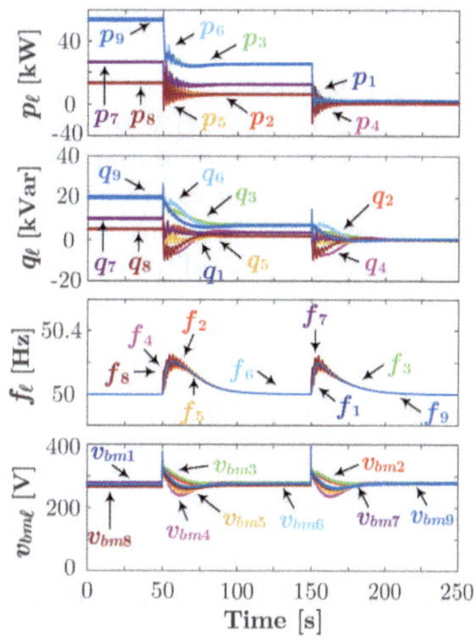

Fig. 9.18 Simulation waveforms of p_ℓ, q_ℓ, f_ℓ, and $v_{bm\ell}$. The impedance of each load is set as $Z_{b\ell}$ = Default Value $\times K_{ld}$. Here, $K_{ld} = 1$ before $t = 50$ s, $K_{ld} = 2$ when $t = 50 \sim 150$ s, and $K_{ld} = 20$ after $t = 150$ s. All parameters adopt their default values with $\chi_{mc} = 0.3$, $\gamma_{mc} = 0.8$, $\chi_{lc} = 0.75$, and $\gamma_{lc} = 4.2 \times 10^{-4}$. Selected communication network and driver controllers are shown in Fig. 9.15

for Fig. 9.19a, while the communication link between the microgrid controller and local controller 4 fails for Fig. 9.19b. After such failure, the load Z_{b1} is switched to test the system stability. From Fig. 9.19, the system operates normally under the load change. Moreover, under the first load change, f_ℓ converges to 50 Hz within 50 s in Fig. 9.19a, while f_ℓ converges to 50 Hz within 100 s in Fig. 9.19b. This is consistent with the analysis towards (9.45), where less driver controllers will reduce the bandwidth of the secondary frequency control loop.

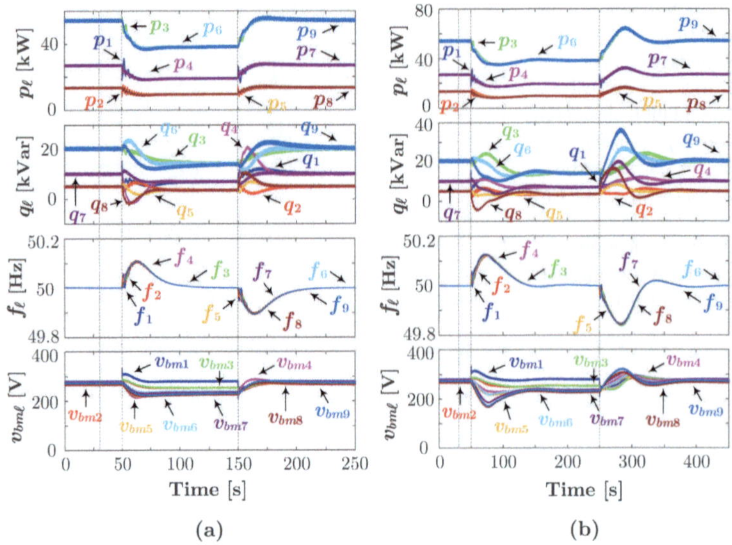

Fig. 9.19 Simulation waveforms of p_ℓ, q_ℓ, f_ℓ, and $v_{bm\ell}$. All parameters adopt their default values with $\chi_{mc} = 0.3$, $\gamma_{mc} = 0.8$, $\chi_{lc} = 0.75$, and $\gamma_{lc} = 4.2 \times 10^{-4}$. Selected communication network and driver controllers are shown in Fig. 9.15. (**a**) Communication link between local controllers 1 and 2 is broken after $t = 30$ s. The load Z_{b1} is switched OFF when $t = 50 \sim 150$ s, but switched ON otherwise. (**b**) Communication link between the microgrid controller and local controller 4 is broken after $t = 30$ s. The load Z_{b1} is switched OFF when $t = 50 \sim 250$ s, but switched ON otherwise

9.5.3 Comparison

Much of the prior work has focused on the stability status of the usual EQP, and the eigenvalues [37, 143, 144, 147], Bode diagrams [148], Lyapunov functions [139, 145, 149], and root locus [146] were used for evaluation. However, of equal importance are the impact of topologies of communication networks and selections of driver controllers on the microgrid stability. In this chapter, the voltage-loop dynamics and communication delay are considered to ensure accuracy of the model for control parameter design. Also, frequency stability and the $N-1$ secure operation of communication network and driver controllers are also incorporated in the design procedure. These considerations have not been incorporated in prior work.

9.6 Experimental Verifications

In this chapter, we adopt hardware-in-loop experiments to verify the theoretical findings reported in the previous sections. The circuit diagram of the three-bus

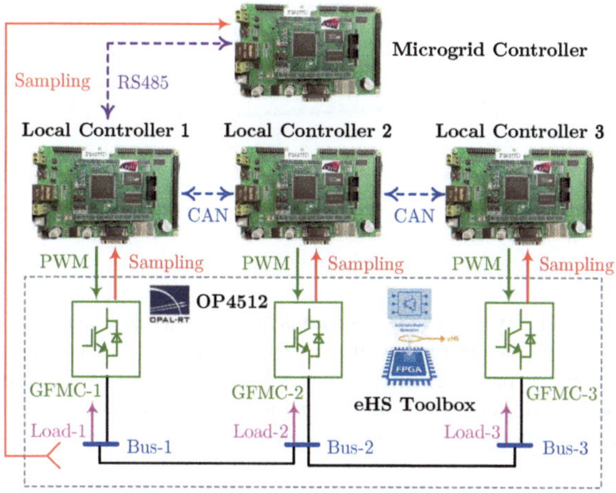

Fig. 9.20 Configuration of the laboratory prototype

microgrid is shown in Fig. 9.4, and the configuration of the laboratory prototype is shown in Fig. 9.20. The power circuit of GFMCs, the microgrid power network, and the loads are emulated in the OPAL-RT equipment OP4512. Here, the eHS toolbox is also adopted to enable the high-precision emulation of a microgrid containing many GFMCs. The DSPs TMS320F28377D are used as the local controllers and the microgrid controller. Each local controller samples the voltages and currents of the corresponding GFMC and sends the PWM signals to the GFMC. These local controllers communicate with each other through the CAN communication. The microgrid controller samples the Bus-1 voltage and sends the pinning-based control signals to a specific local controller through the RS485 communication. For the sake of observation, we define $q_{e\ell} = q_1 - (n_{q\ell}/n_{q1})q_\ell$. Moreover, f_{MG} is the frequency of the microgrid (i.e., $f_{MG} = \omega_{MG}/(2\pi)$), and v_{MG} is the magnitude of Bus-1 voltage.

9.6.1 Virtual-Impedance-Based Secondary Control

For the control strategy $S_s = 2$, z_ℓ is the decrement of the virtual impedance, as shown in (9.7). Based on this control strategy, the experimental waveforms are exhibited in Fig. 9.21. Here, we observe that q_{e2} and q_{e3} converge to zero, and f_{MG} and v_{MG} reach their corresponding target values after the secondary control takes effect. Specifically, corresponding to Fig. 9.21a, the communication network shown in Fig. 9.8a is adopted, and q_{e2} and q_{e3} converge to zero within 145 s. However, corresponding to Fig. 9.21b, the communication network shown in Fig. 9.8b is adopted, and q_{e2} and q_{e3} converge to zero within 80 s, which is much faster than the previous case. These verify the analytical finding about the impact of communication network topologies on the microgrid performance, which is also effective for $S_s = 2$.

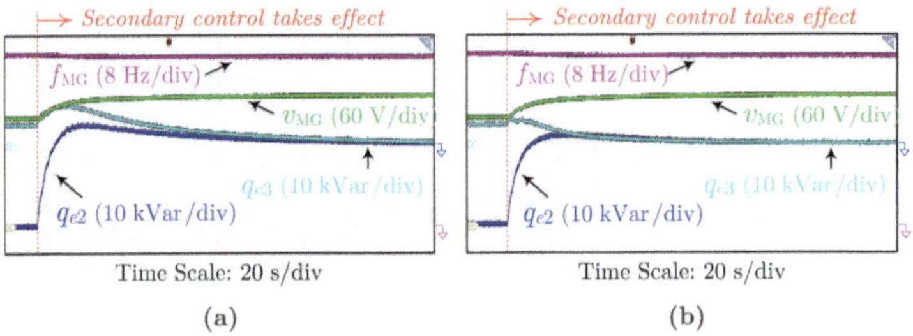

Fig. 9.21 Experimental waveforms of f_{MG}, v_{MG}, q_{e2}, and q_{e3}. Control strategy $S_s = 2$ is adopted, i.e., z_ℓ is the decrement of the virtual impedance. Secondary control does not operate at the beginning, but takes effect later. All parameters adopt their default values. Driver controller is local controller 1. (**a**) Communication network shown in Fig. 9.8a is adopted. (**b**) Communication network shown in Fig. 9.8b is adopted

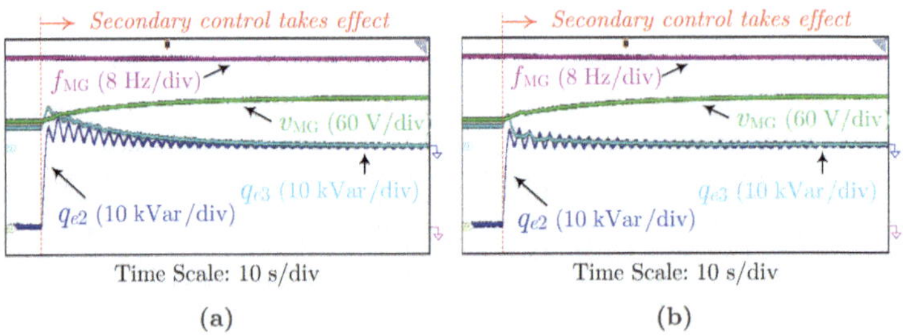

Fig. 9.22 Experimental waveforms of f_{MG}, v_{MG}, q_{e2}, and q_{e3}. Control strategy $S_s = 1$ is adopted, i.e., z_ℓ is $\Delta v_{c\ell}$. Secondary control takes effect from a later time. All parameters adopt their default values. Driver controller is local controller 1. (**a**) Communication network shown in Fig. 9.8a is adopted. (**b**) Communication network shown in Fig. 9.8b is adopted

9.6.2 *Effects of Communication Networks*

For the control strategy $S_s = 1$, z_ℓ is the increment of the d-axis voltage reference (i.e., $\Delta v_{c\ell}$), as shown in (9.6). Based on this control strategy, the experimental waveforms are exhibited in Figs. 9.22 and 9.23. For Fig. 9.22a and b, the driver controller is local controller 1. In Fig. 9.22a, the communication network shown in Fig. 9.8a is adopted. After the secondary control operates, $f_{\mathrm{MG}} = 50$ Hz, and $v_{\mathrm{MG}} = 280$ V eventually. Moreover, q_{e2} and q_{e3} converge to zero within 60 s. For the results given in Fig. 9.22b, the communication network shown in Fig. 9.8b is adopted. After the secondary control takes effect, f_{MG} and v_{MG} finally reach their corresponding target values, and q_{e2} and q_{e3} converge to the zero value within 40 s. Comparing Fig. 9.21a with b and comparing Fig. 9.22a with b, we find that the convergence speed using the network shown in Fig. 9.8a is slower than that using the network shown in Fig. 9.8b, for both types of control strategies. This is consistent with the analytical findings presented in Sect. 9.3.1.

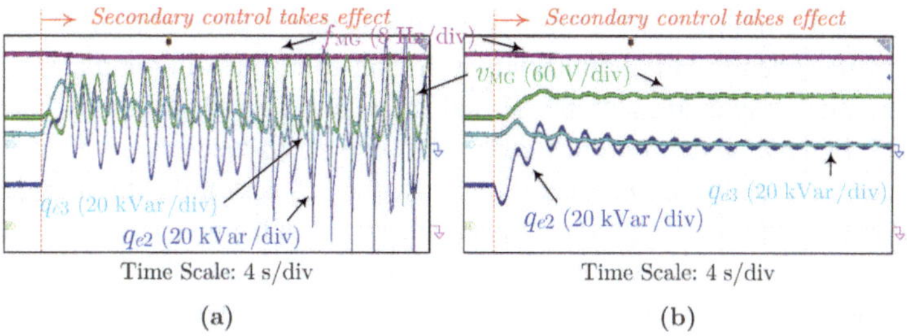

Fig. 9.23 Experimental waveforms of f_{MG}, v_{MG}, q_{e2}, and q_{e3}. Control strategy $S_s = 1$ is adopted, i.e., z_ℓ is $\Delta v_{c\ell}$. Secondary control does not operate at the beginning, but takes effect later. Here, $\chi_{mc} = \gamma_{mc} = 9$ p.u., and other parameters adopt their default values. Communication network shown in Fig. 9.8a is adopted. (**a**) Driver controller is local controller 1. (**b**) Driver controller is local controller 2

9.6.3 *Effects of Driver Controllers*

Corresponding to results shown in Fig. 9.23a and b, the communication network given in Fig. 9.8a is adopted, and the pinning-based control parameters are set as $\chi_{mc} = \gamma_{mc} = 9$ p.u. In Fig. 9.23a, the driver controller is local controller 1. After the secondary control takes effect, v_{MG}, q_{e2}, and q_{e3} exhibit a significantly fierce oscillation. However, for Fig. 9.23b where the driver controller is local controller 2, after the secondary control takes effect, the oscillation in v_{MG}, q_{e2}, and q_{e3} is not fierce and finally subsides. Besides, v_{MG} reaches its target value quickly, and q_{e2} and q_{e3} eventually converge to the zero value, implying the reactive power sharing among the three GFMCs. These experimental results confirm the significant effect of the choice of driver controllers on the system stable operation and are consistent with the analytical findings presented in Fig. 9.9 and Sect. 9.3.2.2.

9.7 Recapitulation

Secondary control plays an important role in shaping the complex dynamics of islanded microgrids comprising multiple GFMCs. In this chapter, we study the impact of distributed secondary control on microgrid's stability. The main findings in this chapter are as follows:

1. *Basic structures for distributed control:* Two basic structures for existing control schemes to achieve the voltage deviation compensation and reactive power sharing are established. For structure that uses a single equation to achieve the above objectives, the control protocol is made up of pinning-based and consensus-based algorithms.
2. *Impact of secondary control on dynamical behavior:* Upon an in-depth study of the saddle-node bifurcation behavior, subcritical Hopf bifurcation behavior,

and hidden attractors of the microgrid, the voltage-increment-based secondary control has been found to contribute to higher robustness in maintaining system stability, in comparison with the virtual-impedance-based secondary control.

3. *Effects of communication networks and driver controllers:* It is found that the topology of the communication network has a significant impact on the convergence speed of reactive power sharing. Such impact can be quantified by the algebraic connectivity of the network. Besides, the choice of driver controllers may have significant effect on the small-signal stability and transient stability. The conditional number of the system's Jacobian is adopted to measure the interaction between different state variables and to estimate the transient oscillation intensity.
4. *Application to design:* A design procedure of the distributed secondary control system is demonstrated by combining the aforementioned analytical results.

Chapter 10
Summary and Extensions

This book offers a comprehensive analysis of the intricate behavior and stability issues associated with grid-connected power electronic systems. These systems play a crucial role in mitigating instability, preventing power blackouts, and facilitating the integration of renewable energy sources into modern power grids. The book begins by examining the complex behavior exhibited by individual grid-connected converters. It delves into various local bifurcations that can arise in grid-following rectifiers, leading to voltage instability in the power grid. Additionally, the book explores the significance of homoclinic bifurcation in grid-connected converters, which fundamentally influences the stable equilibrium point's basin of attraction. This aspect is highly relevant to enhancing multi-swing stability and transient synchronization stability. The research work also considers the impact of outer control loops on grid-connected converters. Subsequently, the book expands its scope to encompass the complex behavior of entire interconnected systems comprising multiple converters. It proposes a methodology for establishing reduced-order models of multi-converter microgrids that effectively describe their synchronization behavior. It is important to note that the interaction between heterogeneous converters can give rise to diverse hidden attractors, which manifest as various sub-synchronous oscillations. These hidden attractors have a significant impact on transient synchronization stability, both in terms of detrimental effects and potential benefits. Importantly, such hidden attractors are absent in single grid-connected converters and microgrids consisting solely of homogeneous converters. Finally, the book explores the influence of distributed secondary control on microgrid stability and presents a systematic approach for designing secondary control systems. By studying these aspects, the book provides a comprehensive understanding of the complex behavior and stability considerations in grid-connected power electronic systems, shedding light on critical factors that impact their performance and reliability.

The analysis of complex behavior and stability in grid-connected power electronic systems remains an important and evolving field. While this book offers

J. Yang, C. K. Tse, *Complex Behavior of Grid-Connected Power Electronics Systems*, https://doi.org/10.1007/978-3-031-69340-3_10

valuable insights, it also points to areas that require further research. First, the models developed in this book effectively capture sustained oscillations with relatively low frequencies, such as sub-synchronous oscillations. However, they are not applicable for studying oscillations with relatively high frequencies, such as subharmonic oscillations. To address this, a future research direction could involve adopting a discrete-time iterative mapping approach to develop suitable models that capture these high-frequency oscillations in both single grid-connected converters and multi-converter microgrids. Second, while the established models in this book have relatively low orders, allowing for the study of global bifurcation behavior, they limit the observation and analysis of diverse nonlinear dynamics within the converter system. Therefore, there is an urgent need to develop appropriate analytical tools to study the global bifurcation behavior of high-order converter systems, striking a balance between complexity and accuracy. Third, as the number of components and system scale increase, distinct nonlinear behaviors may emerge, particularly in converter-dominated systems that evolve into cyber-multitudinal-physical systems. This is especially true in the case of large-scale cyber/physical networks with heterogeneous nodes. These new emerging behaviors can have a significant impact on power system stability. Therefore, future attention should be directed towards understanding the emergence mechanism, physical characteristics, and specific influences of these new behaviors. By addressing these research gaps, future studies can further advance our understanding of complex behavior and stability in grid-connected power electronic systems, enabling improved system design and enhanced stability analysis in the face of evolving power system challenges.

Bibliography

1. C.K. Tse, M. Huang, X. Zhang, D. Liu, X.L. Li, Circuits and systems issues in power electronics penetrated power grid. IEEE Open J. Circ. Syst. **1**, 140–156 (2020)
2. N. Hatziargyriou, J. Milanovic, C. Rahmann, V. Ajjarapu, C.A. Cañizares, I. Erlich, D. Hill, I. Hiskens, I. Kamwa, B. Pal, P. Pourbeik, J. Sanchez-Gasca, A. Stankovic, T.V. Cutsem, V. Vittal, C. Vournas, Definition and classification of power system stability - Revisited and extended. IEEE Trans. Power Syst. **36**(4), 3271–3281 (2021)
3. M. Farrokhabadi, C.A. Cañizares, J.W. Simpson-Porco, E. Nasr, L. Fan, P.A. Mendoza-Araya, R. Tonkoski, U. Tamrakar, N. Hatziargyriou, D. Lagos, R.W. Wies, M. Paolone, M.Liserre, L. Meegahapola, M. Kabalan, A.H. Hajimiragha, D. Peralta, M.A. Elizondo, K.P. Schneider, F.K. Tuffner, J. Reilly, Microgrid stability definitions, analysis, and examples. IEEE Trans. Power Syst. **35**(1),13–29 (2020)
4. Y. Gu, T.C. Green, Power system stability with a high penetration of inverter-based resources. Proc. IEEE **111**(7), 832–853 (2023)
5. C.K. Tse, *Complex Behavior of Switching Power Converters* (CRC Press, Boca Raton, 2003)
6. C.M.F.S. Reza, D.D.-C. Lu, Recent progress and future research direction of nonlinear dynamics and bifurcation analysis of grid-connected power converter circuits and systems. IEEE J. Emerg. Sel. Topics Power Electron. **8**(4), 3193–3203 (2020)
7. F. Blaabjerg, Y. Yang, K.A. Kim, J. Rodriguez, Power electronics technology for large-scale renewable energy generation. Proc. IEEE **111**(4), 335–355 (2023)
8. M. Liserre, M.A. Perez, M. Langwasser, C.A. Rojas, Z. Zhou, Unlocking the hidden capacity of the electrical grid through smart transformer and smart transmission. Proc. IEEE **111**(4), 421–437 (2023)
9. Y. Lin, J.H. Eto, B.B. Johnson, J.D. Flicker, R.H. Lasseter, H.N.V. Pico, G.-S. Seo, B.J. Pierre, A. Ellis, J. Miller, G. Yuan, Pathways to the next-generation power system with inverter-based resources: challenges and recommendations. IEEE Electrific. Mag. **10**(1), 10–21 (2022)
10. S. Rivera, S.M. Goetz, S. Kouro, P.W. Lehn, M. Pathmanathan, P. Bauer, R.A. Mastromauro, Charging infrastructure and grid integration for electromobility. Proc. IEEE **111**(4), 371–396 (2023)
11. S.Y. Hui, C.K. Lee, F.F. Wu, Electric springs—a new smart grid technology. IEEE Trans. Smart Grid **3**(3), 1552–1561 (2012)
12. J. Guo, T. Chen, B. Chaudhuri, S.Y. Hui, Stability of isolated microgrids with renewable generation and smart loads. IEEE Trans. Sustain. Energy **11**(4), 2845–2854 (2020)
13. J.M. Guerrero, J.C. Vasquez, J. Matas, L.G. de Vicuna, M. Castilla, Hierarchical control of droop-controlled AC and DC microgrids—a general approach toward standardization. IEEE Trans. Ind. Electron. **58**(1), 158–172 (2011)

J. Yang, C. K. Tse, *Complex Behavior of Grid-Connected Power Electronics Systems*, https://doi.org/10.1007/978-3-031-69340-3

14. X. Wu, Y. Xu, X. Wu, J. He, J.M. Guerrero, C.-C. Liu, K.P. Schneider, D.T. Ton, A two-layer distributed cooperative control method for islanded networked microgrid systems. IEEE Trans. Smart Grid **11**(2), 942–957 (2020)
15. X. Ruan, X. Wang, D. Pan, D. Yang, W. Li, C. Bao, *Control Techniques for LCL-Type Grid-Connected Inverters* (Science Press, Beijing; Springer, Singapore, 2018)
16. L. Zhang, L. Harnefors, H.-P. Nee, Power-synchronization control of grid-connected voltage-source converters. IEEE Trans. Power Syst. **25**(2), 809–820 (2010)
17. Q.-C. Zhong, G. Weiss, Synchronverters: inverters that mimic synchronous generators. IEEE Trans. Ind. Electron. **58**(4), 1259–1267 (2011)
18. D. Pan, X. Wang, F. Liu, R. Shi, Transient stability of voltage-source converters with grid-forming control: a design-oriented study. IEEE J. Emerg. Sel. Topics Power Electron. **8**(2), 1019–1033 (2020)
19. H. Yu, M.A. Awal, H. Tu, I. Husain, S. Lukic, Comparative transient stability assessment of droop and dispatchable virtual oscillator controlled grid-connected inverters. IEEE Trans. Power Electron. **36**(2), 2119–2130 (2021)
20. M. Lu, Virtual oscillator grid-forming inverters: state of the art, modeling, and stability. IEEE Trans. Power Electron. **37**(10), 11579–11591 (2022)
21. S. Luo, W. Chen, X. Li, Z. Hao, A new virtual inertial strategy for Andronov–Hopf oscillator based grid-forming inverters. IEEE J. Emerg. Sel. Topics in Power Electron. **12**(2), 1995–2005 (2024)
22. K. Strunz, K. Almunem, C. Wulkow, M. Kuschke, M. Valescudero, X. Guillaud, Enabling 100% renewable power systems through power electronic grid-forming converter and control: system integration for security, stability, and application to Europe. Proc. IEEE **111**(7), 891–915 (2023)
23. A.D. Paquette, D.M. Divan, Virtual impedance current limiting for inverters in microgrids with synchronous generators. IEEE Trans. Ind. Appl. **51**(2), 1630–1638 (2015)
24. J. He, Y.W. Li, Analysis, design, and implementation of virtual impedance for power electronics interfaced distributed generation. IEEE Trans. Ind. Appl. **47**(6), 2525–2538 (2011)
25. X. Gao, D. Zhou, A. Anvari-Moghaddam, F. Blaabjerg, Seamless transitions between grid-following and grid-forming control: a novel switching method, in *2023 International Conference on Power Electronics - ECCE Asia* (2023), pp. 1154–1160
26. Y. Gui, X. Wang, F. Blaabjerg, D. Pan, Control of grid-connected voltage-source converters: the relationship between direct-power control and vector-current control. IEEE Ind. Electron. Mag. **13**(2), 31–40 (2019)
27. C.C. Liu, C.K. Tse, J. Yang, Nonlinear behavior and transient stability of grid-following converters using direct power control under weak grid. IEEE Trans. Sustain. Energy To appear [Online]. https://doi.org/10.1109/TSTE.2024.3413343
28. N. Xue, X. Wu, S. Gumussoy, U. Muenz, A. Mesanovic, C. Heyde, Z. Dong, G. Bharati, S. Chakraborty, L. Cockcroft, L. Dangelmaier, Dynamic security optimization for N-1 secure operation of Hawai'i island system with 100% inverter-based resources. IEEE Trans. Smart Grid **13**(5), 4009–4021 (2022)
29. Q. Peng, J. Fang, Y. Yang, T. Liu, F. Blaabjerg, Maximum virtual inertia from DC-link capacitors considering system stability at voltage control timescale. IEEE J. Emerg. Sel. Topics Circ. Syst. **11**(1), 79–89 (2021)
30. J. Yang, C.K. Tse, M. Huang, D. Liu, Bifurcations of grid-following rectifiers and routes to voltage instability in weak AC grids. IEEE Trans. Power Syst. **38**(2), 1702–1713 (2023)
31. Y. Han, H. Li, P. Shen, E.A.A. Coelho, J.M. Guerrero, Review of active and reactive power sharing strategies in hierarchical controlled microgrids. IEEE Trans. Power Electron. **32**(3), 2427–2451 (2017)
32. X. Zhang, J. Wu, A.O. Fapojuwo, Z. Galias, C.K. Tse, Guest editorial complex cyber-multitudinal-physical systems: analysis, decision-making, and AI applications. IEEE J. Emerg. Sel. Topics Circ. Syst. **13**(3), 641–645 (2023)

33. S. D'silva, M. Shadmand, S. Bayhan, H. Abu-Rub, Towards grid of microgrids: seamless transition between grid-connected and islanded modes of operation. IEEE Open Ind. Electron. Soc. **1**, 66–81 (2020)
34. B. Fan, Q. Li, W. Wang, G. Yao, H. Ma, X. Zeng, J.M. Guerrero, A novel droop control strategy of reactive power sharing based on adaptive virtual impedance in microgrids. IEEE Trans. Ind. Electron. **69**(11), 11335–11347 (2022)
35. J. He, Y.W. Li, F. Blaabjerg, An enhanced islanding microgrid reactive power, imbalance power, and harmonic power sharing scheme. IEEE Trans. Power Electron. **30**(6), 3389–3401 (2015)
36. H. Mahmood, D. Michaelson, J. Jiang, Accurate reactive power sharing in an islanded microgrid using adaptive virtual impedances. IEEE Trans. Power Electron. **30**(3), 1605–1617 (2015)
37. X. Wu, Y. Xu, J. He, X. Wang, J.C. Vasquez, J.M. Guerrero, Pinning-based hierarchical and distributed cooperative control for AC microgrid clusters. IEEE Trans. Power Electron. **35**(9), 9865–9885 (2020)
38. J. Yang, C.C. Liu, C.K. Tse, M. Huang, Stability analysis and design of islanded microgrids integrating pinning-based and consensus-based distributed secondary control. IEEE Trans. Power Electron. **39**(8), 9490–9504 (2024)
39. P. Kundur, J. Paserba, V. Ajjarapu, G. Andersson, A. Bose, C.A. Cañizares, N. Hatziargyriou, D. Hill, A. Stankovic, C. Taylor, T.V. Cutsem, V. Vittal, Definition and classification of power system stability. IEEE Trans. Power Syst. **19**(2), 1387–1401 (2004)
40. P. Kundur, *Power System Stability And Control* (McGraw-Hill, New York, 1993)
41. X. Wang, M.G. Taul, H. Wu, Y. Liao, F. Blaabjerg, L. Harnefors, Grid-synchronization stability of converter-based resources—an overview. IEEE Open J. Ind. Appl. **1**, 115–134 (2020)
42. M.J. Li, C.K. Tse, M. Yi, Steady-state cascading failure model with voltage instability event detection. IEEE Trans. Circ. Syst. I, Reg. Papers **71**(1), 463–472 (2024)
43. W. Wang, G. Li, G. He, *Analysis and Mitigation of Broadband Oscillation in Renewable Energy Generation and AC/DC Transmission Systems* (China Electric Power Press, Beijing; Springer Nature, Singapore, 2023)
44. J. Sun, M. Li, Z. Zhang, T. Xu, J. He, H. Wang, G. Li, Renewable energy transmission by HVDC across the continent: system challenges and opportunities. CSEE J. Power Energy Syst. **3**(4), 353–364 (2017)
45. Y. Cheng, L. Fan, J. Rose, F. Huang, J. Schmall, X. Wang, X. Xie, J. Shair, J. Ramamurthy, N. Modi, C. Li, C. Wang, S. Shah, B.C. Pal, Z. Miao, A. Isaacs, J. Mahseredjian, Z.J. Zhou, Real-world subsynchronous oscillation events in power grids with high penetrations of inverter-based resources. IEEE Trans. Power Syst. **38**(1), 316–330 (2023)
46. C. Yoon, H. Bai, R.N. Beres, X. Wang, C.L. Bak, F. Blaabjerg, Harmonic stability assessment for multiparalleled, grid-connected inverters. IEEE Trans. Sustain. Energy **7**(4), 1388–1397 (2016)
47. H.H.C. Iu, Y.F. Zhou, C.K. Tse, Fast-scale instability in a PFC boost converter under average current-mode control. Int. J. Circ. Theory Appl. **31**(6), 611–624 (2003)
48. H. Ji, F. Xie, L. Shen, R. Yang, B. Zhang, Unstable behavior analysis and stabilization of double-loop proportional-integral control H-bridge inverter with inductive impedance load. Int. J. Circ. Theory Appl. **50**(3), 904–925 (2022)
49. L. Yang, L. Yang, F. Yang, X. Ma, Slow-scale and fast-scale instabilities in parallel-connected single-phase H-bridge inverters: a design-oriented study. Int. J. Bifurcat. Chaos **30**(1), 2050005 (2020)
50. H. Cao, F. Wang, J. Liu, Analysis of fast-scale instability in three-level T-type single-phase inverter feeding diode-bridge rectifier with inductive load. IEEE Trans. Power Electron. **37**(12), 15066–15083 (2022)
51. E. Lenz, D.J. Pagano, J. Pou, Bifurcation analysis of parallel-connected voltage-source inverters with constant power loads. IEEE Trans. Smart Grid **9**(6), 5482–5493 (2018)

52. J. Yang, C.K. Tse, D. Liu, Nonlinear behavior and reduced-order models of islanded microgrid. IEEE Trans. Power Electron. **37**(8), 9212–9225 (2022)
53. J. Yang, C.K. Tse, D. Liu, Sub-synchronous oscillations and transient stability of islanded microgrid. IEEE Trans. Power Syst. **38**(4), 3760–3774 (2023)
54. Y.A. Kuznetsov, *Elements of Applied Bifurcation Theory*, 2nd edn. (Springer, New York, 1998)
55. J. Yang, C.K. Tse, M. Huang, X. Fu, Homoclinic bifurcation of a grid-forming voltage source converter. IEEE Trans. Power Electron. **36**(11), 13176–13187 (2021)
56. J. Zhao, M. Huang, H. Yan, C.K. Tse, X. Zha, Nonlinear and transient stability analysis of phase-locked loops in grid-connected converters. IEEE Trans. Power Electron. **36**(1), 1018–1029 (2021)
57. Y. Chen, C.K. Tse, S.-S. Qiu, L. Lindenmuller, W. Schwarz, Coexisting fast-scale and slow-scale instability in current-mode controlled DC/DC converters: analysis, simulation and experimental results. IEEE Trans. Circ. Syst. I, Reg. Papers **55**(10), 3335–3348 (2008)
58. C. Wan, M. Huang, C.K. Tse, X. Ruan, Effects of interaction of power converters coupled via power grid: a design-oriented study. IEEE Trans. Power Electron. **30**(7), 3589–3600 (2015)
59. S. Wang, Z. Liu, J. Liu, D. Boroyevich, R. Burgos, Small-signal modeling and stability prediction of parallel droop-controlled inverters based on terminal characteristics of individual inverters. IEEE Trans. Power Electron. **35**(1), 1045–1063 (2020)
60. Y. Zhu, Y. Gu, Y. Li, T.C. Green, Impedance-based root-cause analysis: comparative study of impedance models and calculation of eigenvalue sensitivity. IEEE Trans. Power Syst. **38**(2), 1642–1654 (2023)
61. C. Yang, L. Huang, H. Xin, P. Ju, Placing grid-forming converters to enhance small signal stability of PLL-integrated power systems. IEEE Trans. Power Syst. **36**(4), 3563–3573 (2021)
62. X. Fu, J. Sun, M. Huang, Z. Tian, H. Yan, H.H.C. Iu, P. Hu, X. Zha, Large-signal stability of grid-forming and grid-following controls in voltage source converter: a comparative study. IEEE Trans. Power Electron. **36**(7), 7832–7840 (2021)
63. Y. Zhang, C. Zhang, X. Cai, Large-signal grid-synchronization stability analysis of PLL-based VSCs using Lyapunov's direct method. IEEE Trans. Power Syst. **37**(1), 788–791 (2022)
64. H. Cheng, Z. Shuai, C. Shen, X. Liu, Z. Li, Z.J. Shen, Transient angle stability of paralleled synchronous and virtual synchronous generators in islanded microgrids. IEEE Trans. Power Electron. **35**(8), 8751–8765 (2020)
65. S. Banerjee, Coexisting attractors, chaotic saddles, and fractal basins in a power electronic circuit. IEEE Trans. Circ. Syst. I **9**(44), 847–849 (1997)
66. X. Geng, F. Xie, B. Zhang, D. Qiu, Y. Chen, R. Wang, Identifying coexisting attractors and quantifying dynamics characteristics of DC–DC converter based on improved variational mode decomposition and wavelet transform. IEEE Trans. Power Electron. **3**(38), 3928–3938 (2023)
67. V. Venkatasubramanian, W. Ji, Coexistence of four different attractors in a fundamental power system model. IEEE Trans. Circ. Syst. I **3**(46), 405–409 (1999)
68. NERC files: Reliability guideline – Forced oscillation monitoring and mitigation. https://www.nerc.com/comm/PC_Reliability_Guidelines_DL/Reliability_Guideline_-_Forced_Oscillations_-_2017-07-31_-_FINAL.pdf. (2017). Accessed 25 May 2022
69. X. Wang, N.V. Kuznetsov, G. Chen, *Chaotic Systems with Multistability and Hidden Attractors* (Springer, Switzerland, 2021)
70. G.A. Leonov, N.V. Kuznetsov, Hidden attractors in dynamical systems. From hidden oscillations in Hilbert-Kolmogorov Aizerman, and Kalman problems to hidden chaotic attractor in Chua circuits. Int. J. Bifurcat. Chaos **23**(1), 1330002 (2013)
71. G.A. Leonov, N.V. Kuznetsovb, V.I. Vagaitsev, Localization of hidden chua's attractors. Phys. Lett. A **375**(23), 2230–2233 (2011)
72. X. Wang, G. Chen, A chaotic system with only one stable equilibrium. Commun. Nonlinear Sci. Numer. Simulat. **17**(3), 1264–1272 (2012)
73. J.C. Sprott, X. Wang, G. Chen, Coexistence of point, periodic and strange attractors. Int. J. Bifurcat. Chaos **23**(5), 1350093 (2013)

74. H.-D. Chiang, *Direct Methods for Stability Analysis of Electric Power Systems: Theoretical Foundation, BCU Methodologies, and Applications* (John Wiley and Sons, Canada, 2011)
75. C.A. Cañizares, On bifurcations, voltage collapse and load modeling. IEEE Trans. Power Syst. **10**(1), 512–522 (1995)
76. C.A. Cañizares, A.C.Z.D. Souza, V.H. Quintana, Comparison of performance indices for detection of proximity to voltage collapse. IEEE Trans. Power Syst. **11**(3), 1441–1450 (1996)
77. J.W. Simpson-Porco, F. Dörfler, F. Bullo, Voltage collapse in complex power grids. Nat. Commun., **7**(10790), 1–8 (2016)
78. D.L.H. Aik, G. Andersson, Fundamental analysis of voltage and power stability of single-infeed voltage-source converter HVDC systems. IEEE Trans. Power Del. **34**(1), 365–375 (2019)
79. J. Chen, F. Milano, T. O'Donnell, Assessment of grid-feeding converter voltage stability. IEEE Trans. Power Syst. **34**(5), 3980–3982 (2019)
80. X. He, H. Geng, Transient stability of power systems integrated with inverter-based generation. IEEE Trans. Power Syst. **36**(1), 553–556 (2021)
81. M. Huang, S. Wong, C.K. Tse, X. Ruan, Catastrophic bifurcation in three-phase voltage-source converters. IEEE Trans. Circ. Syst. I, Reg. Papers **60**(4), 1062–1071 (2013)
82. M. Huang, C.K. Tse, S. Wong, C. Wan, X. Ruan, Low-frequency Hopf bifurcation and its effects on stability margin in three-phase PFC power supplies connected to non-ideal power grid. IEEE Trans. Circ. Syst. I, Reg. Papers **60**(12), 3328–3340 (2013)
83. M. Huang, Y. Peng, C.K. Tse, Y. Liu, J. Sun, X. Zha, Bifurcation and large-signal stability analysis of three-phase voltage source converter under grid voltage dips. IEEE Trans. Power Electron. **32**(11), 8868–8879 (2017)
84. Z. Yang, R. Ma, S. Cheng, M. Zhan, Nonlinear modeling and analysis of grid-connected voltage-source converters under voltage dips. IEEE J. Emerg. Sel. Topics Power Electron. **8**(4), 3281–3292 (2020)
85. M. Liserre, F. Blaabjerg, S. Hansen, Design and control of an LCL-filter-based three-phase active rectifier. IEEE Trans. Ind. Appl. **41**(5), 1281–1291 (2005)
86. E.M. Izhikevich, Subcritical elliptic bursting of Bautin type. SIAM J. Appl. Math. **60**(2), 503–535 (1999)
87. B. Hassard, Y.H. Wan, Bifurcation formulate derived from center manifold theory. J. Math. Anal. Appl. **63**(1), 297–312 (1978)
88. H. Wu, X. Wang, A mode-adaptive power-angle control method for transient stability enhancement of virtual synchronous generators. IEEE J. Emerg. Sel. Topics Power Electron. **8**(2), 1034–1049 (2020)
89. R.C. Robinson, *An Introduction to Dynamical Systems: Continuous and Discrete*, 2nd edn. (American Mathematical Society, Providence, 2012)
90. R. Ma, J. Li, J. Kurths, S. Cheng, M. Zhan, Generalized swing equation and transient synchronous stability with PLL-based VSC. IEEE Trans. Energy Convers. **37**(2), 1428–1441 (2022)
91. Q. Hu, L. Fu, F. Ma, F. Ji, Large signal synchronizing instability of PLL-based VSC connected to weak AC grid. IEEE Trans. Power Syst. **34**(4), 3220–3229 (2019)
92. H. Wu, X. Wang, Design-oriented transient stability analysis of PLL-synchronized voltage-source converters. IEEE Trans. Power Electron. **35**(4), 3573–3589 (2020)
93. X. Xiong, C. Wu, F. Blaabjerg, Effects of virtual resistance on transient stability of virtual synchronous generators under grid voltage sag. IEEE Trans. Ind. Electron. **69**(5), 4754–4764 (2022)
94. M. Chen, D. Zhou, F. Blaabjerg, Enhanced transient angle stability control of grid-forming converter based on virtual synchronous generator. IEEE Trans. Ind. Electron. **69**(9), 9133–9144 (2022)
95. P. Hu, Z. Chen, Y. Yu, D. Jiang, On transient instability mechanism of PLL-based VSC connected to a weak grid. IEEE Trans. Ind. Electron. **70**(4), 3836–3846 (2023)

96. M.G. Taul, C. Wu, S.-F. Chou, F. Blaabjerg, Optimal controller design for transient stability enhancement of grid-following converters under weak-grid conditions. IEEE Trans. Power Electron. **36**(9), 10251–10264 (2021)
97. J. Yang, C.K. Tse, M. Huang, D. Liu, C.C. Liu, X. Fu, Comparison of homoclinic bifurcations between grid-following and grid-forming converters. IEEE Trans. Ind. Electron. **71**(5), 4731–4741 (2024)
98. J. Chen, M. Liu, T. O'Donnell, F. Milano, Impact of current transients on the synchronization stability assessment of grid-feeding converters. IEEE Trans. Power Syst. **35**(5), 4131–4134 (2020)
99. Q. Hu, L. Fu, F. Ma, F. Ji, Y. Zhang, Analogized synchronous-generator model of PLL-based VSC and transient synchronizing stability of converter dominated power system. IEEE Trans. Sustain. Energy **12**(2), 1174–1185 (2021)
100. L. Harnefors, M. Bongiorno, S. Lundberg, Input-admittance calculation and shaping for controlled voltage-source converters. IEEE Trans. Ind. Electron. **54**(6), 3323–3334 (2007)
101. X. Wang, X. Ruan, S. Liu, C.K. Tse, Full feedforward of grid voltage for grid-connected inverter with LCL filter to suppress current distortion due to grid voltage harmonics. IEEE Trans. Power Electron. **25**(12), 3119–3127 (2010)
102. D. Yang, X. Ruan, H. Wu, Impedance shaping of the grid-connected inverter with LCL filter to improve its adaptability to the weak grid condition. IEEE Trans. Power Electron. **29**(11), 5795–5805 (2014)
103. C.C. Liu, J. Yang, C.K. Tse, M. Huang, Transient synchronization stability of grid-following converters considering nonideal current loop. IEEE Trans. Power Electron. **38**(11), 13757–13769 (2023)
104. D. Liu, X. Zhang, C.K. Tse, A tutorial on modeling and analysis of cascading failure in future power grids. IEEE Trans. Circ. Syst. II, Exp. Briefs **68**(1), 49–55 (2021)
105. X. Zhang, C.K. Tse, Assessment of robustness of power systems from a network perspective. IEEE J. Emerg. Sel. Topics Circ. Syst. **5**(3), 456–464 (2015)
106. N. Pogaku, M. Prodanovic, T.C. Green, Modeling, analysis and testing of autonomous operation of an inverter-based microgrid. IEEE Trans. Power Electron. **22**(2), 613–625 (2007)
107. Y. Wang, X. Wang, Z. Chen, F. Blaabjerg, Small-signal stability analysis of inverter-fed power systems using component connection method. IEEE Trans. Smart Grid **9**(5), 5301–5310 (2018)
108. M. Kabalan, P. Singh, D. Niebur, A design and optimization tool for inverter-based microgrids using large-signal nonlinear analysis. IEEE Trans. Smart Grid **10**(4), 4566–4576 (2019)
109. I.P. Nikolakakos, H.H. Zeineldin, M.S. El-Moursi, J.L. Kirtley, Reduced-order model for inter-inverter oscillations in islanded droop-controlled microgrids. IEEE Trans. Smart Grid **9**(5), 4953–4963 (2018)
110. P. Vorobev, P. Huang, M. Al Hosani, J.L. Kirtley, K. Turitsyn, High-fidelity model order reduction for microgrids stability assessment. IEEE Trans. Power Syst. **33**(1), 874–887 (2018)
111. L. Luo, S.V. Dhople, Spatiotemporal model reduction of inverter-based islanded microgrids. IEEE Trans. Energy Conv. **29**(4), 823–832 (2014)
112. M. Rasheduzzaman, J.A. Mueller, J.W. Kimball, Reduced-order small-signal model of microgrid systems. IEEE Trans. Sustain. Energy **6**(4), 1292–1305 (2015)
113. Y. Gu, N. Bottrell, T.C. Green, Reduced-order models for representing converters in power system studies. IEEE Trans. Power Electron. **33**(4), 3644–3654 (2018)
114. Z. Shuai, Y. Peng, X. Liu, Z. Li, J.M. Guerrero, Z.J. Shen, Parameter stability region analysis of islanded microgrid based on bifurcation theory. IEEE Trans. Smart Grid **10**(6), 6580–6591 (2019)
115. J. Rocabert, A. Luna, F. Blaabjerg, P. Rodríguez, Control of power converters in AC microgrids. IEEE Trans. Power Electron. **27**(11), 4734–4749 (2012)
116. Y. Wang, X. Wang, F. Blaabjerg, Z. Chen, Harmonic instability assessment using state-space modeling and participation analysis in inverter-fed power systems. IEEE Trans. Ind. Electron. **64**(1), 806–816 (2017)

117. S. Buso, P. Mattavelli, *Digital Control in Power Electronics* (Morgan and Claypool, San Francisco, 2006)
118. O. Ajala, N. Baeckeland, B. Johnson, S. Dhople, A. Domínguez-García, Model reduction and dynamic aggregation of grid-forming inverter networks. IEEE Trans. Power Syst. **38**(6), 5475–5490 (2023)
119. Y. Liu, H. Geng, C. He, W. Ding, C. Shen, G. Yang, Equivalent aggregated modeling of multi-VSC system for transient synchronization stability analysis. IEEE Trans. Power Syst. **39**(2), 4296–4310 (2024)
120. S. Skogestad, I. Postlethwaite, *Multivariable Feedback Control: Analysis and Design* (John Wiley and Sons, Chichester, 2001)
121. X. Wang, F. Blaabjerg, Harmonic stability in power electronic-based power systems: concept, modeling, and analysis. IEEE Trans. Smart Grid **10**(3), 2858–2870 (2019)
122. H. Wu, X. Wang, Design-oriented transient stability analysis of PLL-synchronized voltage-source converters. IEEE Trans. Power Electron. **35**(4), 3573–3589 (2020)
123. J. Chen, M. Liu, H. Geng, T. O'Donnell, F. Milano, Impact of PLL frequency limiter on synchronization stability of grid feeding converter. IEEE Trans. Power Syst. **37**(3), 2487–2490 (2022)
124. L. Yang, Q. Yang, G. Chen, Hidden attractors, singularly degenerate heteroclinic orbits, multistability and physical realization of a new 6D hyperchaotic system. Commun. Nonlinear Sci. Numer. Simul. **90**, 105362 (2020)
125. J. Ni, L. Liu, C. Liu, X. Hu, S. Li, Fast fixed-time nonsingular terminal sliding mode control and its application to chaos suppression in power system. IEEE Trans. Circ. Syst. II, Exp. Briefs **64**(2), 151–155 (2017)
126. H.-D. Chiang, C.-W. Liu, P.P. Varaiya, F.F. Wu, M.G. Lauby, Chaos in a simple power system. IEEE Trans. Power Syst. **8**(4), 1407–1417 (1993)
127. C.-W. Tan, M. Varghese, P. Varaiya, F.F. Wu, Bifurcation, chaos, and voltage collapse in power systems. Proc. IEEE **83**(11), 1484–1496 (1995)
128. S.H. Strogatz, Human sleep and circadian rhythms: a simple model based on two coupled oscillators. J. Math. Biol. **25**(3), 327–347 (1987)
129. K. Geist, U. Parlitz, W. Lauterborn, Comparison of different methods for computing Lyapunov exponents. Prog. Theor. Phys. **83**(5), 875–893 (1990)
130. H.-O. Peitgen, H. Jürgens, D. Saupe, *Chaos and Fractals: New Frontiers of Science*, 2nd edn. (Springer, New York, 2004)
131. C.P. Silva, Shil'nikov's theorem-A tutorial. IEEE Trans. Circ. Syst. I, Fundam. Theory Appl. **40**(10), 675–682 (1993)
132. NERC files: Multiple solar PV disturbances in CAISO. https://www.nerc.com/pa/rrm/ea/Documents/NERC_2021_California_Solar_PV_Disturbances_Report.pdf. (2021). Accessed 8 Jun 2022
133. S. Sahoo, T. Dragičević, F. Blaabjerg, Cyber security in control of grid-tied power electronic converters—challenges and vulnerabilities. IEEE J. Emerg. Sel. Topics in Power Electron. **9**(5), 5326–5340 (2021)
134. IEEE Working Group, Effects of switching network disturbances on turbine-generator shaft systems. IEEE Trans. Power Appar. Syst. **PAS-101**(9), 3151–3157 (1982)
135. Y. Li, L. Fan, Z. Miao, Wind in weak grids: low-frequency oscillations, subsynchronous oscillations, and torsional interactions. IEEE Trans. Power Syst. **35**(1),109–118 (2020)
136. R. Ma, Y. Zhang, M. Zhan, K. Cao, D. Liu, K. Jiang, S. Cheng, Dominant transient equations of grid-following and grid-forming converters by controlling-unstable-equilibrium-point-based participation factor analysis. IEEE Trans. Power Syst. **39**(3), 4818–4834 (2024)
137. Á. Calsina, S. Cuadrado, B. Vidiella, J. Sardanyés, About ghost transients in spatial continuous media. Chaos Solitons Fractals **166**, 112915 (2023)
138. C.K. Tse, M. Li, Design-oriented bifurcation analysis of power electronics systems. Int. J. Bifurcat. Chaos **21**(6), 1523–1537 (2011)
139. J. Lai, X. Lu, Z. Dong, S. Cheng, Resilient distributed multiagent control for AC microgrid networks subject to disturbances. IEEE Trans. Syst. Man Cybern. Syst. **52**(1), 43–53 (2022)

140. G. Chen, Pinning control of complex dynamical networks. IEEE Trans. Consum. Electron. **68**(4), 336–343 (2022)
141. Y.-Y. Liu, J.-J. Slotine, A.-L. Baraba'si, Controllability of complex networks. Nature **473**, 167–173 (2011)
142. X.F. Wang, G. Chen, Pinning control of scale-free dynamical networks. Phys. A **310**(3–4), 521–531 (2002)
143. G. Chen, Z. Guo, Distributed secondary and optimal active power sharing control for islanded microgrids with communication delays. IEEE Trans. Smart Grid **10**(2), 2002–2014 (2019)
144. J.W. Simpson-Porco, Q. Shafiee, F. Dörfler, J.C. Vasquez, J.M. Guerrero, F. Bullo, Secondary frequency and voltage control of islanded microgrids via distributed averaging. IEEE Trans. Ind. Electron. **62**(11), 7025–7038 (2015)
145. H. Zhang, S. Kim, Q. Sun, J. Zhou, Distributed adaptive virtual impedance control for accurate reactive power sharing based on consensus control in microgrids. IEEE Trans. Smart Grid **8**(4), 1749–1761 (2017)
146. Y. Khayat, R. Heydari, M. Naderi, T. Dragičević, Q. Shafiee, M. Fathi, H. Bevrani, F. Blaabjerg, Decentralized frequency control of AC microgrids: an estimation-based consensus approach. IEEE J. Emerg. Sel. Topics Power Electron. **9**(5), 5183–5191 (2021)
147. Q. Shafiee, Č. Stefanović, T. Dragičević, P. Popovski, J.C. Vasquez, J.M. Guerrero, Robust networked control scheme for distributed secondary control of islanded microgrids. IEEE Trans. Ind. Electron. **61**(10), 5363–5374 (2014)
148. K. Feng, C. Liu, Multi-rate sampling control design and stability analysis for frequency and voltage regulation in islanded microgrids. IEEE Trans. Sustain. Energy **14**(1), 704–716 (2023)
149. A. Bidram, A. Davoudi, F.L. Lewis, J.M. Guerrero, Distributed cooperative secondary control of microgrids using feedback linearization. IEEE Trans. Power Syst. **28**(3), 3462–3470 (2013)
150. S. Rath, D. Pal, P.S. Sharma, B.K. Panigrahi, A cyber-secure distributed control architecture for autonomous AC microgrid. IEEE Syst. J. **15**(3), 3324–3335 (2021)
151. Y.C.C. Wong, C.S. Lim, M.D. Rotaru, A. Cruden, X. Kong, Consensus virtual output impedance control based on the novel droop equivalent impedance concept for a multi-bus radial microgrid. IEEE Trans. Energy Convers. **35**(2), 1078–1087 (2020)
152. Y.C.C. Wong, C.S. Lim, A. Cruden, M.D. Rotaru, P.K. Ray, A consensus-based adaptive virtual output impedance control scheme for reactive power sharing in radial microgrids. IEEE Trans. Ind. Appl. **57**(1), 784–794 (2021)
153. R. Olfati-Saber, J.A. Fax, R.M. Murray, Consensus and cooperation in networked multi-agent systems. Proc. IEEE **95**(1), 215–233 (2007)
154. J. Yang, J. Liu, Y. Shi, N. Zhao, J. Zhang, L. Fu, T.Q. Zheng, Carrier-based digital PWM and multirate technique of a cascaded H-bridge converter for power electronic traction transformers. IEEE J. Emerg. Sel. Topics Power Electron. **7**(2), 1207–1223 (2019)
155. T. Dragičević, Model predictive control of power converters for robust and fast operation of AC microgrids. IEEE Trans. Power Electron. **33**(7), 6304–6317 (2018)
156. A. Rafiee, Y. Batmani, A. Mehrizi-Sani, T. Kato, H. Bevrani, Zero-level control design in AC microgrids: a robust multivariable approach. IEEE Trans. Power Electron. **38**(7), 8394–8405 (2023)
157. J. Rocabert, A. Luna, F. Blaabjerg, P. Rodríguez, Control of power converters in AC microgrids. IEEE Trans. Power Electron. **27**(11), 4734–4749 (2012)
158. G. Chen, X. Wang, X. Li, *Fundamentals of Complex Networks: Models, Structures and Dynamics* (Wiley, Singapore, 2015)
159. E.A.S. Ducoin, Y. Gu, B. Chaudhuri, T.C. Green, Analytical design of contributions of grid-forming and grid-following inverters to frequency stability. IEEE Trans. Power Syst. **39**(5), 6345–6358 (2024)
160. Z. Yuan, C. Zhao, Z. Di, W.-X. Wang, Y.-C. Lai, Exact controllability of complex networks. Nat. Commun. **4**(2447), 1–9 (2013)

Index

J. Yang, C. K. Tse, *Complex Behavior of Grid-Connected Power Electronics Systems*, https://doi.org/10.1007/978-3-031-69340-3

MIX
Papier aus verantwortungsvollen Quellen
Paper from responsible sources
FSC® C105338

If you have any concerns about our products,
you can contact us on
ProductSafety@springernature.com

In case Publisher is established outside the EU,
the EU authorized representative is:
Springer Nature Customer Service Center GmbH
Europaplatz 3, 69115 Heidelberg, Germany

Printed by Libri Plureos GmbH
in Hamburg, Germany